Emerging Carbyne

Emerging Carbyne

Truly One-Dimensional Form of Carbon

edited by
Yahachi Saito

JENNY STANFORD
PUBLISHING

Published by

Jenny Stanford Publishing Pte. Ltd.
101 Thomson Road
#06-01, United Square
Singapore 307591

Email: editorial@jennystanford.com
Web: www.jennystanford.com

British Library Cataloguing-in-Publication Data
A catalogue record for this book is available from the British Library.

Emerging Carbyne: Truly One-Dimensional Form of Carbon

ISBN 978-981-5129-15-1 (Hardcover)
ISBN 978-1-003-51036-9 (eBook)

Contents

Preface

Carbon materials have supported the development of human culture and civilization from ancient times. Carbon materials are used in various forms (charcoal, soot, graphite, coke, carbon fiber, diamond, etc.) in a variety of fields ranging from our daily lives and manufacturing industry to cutting-edge science and aerospace technology. It can be said that our modern life would not be possible without carbon materials. Although carbon is a simple substance (substance consisting of one element), it has a variety of structures depending on how it is made, and it forms materials with multifaceted properties due to the difference in structure. It has been known that the carbon produced by nature as a simple substance is graphite and diamond, but in the late 1960s and mid-1980s, the possibilities of the existence of new carbon allotropes were reported, and the new structures of carbon, carbyne and fullerene, attracted a lot of attention. News on the discovery of a new carbon allotrope called carbyne appeared in 1968, which reported that the allotrope occurred in the mineral "chaoite" formed in the crater of a meteorite impact and consisted of linear chains of carbon atoms. Another new allotrope, C_{60} fullerene, with a spherically closed structure made of carbon pentagons and hexagons was produced by laser ablation of graphite in 1985. Long before the discovery of chaoite, A. Baeyer, an eminent organic chemist, had attempted to synthesize linear carbon chains in 1885, but the techniques of the time were not successful. In addition, carbyne that was insisted to exist in geological materials including chaoite has not been fully evaluated in terms of structure and is not convincing from the current point of view. However, at the beginning of this century, an unmistakable existence of carbyne that was protected inside carbon nanotubes was revealed by X. Zhao (one of the authors of this book), which brought this one-dimensional material to the forefront of carbon science. This book summarizes the results of research on the generation and properties of carbyne since 2003, which have been confirmed by the latest analytical evaluation techniques and theoretical support. I hope that the book will be useful in the development of related disciplines in the future.

Yahachi Saito
Nagakute, Japan
May 2024

Chapter 1

Historical Survey of "Carbyne" Research Leading to This Book

Yahachi Saito
Toyota Physical and Chemical Research Institute,
Nagakute, Aichi 480-1192, Japan
yysaito@sky.plala.or.jp

Among a variety of forms of elemental carbon, carbon monoatomic chains (so called "carbyne") were elusive to obtain. At the beginning of the 21st century, various chemical and physical methods to synthesize carbyne were devised and enabled to characterize the structure and properties of carbyne. In this chapter, a history of carbyne research is briefly reviewed, and the recent development of synthesis method and expected properties of carbyne are surveyed.

1.1 Introduction

Although carbon is a simple substance, it takes on a wide variety of structures depending on how it is made, and the difference in structure makes it a substance with multifaceted properties. It

Emerging Carbyne: Truly One-Dimensional Form of Carbon
Edited by Yahachi Saito

ISBN 978-981-5129-15-1 (Hardcover), 978-1-003-51036-9 (eBook)
www.jennystanford.com

is well known that graphite and diamond are produced as simple substances from the earth's crust over a long period of geological time. In the late 1960s and mid-1980s, the possibilities of the existence of new carbon allotropes were reported. C_{60} fullerenes family [1] and graphene [2] were successfully produced and their exotic properties attracted great attentions. On the other hand, carbon monoatomic chains (so called "carbyne"), being elusive for a long time since the first trial of their production in 1885 [3], have been research subjects of chemical synthesis [4, 5] and interstellar substances [6]. Owing to the progress of synthesis methods and characterization techniques of products, long carbon monoatomic chain became the real entities: Carbynes are now synthesized by both chemical and physical methods and can exist stably, though the amount is still microscopic or semi-macroscopic, as described in this review book. Studies on the truly one-dimensional form of carbon, carbyne, are now attracting researchers of materials scientists because of its unique structure and unexpectedly extreme properties. This chapter reviews the structural diversity of carbon materials, the history of the discovery and synthesis of linear carbon chains (carbynes), and their predicted properties.

1.2 Versatile Allotropes of Carbon

Elemental carbon appears in versatile forms from the well-known diamond and graphite, and newly synthesized fullerenes, nanotubes and graphene to amorphous carbon whose structure approaches that of graphite in continuous degrees. Three-dimensional sp^3 diamond and pseudo-two-dimensional sp^2 graphite are stable allotropes that have been known for a long time, but since 1985, new carbons such as pseudo-0-dimensional $sp^{2+\delta}$ C_{60} fullerene [1], pseudo-1-dimensional $sp^{2+\delta}$ nanotubes [7], and two-dimensional sp^2 graphene [2] have been discovered and synthesized one after another (Fig. 1.1). However, the so-called sp^1 carbyne has been elusive for more than 100 years, though the synthesis of carbyne was keenly attempted. The authenticity of its existence was a subject of discussion and sometimes cast doubts. The status of scientific researches on carbyne up to the last century can be found in the review paper by Kudryavtsev *et al.* [4]. After a time passed without major progress, it is only recent that its synthesis has been successfully demonstrated and its existence has actually been confirmed. The new form of linear all-carbon chains is named "carbyne" from the word carbon with the ending "yne" which is customary used in organic chemistry to indicate an acetylenic linkage [4].

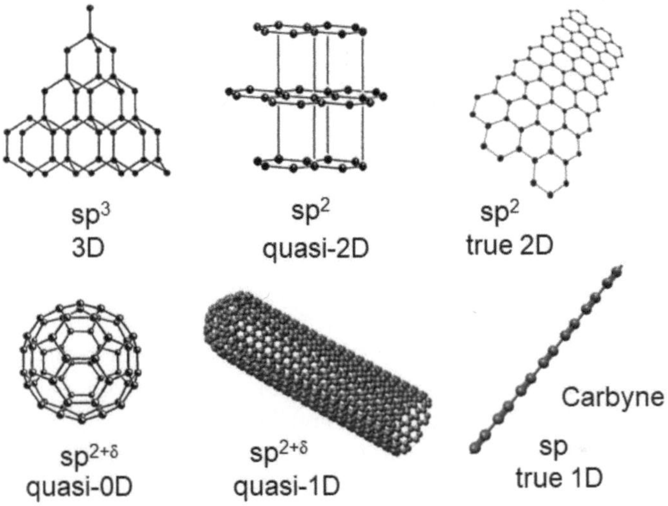

Figure 1.1 Allotropes of carbon.

1.3 Carbyne in Nature

It has been pointed out that one-dimensional allotropes of carbon exist in nature, including unconfirmed ones, for example, the report published in 1968 on carbon allotropes called "chaoite" found in a crater after a meteorite fall (Ries Crater in Bavaria, Germany) [8] has been controversial. Bearing on the chaoite discovery, Whittaker carried out a study of several natural graphitic ores from a wide geographic area using X-ray diffraction and ion microprobe analysis, and indicated that many graphites contain small amounts of carbyne forms of carbon [9]. He concluded that chaoite existed in the gneiss before the meteorite impact; hence, the impact would have had nothing to do with chaoite formation. These claims, lacking compelling evidence of the presence of carbyne in the ores, have been questioned [10].

Radio astronomy has shown that short one-dimensional carbon chains such as polyynylcyanides, H-$(C\equiv C)_n$-CN (n = 2–4), are floating in interstellar space [6]. It is well known that C_{60} fullerenes were discovered in ambitious research to produce the carbon chains in laboratories on Earth [1]. In addition, some plants and fungi reportedly contain polyyne derivatives [4].

1.4 Chemical Synthesis

The synthesis of polyacetylene (tetraacetylene) was first attempted by Adolf von Baeyer in 1885 [3], but it was shown that the synthesis was difficult even by famous chemists, so synthesis research was for a while subsided.

The synthesis of polyyne H–C_n–H by dehydrogenation polymerization of acetylene has been carried out by Russian research groups and others since 1959 [4]. Czech researchers, using poly(tetrafluoroethylene) (PTFE) as a carbyne precursor, defluorinated PTFE with alkali metal (mainly Li) amalgams under the moderate condition to obtain carbyne chains by the reaction [5]:

$$\begin{array}{c} F\quad F \\ |\quad\ | \\ (-C-C-)_n \\ |\quad\ | \\ F\quad F \end{array} + 4n(e^- + Li^+) \rightarrow (-C\equiv C-)_n + 4nLiF \qquad (1.1)$$

This reaction also produces graphite-based carbon, and a pure carbyne solid has not yet been formed. Moreover, when stored for a long time at room temperature, the proportion of graphite-based components increases. This is said to be due to the fact that when the carbyne chains are closely spaced, cross-linking reactions between molecules occur, as shown in Equation (1.2), and they change to graphite.

$$\begin{array}{c} -C\equiv C-C\equiv C- \\ + \\ -C\equiv C-C\equiv C- \end{array} \rightarrow \begin{array}{c} |\quad\ | \\ -C=C-C=C- \\ |\quad\ | \\ -C=C-C=C- \\ |\quad\ | \end{array} \qquad (1.2)$$

Therefore, they argue that the presence of impurities and terminal functional groups that play a role in increasing the spacing between molecules is indispensable for the production of stable carbyne.

When the number of carbon atoms exceeds around 10, the one-dimensional structure is unstable and breaks due to the cross-linking. In order to stabilize the long one-dimensional structure, spatially bulky atomic groups are tied to both ends of the chain [11, 12], and polyyne consisting of 44 sp-carbon atoms has been synthesized so far [12].

1.5 Physical Synthesis

Physical methods, based on modification of carbon using a high-energy process, include several techniques such as resistive heating of pyrolytic graphite at 2700–3000 K, laser irradiation onto a graphite, a short-duration, high-current arc discharge between graphitic carbon electrodes [13], ion sputtering of graphite in combination with a bombardment of the growing carbon film by Ar^+ ions, evaporation of carbon by an arc discharge with subsequent condensation of the carbon vapor on a cold substrate [14] and so on, but the identification of the resulting products seemed unconvincing. These experiments aiming at the production of carbynes in solid state have been tried until the middle of the 1990s.

Going down in time, in 2003 a report was published by Zhao *et al.*, demonstrating without doubt the existence of carbyne being stably encapsulated inside multiwall carbon nanotubes prepared by arc discharge [15]. Eight years later, in 2011, Shi *et al.* reported that long linear carbon chains isolated inside double-wall carbon nanotubes were formed by high-temperature heat treatment of the carbon nanotubes in vacuum [16]. In 2017, carbynes encapsulated in single-wall and double-wall carbon nanotubes were unexpectedly discovered in a field emission cathode of carbon nanotube film after suffering discharge damage [17]. These discoveries rely on the accumulation of studies on carbon nanotubes (CNTs) and the recent development of scientific analysis techniques, especially high-resolution transmission electron microscopy and Raman microscopy. Newly emerging studies on carbynes encapsulated in carbon nanotubes are described in Chapters from 4 to 8 of this book.

Sophisticated techniques to prepare short carbon atomic chains under transmission electron microscopy (TEM) observations have been demonstrated since the 2000s. For example, an atomic chain crossing a hole formed in a thin carbon film/graphene by irradiating an electron beam of TEM [18–20], a carbon chain striped from graphene or CNTs by manipulating the carbon substrates with application of electric current during TEM observation [21] were reported. The atomic chains are formed *via* self-organization process induced by electron radiation, mechanical deformation, and Joule heating. The details of these subjects are reviewed in Chapters 9 and 10.

The presence of free-standing carbon atomic chains (consisting more than ten atoms) at the surface of sharpened carbon fibers treated by high electric fields on the order of 10^{11} V/m has been claimed by a research group of Ukraine in 2007 [22]. The unraveling of a carbon atomic wire from an open-ended multiwall carbon nanotube, resembling an unraveling of a thread from a knit sleeve, under the high electric field was first postulated by Smalley' group in 1995 [23]; the hypothesis was devised to explain an enhanced current of electrons emitted from the open-ended nanotube even under moderate bias voltages. Mikhailovskij *et al.*, using field ion microscopy, also tried to measure the tensile strength of a carbon atomic chain under electrostatic forces, and estimated it to be 245 GPa [24], being close to the calculated value of about 200 GPa if 0.2 nm was adapted as the chain diameter [25].

1.6 Physicochemical Synthesis

In 2002, a new technique based on laser ablation in liquid was devised as physicochemical method [26], and a series of polyynes with a variety of length ranging from 6 to ~30 carbon atoms terminated by not only hydrogen but also methyl (CH_3) and cyano (CN) groups were synthesized in liquid phase. Using high-performance liquid chromatography (HPLC), polyynes were separated according to the length and termination. The synthesis method called pulsed laser ablation in liquid (PLAL) and spectroscopic characterizations of a variety of polyynes prepared the method are intensively reviewed in Chapters 2 and 3.

1.7 Expected Physical Properties

Besides the fundamental importance as the true entity of one-dimensional nanowires with an atom-sized diameter, carbyne attracts much interests because of the predicted outstanding physical properties; Extreme mechanical strengths are estimated, *e.g.*, high breaking strength (67–200 GPa) [25] and stiffness (Young's modulus ~32 TPa [27]) exceeding those of quasi-one dimensional CNTs (45 GPa) [28]. Thermal conductivity is also expected to reach extremely high value ~80 kW/K·m at 200–300 K [29], which is attributed to the high frequency of lattice vibration and the long mean-free-path of phonon, being larger than that of CNTs (3.5 kW/K·m) [30]. It has also been shown that semiconductors with direct transition bandgaps can be controlled by the length [31]. Moreover, the flexibility of carbyne is shown to be equivalent to that of graphene nanoribbons and conductive polymer polyacetylene, and the effective surface area for hydrogen is four times that of graphene [27, 32]. Applications of carbyne that are expected to have excellent physical properties include nanoelectronics, spintronic devices [33, 34], high-strength composite materials, and hydrogen storage materials. Possible superconductivity of one-dimensional carbon is presented in Chapter 11.

References

1. Kroto, H. W., Heath, J. R., O'Brien, S. C., Curl, R. F., and Smalley, R. E. (1985). C60: buckminsterfullerene, *Nature* **318**, 162–163.
2. Novoselov, K. S., Geim, A. K., Morozov, S. V., Jiang, D., Zhang, Y., Dubonos, S. V., *et al.* (2004). Electric field effect in atomically thin carbon films, *Science* **306**, 666–669.
3. Baeyer, A. (1885). Uber polyacetylenverbindungen, *Berichte der Deutschen Chemischen Gesellschaft* **18**(1), 674–681.
4. Kudryavtsev, Y. P., Evsyukov, S., Guseva, M., Babaev, V., and Khvostov, V. (1996). Carbyne – a linear chainlike carbon allotrope, in *Chemistry and Physics of Carbon* (Thrower, P. A., ed), vol. 25, Marcel Dekker, New York, Chapter 1.
5. Kavan, L., and Dousek, F. P. (1993). Carbynoid species in electrochemical polymeric carbon, *Synth. Metals* **58**, 63–72.

6. Kroto, H. W. (1992). C60: buckminsterfullerene, the celestial sphere that fell to earth, *Angew. Chem. Int. Edn. Engl.* **31**, 111–129.

7. Iijima, S. (1991). Helical microtubules of graphitic carbon, *Nature* **354**, 56–58.

8. Goresy, A. E., and Donnay, G. (1968). A new allotropic form of carbon from the Ries crater, *Science* **161**, 363–364.

9. Whittaker, A. G. (1979). Carbon: occurrence of carbyne forms of carbon in natural graphites, *Carbon* **17**, 21–24.

10. Smith, P. P. K., and Buseck, P. R. (1982). Carbyne forms of carbon: do they exist? *Science* **216**, 984–986.

11. Eastmond, R., Johnson, T. R., and Walto, D. R. M. (1972). Silylation as a protective method for terminal alkynes in oxidative couplings; A general synthesis of the parent polyynes $H(C{\equiv}C)_nH$ (n = 4–10, 12), *Tetrahedron* **28**, 4601–4616.

12. Chalifoux, W. A., and Tykwinski, R. R. (2010). Synthesis of polyynes to model the sp-carbon allotrope carbyne, *Nat. Chem.* **2**, 967–971.

13. Whittaker, A. G., and Waites, G. M. (1972). Carbon: a suggested new hexagonal crystal form, *Science* **178**, 54–56.

14. Kudryavtsev, Y. P., Esyukov, S. E., Babaev, V. G., Guscva, M. B., Khvostov, V. V., and Krechko, L. M. (1992). Oriented carbyne layers, *Carbon* **30**, 213–221.

15. Zhao, X., Ando, Y., Liu, Y., Jinno, M., and Suzuki, T. (2003). Carbon nanowire made of a long linear carbon chain inserted inside a multiwalled carbon nanotube, *Phys. Rev. Lett.* **90**, 187401.

16. Shi, L., Sheng, L., Yu, L., Am, K., Ando, Y., and Zhao, X. (2011). Ultra-thin double-walled carbon nanotubes: a novel nanocontainer for preparing atomic wires, *Nano Res.* **4**, 759–766.

17. Toma, S., Asaka, K., Irita, M., and Saito, Y. (2019). Bulk synthesis of linear carbon chains confined inside single-wall carbon nanotubes by vacuum discharge, *Surf. Interface Anal.* **51**, 131–135. (Special Issue of 11th Inter. Symp. Atomic Level Characterizations for New Materials and Devices 2017)

18. Troiani, H. E., Miki-Yoshida, M., Camacho-Bragado, G. A., Marques, M. A. L., Rubio, A., Ascencio, J. A., and Jose-Yacaman, M. (2003). Direct observation of the mechanical properties of single-walled carbon nanotubes and their junctions at the atomic level, *Nano Lett.* **3**, 751–755.

19. Chuvilin, A., Meyer, J. C., Algara-Siller, G., and Kaiser, U. (2009). From graphene constrictions to single carbon chains, *New J. Phys.* **11**, 083019.

20. Jin, C., Lan, H., Peng, L., Suenaga, K., and Iijima, S. (2009). Deriving carbon atomic chains from graphene, *Phys. Rev. Lett.* **102**, 205501.

21. Asaka, K., and Kizuka, T. (2005). Atomistic dynamics of deformation, fracture, and joining of individual single-walled carbon nanotubes, *Phys. Rev. B* **72**, 115431.

22. Mikhailovskij, I. M., Wanderka, N., Ksenofontov, V. A., Mazilova, T. I., Sadanov, E. V., and Velicodnaja, O. A. (2007). Preparation and characterization of monoatomic C-chains: unraveling and field emission, *Nanotechnology* **18**, 475705 (6pp).

23. Rinzler, A. G., Hafner, J. H., Nikolaev, P., Lou, L., Kim, S. G., Tomanek, D., Nordlander, P., Colbert, D. T., and Smalley, R. E. (1995). Unraveling nanotubes: field emission from an atomic wire, *Science* **269**, 1550–1553.

24. Mikhailovskij, I. M., Sadanov, E. V., Kotrechko, S., Ksenofontov, V. A., and Mazilova, T. I., (2013). Measurement of the inherent strength of carbon atomic chains, *Phys. Rev. B* **87**, 045410.

25. Liu, X., Zhang, G., and Zhang, Y.-W. (2015). Tunable mechanical and thermal properties of one-dimensional carbyne chain: phase transition and microscopic dynamics, *J. Phys. Chem. C* **119**, 24156–24164.

26. Tsuji, M., Tsuji, T., Kuboyama, S., Yoon, S.-H., Korai, Y., Tsujimoto, T., Kubo, K., Mori, A., and Mochida, I. (2002). Formation of hydrogen-capped polyynes by laser ablation of graphite particles suspended in solution, *Chem. Phys. Lett.* **355**, 101–108.

27. Liu, M., Artyukhov, V. I., Lee, H., Xu, F., and Yakobson, B. I. (2013). Carbyne from first principles: chain of C atoms, a nanorod or a nanorope, *ACS Nano* **7**, 10075–10082.

28. Walters, D. A., Ericson, L. M., Casavant, M. J., Liu, J., Colbert, D. T., Smith, K. A., and Smalley, R. E. (1999). Elastic strain of freely suspended single-wall carbon nanotube ropes, *Appl. Phys. Lett.* **74**, 3803–3805.

29. Wang, M., and Lin, S. (2015). Ballistic thermal transport in carbyne and cumulene with micron-scale spectral acoustic phonon mean free path, *Sci. Rep.* **5**, 18122.

30. Pop, E., Mann, D., Wang, Q., Goodson, K., and Dai, H. (2006). Thermal conductance of an individual single-wall carbon nanotube above room temperature, *Nano Lett.* **6**, 96–100.

31. Yang, S., and Kertesz, M. (2006). Bond length alternation and energy band gap of polyyne, *J. Phys. Chem. A* **110**, 9771–9774.

32. Sorokin, P. B., Lee, H., Antipina, L. Y., Singh, A. K., and Yakobson, B. I. (2011). Calcium-decorated carbyne networks as hydrogen storage media, *Nano Lett.* **11**, 2660–2665.

33. Dong, Y.-J., Wang, X.-F., Yang, S.-W., and Wu, X.-M. (2014). High performance current and spin diode of atomic carbon chain between transversely symmetric ribbon electrodes, *Sci. Rep.* **4**, 06157 (7pp).

34. Romdhane, F. B., Adjizian, J. J., Charlier, J.-C., and Banhart, F. (2017). Electrical transport through atomic carbon chains: the role of contacts, *Carbon* **122**, 92–97.

Chapter 2

Formation and Characterization of Polyynes in Liquid and Solid Media

Tomonari Wakabayashi
Department of Chemistry, Kindai University, Higashi-Osaka, 577-8502, Japan
wakaba@chem.kindai.ac.jp

One-dimensional linear chain molecules of $2n$ = 8–16 sp-hybridized carbon atoms, namely polyynes, are produced by laser ablation of graphite particles suspended in liquid organic solvents. Hydrogen-caped polyynes and their derivatives, i.e., cyanopolyynes, methylpolyynes, and dimethylpolyyens, are identified by ultraviolet (UV) absorption spectra with the help of mass spectrometry. Laser induced fluorescence (LIF), phosphorescence, infrared (IR) absorption, and Raman scattering spectroscopy are performed for size-separated polyynes. The reactive polyyne molecules are dispersed in solid media such as single-wall carbon nanotubes (SWNTs), cyclodextrin (CD), and polyvinyl alcohol (PVA) which is exploited to create UV-polarizing films. As a model compound of carbyne, a condensed form of polyyne molecules remains a challenge.

Emerging Carbyne: Truly One-Dimensional Form of Carbon
Edited by Yahachi Saito

ISBN 978-981-5129-15-1 (Hardcover), 978-1-003-51036-9 (eBook)
www.jennystanford.com

2.1 Introduction

Half a century has passed since cyanopolyyne molecules were detected in space [1]. Researches stimulated by the event in the fields of interstellar chemistry and molecular spectroscopy lead eventually to the discovery of fullerene C_{60} [2], carbon nanotubes [3], and graphene [4], which expanded further into a wide variety of research fields in physics, chemistry, and materials science. It was accidental coincidence in the '70s that the prediction of aromatic stability of the C_{60} molecule [5] paralleled with the detection of cyanopolyyne molecules in molecular clouds and circumstellar shells of red giants. Recent discovery of C_{60} and C_{70} in a planetary nebula [6] represents excellent feedback of detailed physical and chemical analyses in the laboratory to the study of molecular evolution in space. Connection between the motivation for reproduction of interstellar conditions in the laboratory and the idea for combination of laser-vaporization and mass-spectrometry lead to the discovery of fullerene C_{60} [2]. The laboratory reference of IR spectra [7] must have accelerated confirmation for the presence of C_{60} and C_{70} in space during the movement from radio astronomy to the infrared space telescope [6]. An encounter between a new method and a motivation from a bit remote area can promote novel finding in an interdisciplinary area. Cyanopolyyne represents one of the rare cases that the molecule was first detected in space before identification in the laboratory. For a property of cross-linking polymerization at high concentration, thus difficult to handle these molecules in a crystalline form, the research on polyynes still remains state-of-the-art.

The early polyyne research dates back to the organic synthesis aiming at elongation of sp-hybridized conjugate carbon chains in the '70s [8, 9]. Based on the flexible hybridization of carbon, interests in a variety of molecular structures of the element have inspired following studies for chemical syntheses of macrocyclic molecules containing bent sp-carbon chains in the molecular framework, including the formation of monocyclic carbon clusters, i.e., cyclic C_n, from these precursors [10] and characterization of the 2-dimensional carbon-ring structure on clean surfaces [11]. The elegant synthetic route enables longer polyyne molecules to be stabilized by bulky end-capping groups to avoid polymerization with adjacent molecules, then to form crystalline solid subjected to X-ray diffraction analysis [12].

A simple method by laser ablation was found for production of polyyne molecules in 2002 [13]. Another simple method by arc discharge in liquids was also reported during the same period of time [14]. A newly developed method at that time, in which a pulsed laser beam was focused onto a sunken metal target in a liquid to produce metal nanoparticles, was applied to graphite particles suspended in organic solvents to identify hydrogen-capped polyyne molecules remained in solutions after laser ablation. The molecules were identified by UV absorption spectra [13]. With this simpler method of laser ablation of particulate carbon in organic solvents, thorough investigation of polyyne molecules has become feasible even by researchers unfamiliar to organic synthesis.

In Chapter 2, spectroscopic characterization of polyynes in liquid and solid media is introduced [15]. Having characteristic molecular structures and UV absorption features, the series of polyyne molecules depicted in Fig. 2.1 are promising materials for various applications, such as UV polarizer by exploiting their anisotropic properties in the electronic transition. The analysis of the order parameter for orientation of linear polyyne molecules trapped in solid media enables us to study microscopic molecular alignment in the optically anisotropic sample. The research of polyynes is expected to develop further not only as interstellar molecules and condensed-matter materials for optical applications but also as model compounds for carbyne, the truly one-dimensional form of carbon.

Series of polyynes have been produced, separated, and purified for characterization of optical properties [16–41]. Laser ablated polyyne molecules, $X–(C{\equiv}C)_n–Y$ ($n \geq 3$), identified so far in our laboratory are relatively simple ones as listed in Figs. 2.1(a)–(c), i.e., the ones with end-capping groups, X and Y, of a hydrogen –H, a cyano group –CN, a methyl group $–CH_3$, and their combinations. Using these relatively simple polyynes, detailed spectroscopic analysis is possible for molecular transitions depicted in Fig. 2.1(d). In the following sections, topics outlined are *formation and separation of polyyne molecules* in Section 2.2 [16–24], *spectroscopic properties* summarized in Section 2.3 [25–33], including photo-induced formation of a unique complex of polyynes with iodine [34–36]. In Section 2.4, examples are exhibited for dispersion of polyyne molecules in solid media, i.e., *polyyne-doped SWNTs* [37–39], *crystallization of inclusion compounds of polyynes by using cyclodextrin* [40], and *polyyne molecules aligned in PVA films* [41]. Concerning the last topic, *analysis for the order parameter of molecular orientation in UV-polarizing films* will be discussed in detail [15]. Finally, formation of *aggregates of hydrogen-capped polyynes in a thread-like solid form* is briefly mentioned in Section 2.5.

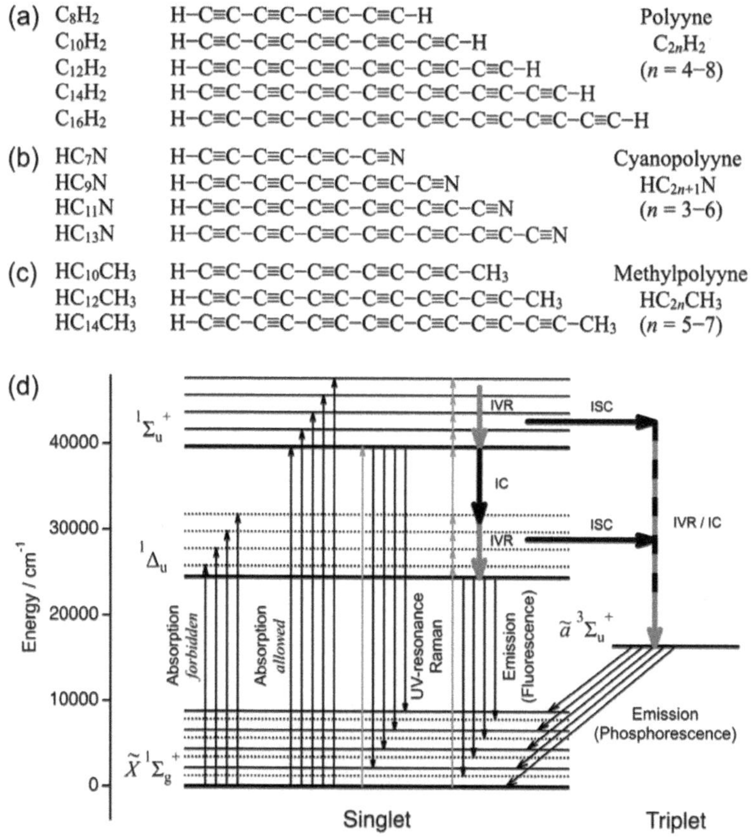

Figure 2.1 Series of polyynes: (a) hydrogen-capped polyyne, $C_{2n}H_2$ (n = 4–8), (b) cyanopolyyne, $HC_{2n+1}N$ (n = 3–6), and (c) methylpolyyne, $HC_{2n}CH_3$ (n = 5–7). (d) Electronic and vibrational energy level schemes of a typical polyyne molecule and its optical transitions: absorption (*allowed/forbidden*), UV-resonance Raman scattering, and emission (fluorescence/phosphorescence) following excitation (upward arrows in gray) and non-radiative relaxation *via* internal vibrational energy redistribution (IVR), internal conversion (IC), and intersystem crossing (ISC) [15].

2.2 Formation and Characterization of Polyynes

2.2.1 Laser Ablation of Graphite Particles in Organic Solvents

Carbon particles suspended in an organic solvent such as hexane are instantaneously heated above 4000 K upon irradiation of an intense laser pulse as depicted in Fig. 2.2. Nascent linear carbon chain molecules fragmented from the surface of hot carbon particles transform in their bonding scheme from sp- to sp^2-hybridization for stabilization upon condensation into larger molecules. There is a report for experimental evidence that, during coagulation of sp-carbon molecules at cryogenic temperature under vacuum, the condensate heats itself up to 2500 K by accumulation of the heat of reaction in its vibrational degrees of freedom [16, 17]. On the other hand, upon laser ablation in a liquid organic solvent, bubbles are formed by evaporation of gas-phase carbon chain molecules, namely linear C_n, emanated from the surface of laser-ablated hot carbon particles. Part of the nascent carbon chain molecules react with solvent molecules at interface between the gas bubble and the liquid phase to extract hydrogen to form polyynes, $C_{2n}H_2$. The resulting polyyne molecules are soluble in the organic solvent, and hence neither adsorbed nor condensed on the carbon particles but dispersed in the liquid phase. Since the laser pulses used for ablation have an infrared or a visible wavelength, *e.g.*, Nd:YAG fundamental at 1064 nm or second harmonics at 532 nm, polyyne molecules having strong absorption only in the UV do not absorb photons from subsequent laser pulses and remain in the solution without decomposition.

Both of nanosecond, ns, [18–21] and femtosecond, fs, [20–23] laser pulses are applicable to the formation of polyynes, both in the gas phase [18–21] and in the liquid phase [22, 23]. Nanosecond lasers are operated at a repetition rate of 10–30 Hz and deliver relatively high pulse energies of 20–500 mJ/pulse. For laser ablation in the liquid phase for which sufficiently high-concentration conditions of carbon particles are easily attainable, the laser beam is not necessarily focused in order to obtain several milligrams of polyyne molecules an hour. Femtosecond lasers, with an advantage of extremely high peak intensities to realize filamentation upon focusing in liquids, can provide polyyne molecules solely from the solvent without supplemental carbon particles [22, 23]. Since the pulse energies of typical fs lasers are at most 0.2–2 mJ/pulse, high repetition rates, *e.g.*, 1 kHz or higher, will be required for bulk production of polyynes. In comparison with the scale of organic synthesis where at least 10–100 milligrams of products are requested, further improvements are needed for the production of polyynes by laser ablation, at least an order of magnitude in its scale and efficiency.

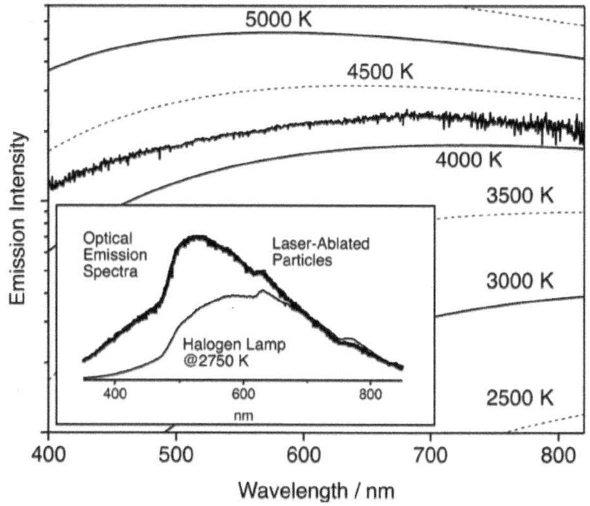

Figure 2.2 Optical spectra of incandescent light emission from laser-ablated carbon particles in liquid hexane, showing blackbody radiation above 4000 K. Inset shows spectra prior to calibration of laser-ablated particles (thick line) and of thermal emission from the filament at 2750 K of a halogen lamp (thin line).

2.2.2 High Performance Liquid Chromatography (HPLC) for Separation and Purification

Among the polyyne molecules formed by laser ablation in *n*-hexane, *n*-C_6H_{14}, major products are the series of hydrogen-capped polyynes, $H(C{\equiv}C)_nH$ ($n \geq 3$) depicted in Fig. 2.1(a). When the solvent is replaced by nitrogen-containing one such as acetonitrile, CH_3CN, the series of cyanopolyynes, $H(C{\equiv}C)_nC{\equiv}N$ ($n \geq 3$) depicted in Fig. 2.1(b), are also formed in addition to the regular polyynes, $H(C{\equiv}C)_nH$ [24]. In both cases, a variety of sizes of polyyne and cyanopolyyne molecules are formed at the same time in one pot.

In addition, unidentified polyyne derivatives are also present in solutions after laser ablation. Therefore, separation and purification are necessary to obtain targeted samples of size-separated polyyne and cyanopolyyne molecules.

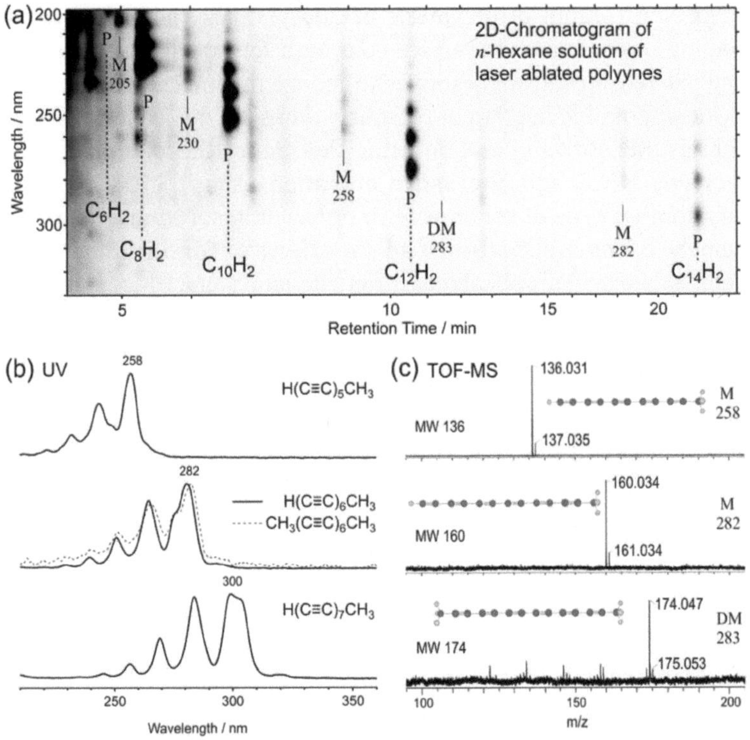

Figure 2.3 Separation and identification of polyyne molecules. (a) 2-dimensional chromatogram of analytical HPLC operation for a laser ablated solution of *n*-hexane after removal of carbon particles by filtration. The vertical cross section corresponds to the UV absorption spectrum of eluting molecular species at the retention time. In addition to the major series of hydrogen-capped polyynes, $C_{2n}H_2$ (n = 3–7), a minor series of methylpolyynes (M), $H(C{\equiv}C)_nCH_3$ (n = 3–6), and a dimethypolyyne (DM), $CH_3(C{\equiv}C)_6CH_3$, are discernible. (b) UV absorption spectra of methylpolyynes (solid line) and a dimethypolyyne (dashed line). (c) Time-of-flight mass spectra (TOF-MS) of some fractions from HPLC, corresponding to M258, M282, and DM283 in (a), recorded by using GC-TOFMS (JEOL JMS-T100GCV), to provide molecular weight, m/z 136, 160, and 174, respectively. The UV spectra of methyl-$C_{12}H$ and dimethyl-C_{12} are almost identical but slightly redshifted by ~1 nm for the latter as in the middle panel of (b), while the retention time of the heavier dimethyl-C_{12} (m/z 174) is distinctly shorter than that of polar methyl-$C_{12}H$ (m/z 160), and rather close to the same centrosymmetric, non-polar $C_{12}H_2$ (m/z 146) as in (a).

High performance liquid chromatography (HPLC) is a necessary tool for separation and purification as illustrated in Fig. 2.3. Experimental procedures have been established for preparative use of HPLC to obtain a milligram order of size-separated polyyne molecules from the laser ablated mixture, *e.g.*, use of a strongly adsorbing column of polymeric octadecyl silica (polymeric-ODS), cooling of columns with an ice-cold bath for stronger adsorption, and use of concentrated samples to reduce the injection volume, all to avoid possible overlap of retention times for different molecular species. Repetitive operation using the same ODS column, namely *recycling* HPLC, and alternative operation using the ODS column and another type of columns with different adsorption properties, namely *2-step* HPLC, turned to be effective. For the final stage before a particular optical measurement, additional HPLC operation is performed to get rid of possible impurities which may affect the relevant spectroscopic measurement. Part of laser ablated polyynes ($C_{2n}H_2$), methylpolyynes (M), and a dimethylpolyyne (DM), identified so far by UV absorption and tandem operation of gas chromatography and time-of-flight mass spectrometry (GC-TOFMS) are listed in Table 2.1.

Table 2.1 Typical polyynes formed by laser ablation of carbon particles in *n*-hexane

Molecule	Category	Structure	λ_{max} (nm)	M (m/z)	Nomenclature
C_6H_2	P	$H(C{\equiv}C)_3H$	199	-	Hexatriyne
C_7H_4	M	$H(C{\equiv}C)_3CH_3$	205	-	Methylhexatriyne
C_8H_2	P	$H(C{\equiv}C)_4H$	227	98	Octatetrayne
C_9H_4	M	$H(C{\equiv}C)_4CH_3$	230	-	Methyloctatetrayne
$C_{10}H_2$	P	$H(C{\equiv}C)_5H$	252	122	Decapentayne
$C_{11}H_4$	M	$H(C{\equiv}C)_5CH_3$	258	136	Methyldecapentayne
$C_{12}H_2$	P	$H(C{\equiv}C)_6H$	276	146	Dodecahexayne
$C_{14}H_6$	DM	$CH_3(C{\equiv}C)_6CH_3$	283	174	Dimethyldodecahexayne
$C_{13}H_4$	M	$H(C{\equiv}C)_6CH_3$	282	160	Methyldodecahexayne
$C_{14}H_2$	P	$H(C{\equiv}C)_7H$	297	-	Tetradecaheptayne
$C_{15}H_4$	M	$H(C{\equiv}C)_7CH_3$	300	-	Methyltetradecaheptayne
$C_{16}H_2$	P	$H(C{\equiv}C)_8H$	316	-	Hexadecaoctayne

Molecules are categorized as P, M, and DM according to the behavior in HPLC, UV, and molecular mass in Fig. 2.3. Note that the wavelength at the UV-absorption maximum, λ_{max}, is unique for each species.

2.3 Optical Properties of Polyynes

2.3.1 Electronic States of Polyyne Molecules

The centrosymmetric linear polyyne molecule, $H(C{\equiv}C)_nH$, comprises $2n$ carbon atoms of sp-hybridization each of which is connected with two adjacent ones by σ bonds to form an unbranched carbon chain. Unhybridized two 2p orbitals remaining on each C atom which extend perpendicularly to the molecular axis overlap with the same 2p orbitals on adjacent C atoms to form two π-bonding systems orthogonal to each other (Fig. 2.4(a) left). For the π orbitals, the number of nodes for the wavefunction increases one by one as the orbital energy increases stepwise, and antisymmetric (u: *ungerade*) and symmetric (g: *gerade*) orbitals appear alternatively (Fig. 2.4(a) right). The ground state has closed electronic structure having doubly degenerate highest occupied molecular orbital (HOMO) filled with four electrons (Fig. 2.4(b) left). Because of the degeneracy of HOMO and lowest unoccupied molecular orbital (LUMO), four patterns are possible for the single-electron excitation from HOMO to LUMO, bearing three electronically excited states (Fig. 2.4(b) middle). As symmetry considerations read, the direct product of two irreducible representations for half-filled doubly degenerate orbitals of π_g and π_u includes three excited states, i.e., $\pi_g \times \pi_u = \Sigma_u^+ + \Sigma_u^- + \Delta_u$. Among the three, the Δ_u state is doubly degenerate, while the Σ_u^+ and Σ_u^- states are not degenerate. In other words, the *four* eigenstates, Σ_u^+, Σ_u^-, and Δ_u (doubly degenerate), stem from superposition of the *four* election configurations for the HOMO-LUMO excitation (Fig. 2.4(b) right). Furthermore, depending on the spin states of the two unpaired electrons, one in HOMO and the other in LUMO, there are singlet and triplet states for each of the three excited states, i.e., ${}^1\Sigma_u^+$, ${}^1\Sigma_u^-$, and ${}^1\Delta_u$ for singlets and ${}^3\Sigma_u^+$, ${}^3\Sigma_u^-$, and ${}^3\Delta_u$ for triplets.

A half of the total $8n$ valence electrons of C atoms in polyyne $C_{2n}H_2$, namely $4n$ electrons, are accommodated in molecular orbitals of the doubly degenerate π-electron system (Fig. 2.4(c)). The optically allowed transition from the ground state, $\tilde{X}\,{}^1\Sigma_g^+$, lifts the molecule to the uppermost state of ${}^1\Sigma_u^+$ among the three states stemming from the HOMO-LUMO excitation (Fig. 2.4(c)) [42]. The optical transitions to the other two states, ${}^1\Sigma_u^-$ and ${}^1\Delta_u$, which are two close-lying lowest spin-singlet excited states, are both forbidden

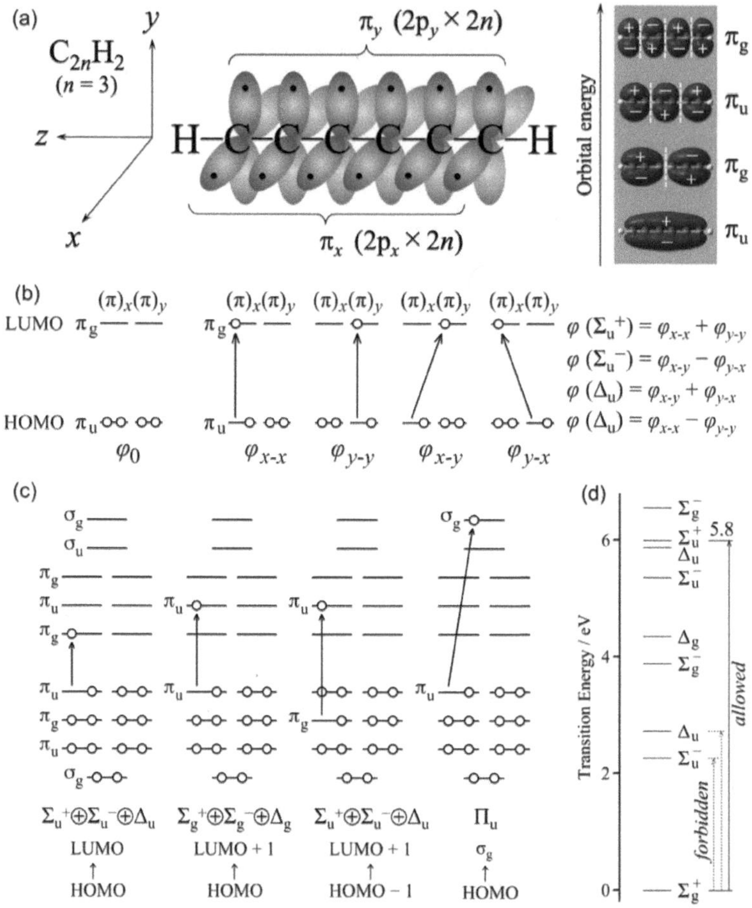

Figure 2.4 (a) Superposition of atomic $2p_x$ and $2p_y$ orbitals constituting doubly degenerate π orbitals of a polyyne molecule. (b) The electron configuration of the ground state (left) and those of singly excited states for the HOMO-LUMO excitation, i.e., four configurations (middle) and four wavefunctions of resulting electronic states (right). (c) Examples of different single-electron excitation schemes and derived electronic states for C_6H_2. (d) Electronic states and excitation energies in the spin-singlet manifold for $C_{10}H_2$ calculated by time-dependent density functional theory (TD-DFT) [25].

by symmetry (Figs. 2.4(c) and (d)) [25]. The allowed transition is observed in the UV region with the excitation energy of 4–5 eV, while the forbidden transitions become weakly allowed by vibronic interaction and are observed in the visible to near-UV (VIS–NUV) region with the transition energy of 2–4 eV, as weak as one seven hundredths of the intensity of the allowed transition in solutions of *n*-hexane [25–27, 31]. The difference of the excitation energies between the allowed (UV) and forbidden (VIS–NUV) transitions for which all the upper electronic states stem from the same electron configurations of the HOMO-LUMO excitation can be more than 2 eV. According to molecular orbital calculations for the spin-singlet single-electron excitation, i.e., time-dependent density functional theory (TD-DFT) or configuration interaction singles (CIS), several electronic states stemming from electron configurations other than the HOMO-LUMO excitation lie in-between the uppermost ${}^1\Sigma_u^+$ and the lowest ${}^1\Sigma_u^-$ and ${}^1\Delta_u$ of the HOMO-LUMO excitation, for all of which the optical transitions are forbidden by symmetry reasons (Figs. 2.4(c) and (d)).

The spin-triplet state, $\tilde{a}\,{}^3\Sigma_u^+$, lies slightly below the lowest spin-singlet excited states of ${}^1\Sigma_u^-$ and ${}^1\Delta_u$ (Fig. 2.1(d)) [42]. The optical transition down to the ground state is spin-forbidden, but observable at low temperature by phosphorescence with lifetimes from some milliseconds to a few tens of milliseconds of the photon energies corresponding to the near-IR to visible (NIR–VIS) region with the transition energy of 1–2 eV.

2.3.2 UV Absorption Spectra

Polyynes are detectable in solutions by UV absorption spectra [8, 9]. In some solvents transparent in the UV such as hexane, acetonitrile, and methanol, polyynes show strong UV absorption with molar absorption coefficients as large as 10^5 L mol^{-1} cm^{-1} at the absorption maximum [8], and thus the concentration as dilute as 10^{-6} mol L^{-1} is sufficient for the observation. The wavelength at the absorption maximum for the series of polyynes, $C_{2n}H_2$, tends to shift toward longer wavelength proportionally to the molecular size, $2n$, thus the size of the π-electron system is identified from the wavelength at the peak maximum (Fig. 2.5(a) and Table 2.1). A series of bands appearing next to the absorption maximum toward shorter wavelength with increments of ~2000 cm^{-1} are associated with vibrational progression of the symmetric stretching vibrational mode excited upon the HOMO-LUMO transition. The relative intensity of peaks in the vibrational progression is explainable by the Franck-

Condon factor between the upper and lower states of the electronic transition. Formation of laser-ablated polyynic derivatives, such as methylpolyynes, H(C≡C)$_n$CH$_3$ (Fig. 2.1(c)), whose symmetry is lower than hydrogen-capped polyynes, H(C≡C)$_n$H (Fig. 2.1(a)), is confirmed by distorted UV absorption spectra (Fig. 2.3(b)) but with more or less the same wavelength at the absorption maximum as the one for the species having a common length of the conjugate π-electron system (see Table 2.1) [36].

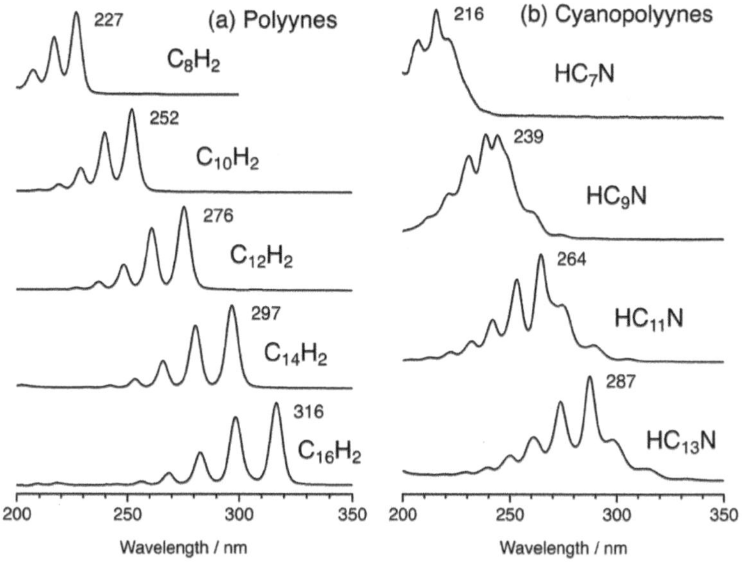

Figure 2.5 UV absorption spectra of (a) polyynes, $C_{2n}H_2$ (n = 4–8), in n-hexane and (b) cyanopolyynes, $HC_{2n+1}N$ (n = 3–6), in acetonitrile. The absorption maximum, λ_{max}, is indicated by wavelength in nm.

Cyanopolyynes, H(C≡C)$_n$C≡N (Fig. 2.1(b)), show absorption bands towards shorter and longer wavelengths of the absorption maximum (Fig. 2.5(b)). The latter features are intrinsically forbidden but weakly allowed through the mechanism of intensity borrowing. Due to the character of the electronic transition, namely configuration mixing of the multiple single-electron excitations, the spectral features are not simply accounted for by the combination of harmonic vibrational frequencies [24, 32]. The intensity of bands in this vibronically allowed transition decreases as the transition energy is separated from the peak maximum of the allowed transition from which the intensity is borrowed. In addition, the 0–0 band without any vibrational activation or deactivation is completely forbidden, showing no spectral signature. Therefore, it is difficult to identify the origin band of the electronic transition locating at the lowest-energy edge in the spectrum [25, 43].

2.3.3 Raman Spectra

Electronic transition which activates Raman scattering of polyyne molecules is the allowed transition in the UV, ${}^1\Sigma_u^+ \leftarrow \tilde{X}\,{}^1\Sigma_g^+$, thus the same vibrational modes are observed both in normal Raman scattering (or non-resonant Raman scattering) where visible light is used for the excitation and in resonance Raman scattering where UV light is used for the excitation (Fig. 2.6(a)) [26–28]. As illustrated from molecular orbital calculations on $C_{14}H_2$ (Fig. 2.6(b)), among the symmetric stretching σ_g modes, the mode having alternative displacement directions for adjacent carbon atoms, i.e., all the triple bonds, C≡C, expand and all the single bond, C–C, shrink in phase, is strongly excited (Fig. 2.6(a)).

In addition to this characteristic mode showing a systematic low-frequency shift with increasing molecular size, signals are observed for some other symmetric stretching σ_g modes as finger prints for identification of the molecular species (Fig. 2.6(a)). In case that the sum frequency of an overtone or a combination of lower frequency vibrational modes lies close to the frequency of the Raman-active fundamental mode, a pair of peaks split by Fermi resonance can be seen, *e.g.*, as observed for the Raman scattering signal of ν_2 mode of $C_{10}H_2$ (Fig. 2.6(a)) [27].

When the excitation-laser frequency is set close to the excitation energy of the allowed transition, the Raman scattering intensity increases by the resonance effect and the vibrational progression becomes conspicuous for overtones, $n \geq 2$, of the same symmetric stretching mode by the mechanism of stimulated Raman scattering (SRS) (Fig. 2.7). Concerning the vibration of polar C–H bonds, the antisymmetric stretching vibrational mode, σ_u at 3305 cm^{-1}, for which the two H atoms at opposite ends of the carbon chain have displacements of the same direction, is a major spectral feature in the IR absorption spectrum, whereas the Raman-active symmetric stretching vibrational mode, σ_g, for which the two H atoms at opposite ends of the carbon chain have displacements of opposite directions, is orders of magnitude weaker than the Raman signal of stretching vibrations of the carbon chain in the Raman spectrum. Polyyne molecules adsorbed on the surface of silver particles are observed by intensified signals of surface enhanced Raman scattering (SERS) [28–30].

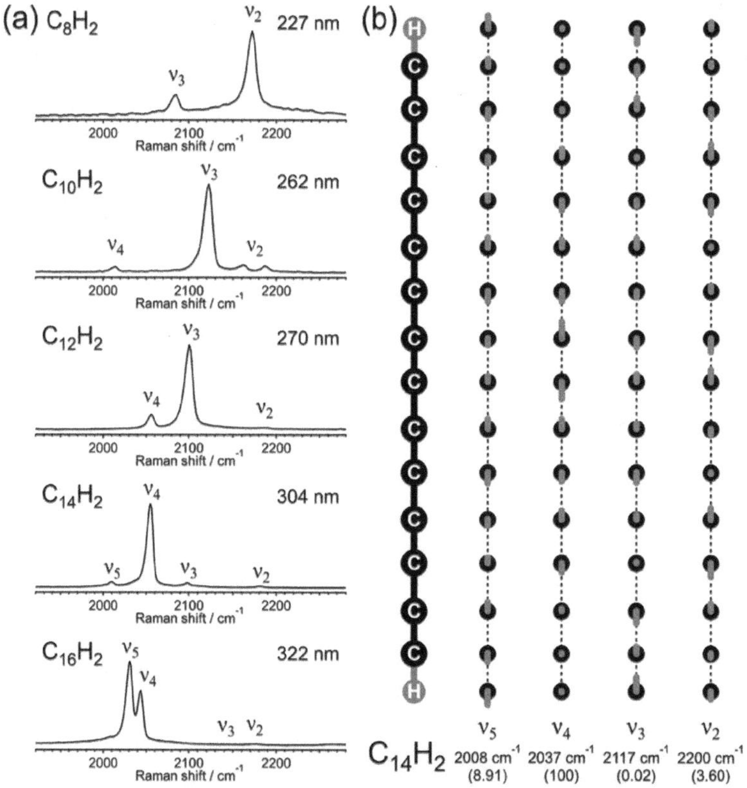

Figure 2.6 (a) UV-resonance Raman spectra of polyyne molecules, $C_{2n}H_2$ (n = 4–8), in n-hexane. The excitation wavelength in nm is indicated in each panel. Peaks indicated are the symmetric stretching $\nu_n\ \sigma_g$ modes of the linear carbon chain. With the increasing molecular size, the peak of the most intense mode shifts to lower frequency and the number of vibrational modes increases. The Raman-active CH-stretching $\nu_1\ \sigma_g$ mode has a higher frequency (outside the range), whereas its scattering intensity is fairly low. (b) Displacements of atoms for the four symmetric stretching $\nu_n\ \sigma_g$ modes (n = 2–5) of $C_{14}H_2$ with harmonic frequencies scaled by 0.9575 for anharmonicity correction and relative Raman activities in parenthesis calculated at the B3LYP/6-31G(d,p) level of density functional theory (DFT) [15].

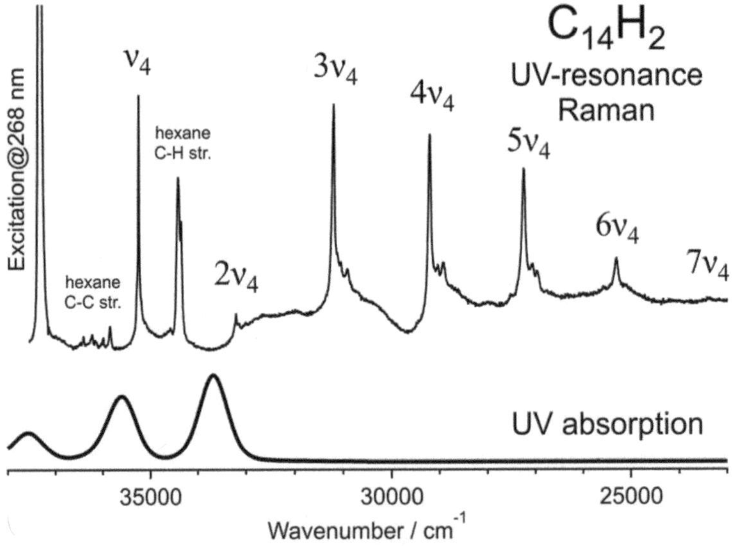

Figure 2.7 UV-resonance Raman spectrum (top) and UV absorption spectrum (bottom) of polyyne, $C_{14}H_2$, in *n*-hexane. The vibrational progression up to the 7th overtone of the ν_4 σ_g mode is conspicuous by the deep-UV excitation at 268 nm. The intensity is substantially diminished for the fundamental (ν_4) and the first overtone ($2\nu_4$) due to self-absorption by the $C_{14}H_2$ molecule [15].

2.3.4 IR Absorption Spectra

IR absorption and Raman scattering are different in their selection rules, thus provide complementary information on molecular vibrational modes. For centrosymmetric linear polyyne molecules, namely $H(C{\equiv}C)_nH$, having the center of symmetry or inversion center, *i*, there is no mode which is active both for the IR and Raman transitions, thus the signal of a vibrational mode appears either in the IR spectrum or in the Raman spectrum. Since the IR absorption is active for a vibrational mode for which the electric dipole moment of the molecule changes upon displacements of the nuclei, signals of stretching and bending modes of polar CH bonds appear strongly, i.e., the antisymmetric stretching σ_u mode at 3305 cm^{-1}, the antisymmetric bending π_u mode at 625 cm^{-1}, and the combination band of $\pi_g + \pi_u$ at 1237 cm^{-1} [35], whereas the IR absorption of the antisymmetric stretching σ_u modes of the carbon chain in 2150–2200 cm^{-1} is relatively weak [44].

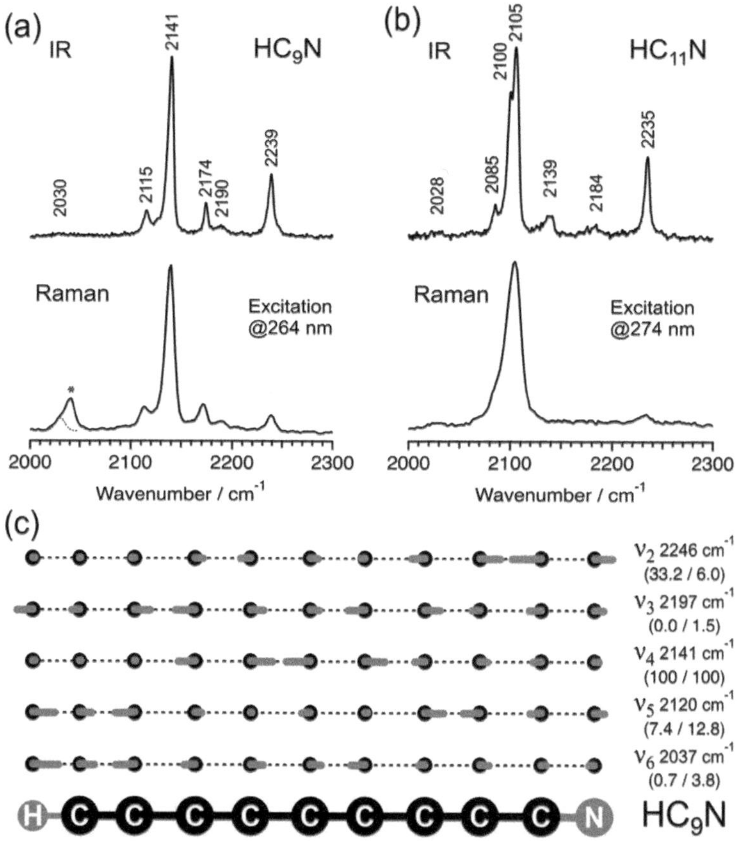

Figure 2.8 IR absorption spectra in carbon tetrachloride (upper trace) and UV-resonance Raman spectra in *n*-hexane (lower trace) of cyanopolyynes, (a) HC_9N and (b) $HC_{11}N$. The peak with an asterisk is due to impurity or fragment molecules produced upon irradiation of the UV-laser pulse [32]. (c) Displacements of atoms for the five stretching ν_n σ modes (n = 2–6) of HC_9N with harmonic frequencies scaled by 0.9575 for anharmonicity correction and relative IR/Raman intensities in parenthesis calculated at the B3LYP/6-31G(d,p) level of density functional theory (DFT) [15].

For the asymmetric linear polyynic molecules without inversion center such as cyanopolyynes, $H(C{\equiv}C)_nC{\equiv}N$, all the vibrational modes, σ and π, are both IR active and Raman active, and their relative intensities are compared in Fig. 2.8 [32]. For HC_9N, the stretching vibration of the polar CN bond, namely ν_2 σ at 2239 cm^{-1}, is observed strongly in the IR absorption, while it is relatively weak in the Raman scattering as in Figs. 2.8(a) and (c). The same tendency applies also to the CN stretching vibration of $HC_{11}N$, namely ν_2 σ at 2235 cm^{-1} in Fig. 2.8(b).

There is a tradeoff between experimental conditions for the IR and Raman spectral measurements. For IR spectra, a commercially available FTIR spectrometer is applicable, but a concentrated solution containing at least a few milligrams of size-separated polyyne molecules in a small volume of carbon tetrachloride, CCl_4, should be carefully prepared [32, 35]. For UV-resonance Raman spectra, on the other hand, a dilute solution containing only some tens of micrograms of purified polyyne molecules in *n*-hexane or acetonitrile is sufficient, *e.g.*, a solution showing 0.6–1.2 absorbance units at the excitation wavelength, but a monochromatic light source or a tunable laser system operated in the UV wavelength range is required in addition to a polychromator equipped with a CCD array detector [23, 26, 27].

2.3.5 Fluorescence Spectra

Among the three electronically excited states stemming from the single-electron excitation between HOMO and LUMO, two excited states of $^1\Sigma_u^-$ and $^1\Delta_u$ lie closely as the first excited states with low excitation energies within the spin-singlet manifold [42]. Following UV excitation to the $^1\Sigma_u^+$ state *via* the allowed transition, the electronically excited polyyne molecule is deactivated through internal conversion (IC) and vibrational relaxation *via* internal vibrational energy redistribution (IVR) to the lowest vibrational level in the $^1\Sigma_u^-$ and $^1\Delta_u$ states from which the molecule is deactivated further down to the ground state, $\tilde{X}\,^1\Sigma_g^+$, by emitting a photon as fluorescence as depicted in Fig. 2.1(d). The fluorescence exhibits a broad-band emission spectrum with vibrational structures spanning a whole range of visible wavelengths as shown in Fig. 2.9(a) corresponding to photon energies of 1.5–4 eV [31].

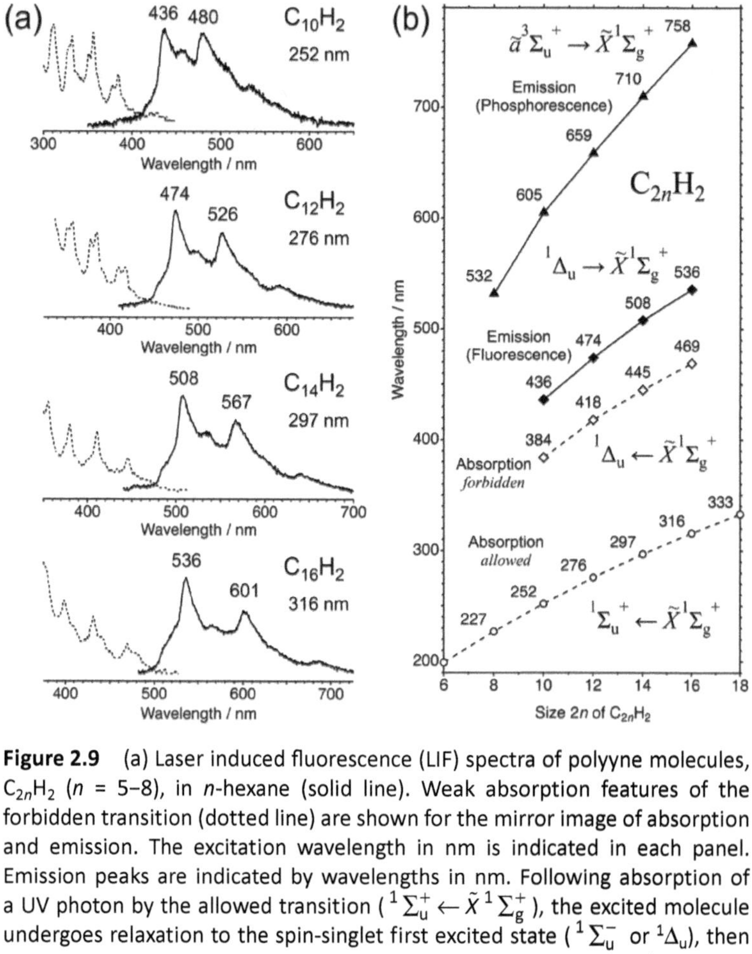

Figure 2.9 (a) Laser induced fluorescence (LIF) spectra of polyyne molecules, $C_{2n}H_2$ (n = 5–8), in n-hexane (solid line). Weak absorption features of the forbidden transition (dotted line) are shown for the mirror image of absorption and emission. The excitation wavelength in nm is indicated in each panel. Emission peaks are indicated by wavelengths in nm. Following absorption of a UV photon by the allowed transition (${}^1\Sigma_u^+ \leftarrow \tilde{X}\,{}^1\Sigma_g^+$), the excited molecule undergoes relaxation to the spin-singlet first excited state (${}^1\Sigma_u^-$ or ${}^1\Delta_u$), then emits a visible photon (400–700 nm) down to the ground state ($\tilde{X}\,{}^1\Sigma_g^+$) [31]. (b) Plots of the observed peak wavelength for optical transitions for the series of $C_{2n}H_2$ polyyne molecules as a function of molecular size, $2n$ [15].

For all the electronic transitions of polyynes, i.e., the allowed transition observed by UV absorption, the forbidden transition activated by vibronic interaction, the laser induced fluorescence (LIF), and phosphorescence as noted in the next section, a common tendency of the increasing transition wavelength or the decreasing HOMO-LUMO gap is confirmed for the increasing length of the π-electron system or the molecular size as plotted in Fig. 2.9(b), which is rationalized by a simple particle-in-a-box model in quantum physics.

2.3.6 Phosphorescence Spectra

Phosphorescence is observed when polyyne molecules are isolated in a solid matrix at cryogenic temperature as depicted in Fig. 2.1(d). In our laboratory, a solution containing size-separated polyyne molecules is sprayed onto a cold copper substrate under vacuum for condensation to form a solid matrix sample [32, 33]. Figure 2.10(a) shows dispersed phosphorescence spectra of polyyne molecules, C_8H_2, $C_{10}H_2$, and $C_{12}H_2$, trapped in solid hexane at 20 K [33]. The excitation wavelength is set close to the absorption maximum of the molecules trapped in the solid matrix hosts, for which a redshift is observed compared to the band in solutions at ambient temperature. The origin band of phosphorescence, namely the 0–0 band, shifts systematically with the molecular size, i.e., 532 nm for C_8H_2, 605 nm for $C_{10}H_2$, and 659 nm for $C_{12}H_2$, respectively, whereas the conspicuous vibrational progression of ~2100 cm^{-1} for the stretching σ_g mode is observed commonly. The lifetime of the phosphorescence decay decreases from 31.3 ms for C_8H_2, 8.3 ms for $C_{10}H_2$, to 3.9 ms for $C_{12}H_2$ as the molecular size increases.

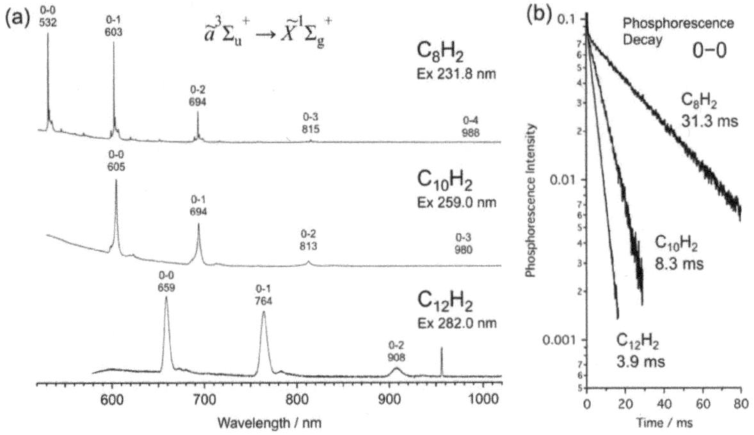

Figure 2.10 (a) Dispersed phosphorescence spectra of polyynes, $C_{2n}H_2$ (n = 4–6), in solid hexane at 20 K. (b) Phosphorescence decay of the 0–0 band peaking at 532, 605, and 659 nm for C_8H_2, $C_{10}H_2$, and $C_{12}H_2$ for showing triplet lifetimes of 31.3, 8.3, and 3.9 milliseconds, respectively [33].

A two-dimensional map is created as shown in Fig. 2.11 from the series of dispersed phosphorescence spectra recorded in the VIS–NIR range by scanning the excitation wavelength in the UV using the frequency-doubled or sum-frequency output from an optical parametric oscillation (OPO) laser system. Besides the typical advantage for low temperature matrix samples, i.e., narrowing of the band in phosphorescence excitation spectra, substantial redshifts and splitting of peaks are observed as a result of tight packing of surrounding solvent molecules which affect electronic states of the polyyne molecule. Another advantage for the measurement of phosphorescence excitation spectra, namely high sensitivity, is enjoyed to observe vibronic bands in the symmetry-forbidden transition in 300–400 nm more clearly for molecules embedded in the low temperature solid matrix host at 20 K in Fig. 2.11 [33].

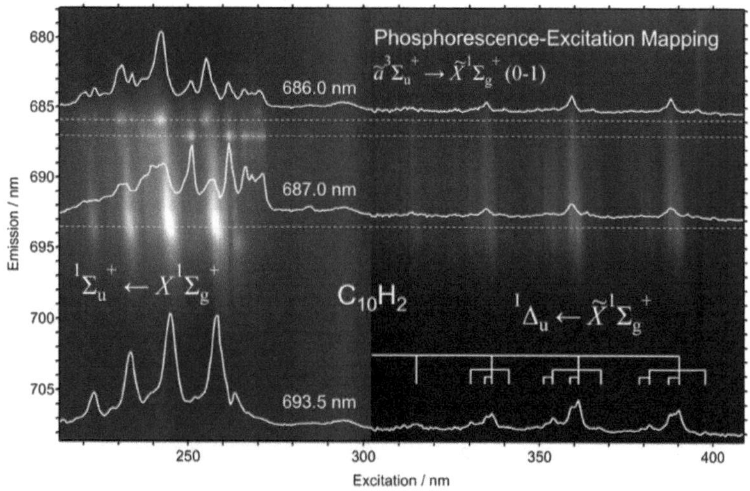

Figure 2.11 Phosphorescence-excitation mapping of polyyne, $C_{10}H_2$, in solid hexane at 20 K. The excitation spectrum for the emission at 693.5 nm (solid line spectrum at bottom) mimics the UV absorption spectrum in solutions but downshifted by ~1040 cm^{-1} (shifted from 252 nm in the solution to 258.8 nm in the solid matrix). Bands in the forbidden transition are intensified in 300–400 nm. Excitation spectra at different emission wavelengths, 687.0 nm and 686.0 nm, show distinctly different vibronic patterns in the same spectral regions of $C_{10}H_2$ [33].

2.3.7 Photoinduced Formation of a Polyyne-Iodine Complex

An experimental trial aiming at the formation of radical species having unpaired electrons in a polyyne molecule lead to the discovery of a novel molecular complex [34–36]. The reaction between polyyne and iodine molecules proceeds upon irradiation of visible light as confirmed by the change in the UV absorption spectrum in Fig. 2.12 [34]. Although the detection of radical species by electron spin resonance (ESR) was negative, the formation of a series of molecular complexes was evidenced by the analysis of systematic

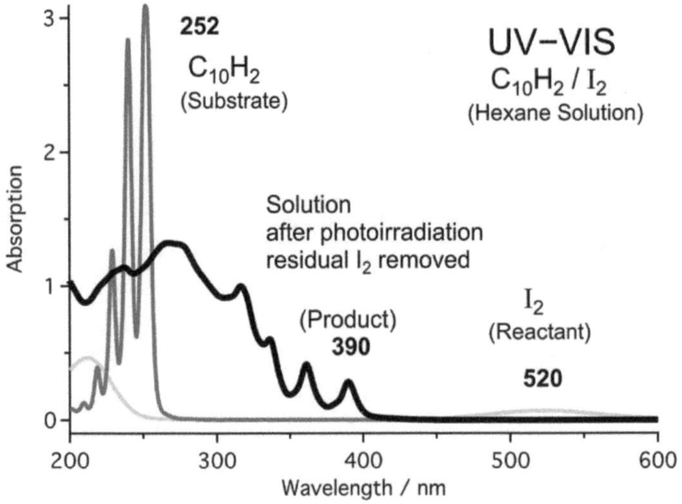

Figure 2.12 Spectral changes upon photoirradiation of polyyne, $C_{10}H_2$, with iodine molecules in *n*-hexane. Under the visible light in the presence of I_2 molecules (light gray line), absorption bands of $C_{10}H_2$ below 260 nm (gray line) diminished and new absorption features appeared in a range of 210–400 nm (thick black line) [34].

changes in the UV absorption intensity under different reaction conditions, where three I_2 molecules react with a polyyne molecule to form a stable complex of the stoichiometry, $C_{2n}H_2I_6$ (n = 5–7). In the absorption spectrum of product species, a group of bands are intensified in the near-UV region which belong to the forbidden transition of the relevant polyyne molecule. The observation of a vibrational increment of ~2000 cm^{-1} which is characteristic for the sp-hybridized carbon chain implies that the resulting complex is not a sequential adduct of three I_2 molecules to three triple bonds in a polyyne molecule but a charge-transfer complex by simultaneous formation of two counter anions of I_3^- nearby one cationic polyyne, $C_{2n}H_2^{2+}$. The formation of an analogous series of complexes using methylpolyynes, namely $HC_{2n}CH_3I_6$ (n = 5–7), was confirmed [36]. After removing molecular iodine from the hexane phase to the aqueous phase by reduction of molecular iodine to iodide, recurrent reaction was observed in the hexane phase to resume the series of UV absorption bands of precursory polyyne molecules under the presence of ambient visible light.

2.4 Polyynes in Solid Media

Polymerization is a major property of polyyne molecules composed of conjugate unsaturated chemical bonds. More than 98% of the mass of hydrogen-capped polyyne molecules, $C_{2n}H_2$ ($n \geq 4$), is carbon. Once the soot-like material is formed upon evaporation of the solvent, it cannot be dissolved again in a solvent to recover the polyyne molecule as its original form. Polyyne molecules should be reserved in dilute solutions of concentrations on the order of 10^{-3}–10^{-4} mol L^{-1}. To exploit optical properties of polyyne molecules, it is natural to utilize appropriate solid matrix hosts as dispersion media to avoid polymerization. Attempts have been made so far to disperse a small amount of polyyne molecules in some solid media.

2.4.1 Polyynes in Carbon Nanotubes

Single-wall carbon nanotubes (SWNTs) are known to stabilize various molecules trapped inside, including fullerene C_{60} molecules [45]. Polyyne molecules can also be accommodated by soaking SWNT films in a solution containing polyyne molecules [37–39]. Raman spectroscopy is suitable for the detection, *e.g.*, polyyne $C_{10}H_2$ trapped in SWNT resonates to visible light to exhibit characteristic Raman signals with comparable intensities to the Raman signal of SWNT (Fig. 2.13). In between the G band at 1590 cm^{-1} and the 2D band at 2680 cm^{-1} of SWNT, Raman bands for symmetric stretching vibrational modes of the sp-hybridized carbon chain, namely P bands, are observed [37–39]. From the observation for decrease in intensity of the P band at 2064 cm^{-1} upon heating of the sample, $C_{10}H_2$@SWNTs, detachment of $C_{10}H_2$ molecules from carbon nanotubes is evidenced [37].

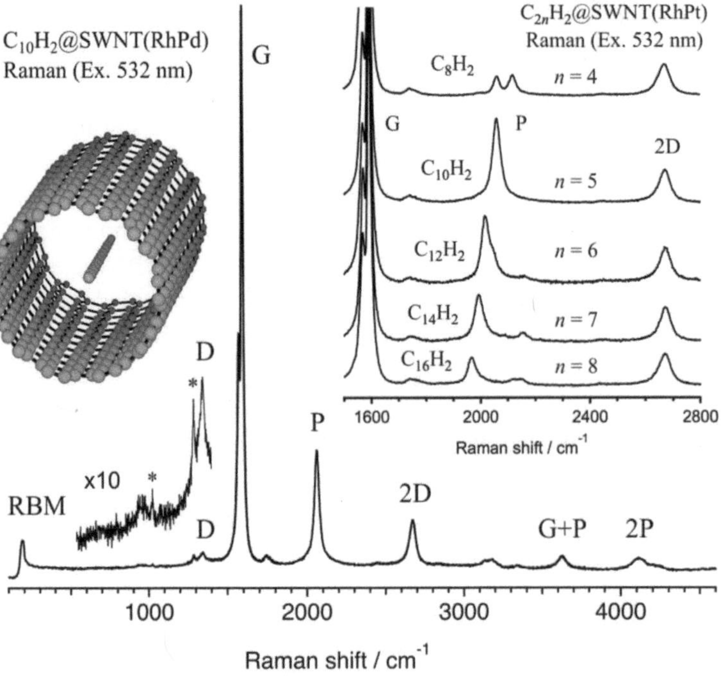

Figure 2.13 Raman spectrum of SWNTs encaging size-selected polyyne molecules, $C_{10}H_2$@SWNT(RhPd). Besides the conspicuous polyynic band, P, the combination, G + P, and the overtone, 2P, are noticeable [38]. Weaker signals with an asterisk are noted for low-frequency vibrational modes of the polyyne molecule (main panel). Polyynic bands for the series of $C_{2n}H_2$@SWNT(RhPt) (n = 4–8) downshift in frequency as the molecular size increases (inset) [39].

Accommodated in SWNTs having a wide distribution of diameters, inhomogeneous broadening of the Raman band is observed for polyynes in SWNTs, in contrast to the relatively narrow Raman band observed for homogeneously dispersed polyynes in solutions. Even alternation of the scattering intensities between nearby Raman bands is conspicuous [38]. Nonetheless, the systematic downshift of vibrational frequencies as a function of molecular size and the same number of nearby Raman bands characteristic for each polyyne molecule are observed as shown in the inset of Fig. 2.13 [39], commonly as in the Raman spectra in solutions in Fig. 2.6(a) [26]. In addition, weaker Raman signals of polyyne molecules, whose frequencies can be strongly dependent or almost independent on the molecular size, are discernible in a spectral range between the radial breathing mode (RBM) in 150–250 cm^{-1} and the D band at 1350 cm^{-1}. Therefore, SWNTs indeed offer suitable environments as spectroscopic capsules for isolating the reactive molecules.

In addition to the fundamental excitation of the normal modes of polyyne molecules and SWNTs, a combination band is observed at 3630 cm^{-1} for which the G band at 1590 cm^{-1} of SWNTs and the P band of polyynes, *e.g.*, 2064 cm^{-1} for $C_{10}H_2$, are simultaneously excited, namely G + P, and an overtone of the polyynic P band, namely 2P, is observed at 4120 cm^{-1}, as in Fig. 2.13 [38]. In considerations of the resonance effect observed for $C_{10}H_2$@SWNTs and $C_{12}H_2$@SWNTs [46], where the $C_{10}H_2$ or $C_{12}H_2$ molecule alone does not have strong absorption in the visible wavelength region, electronic coupling is implied between carbon nanotubes and polyyne molecules.

2.4.2 Crystalline Forms of Polyyne-Cyclodextrin Inclusion Compounds

Cyclodextrin (CD) is a cyclic oligomer of several D-glucose molecules. Depending on the number of D-glucose molecules, α–CD (n = 6), β–CD (n = 7), and γ–CD (n = 8) are available. The CD molecule is known for the property of inclusion as hosting another molecule in its barrel-shaped empty space inside, *e.g.*, α–CD molecules including an acetonitrile molecule for each crystalize with six water molecules for crystallization into colorless transparent crystals [47]. An attempt was made in the laboratory to obtain colorless transparent crystals growing in the concentration-gradient region between polyyne/acetonitrile and α–CD/water solutions shown in Fig. 2.14. In these crystals, a small fraction of the acetonitrile-inclusion compounds of α–CD as in the literature [47] is replaced by polyyne-inclusion compounds of α–CD upon crystallization. A typical size of the crystals was at most 2 mm in length of a cube. Some crystals were pressed to form a thin disc for the measurement of UV–VIS absorption spectra, showing characteristic absorption bands of polyyne molecules as plotted in Fig. 2.14. Compared to the spectra in solutions of acetonitrile and hexane, the peak maximum in the solid α–CD was shifted to longer wavelength and intrinsically weak absorption bands of the forbidden transition in the near-UV region were intensified significantly. The observation of the near-UV absorption bands indicates strength of crystal fields for trapped cyanopolyyne molecules.

The series of crystals for polyyne–CD inclusion compounds have been obtained for various combinations of size-separated polyyne or cyanopolyyne with cyclodextrin of α–CD or γ–CD using acetonitrile or hexane as a solvent, except for the combination with β–CD which is less-soluble in water [40]. For the measurement of the property for molecular alignment along the crystal axis, analysis of the polarized absorption was attempted using a home-built polarized-UV microscope (see Section 2.4.3), but failed for a reason that the transparent crystals in solutions gradually changed into opaque ones in the air by evaporation of the crystal water molecules for drying. For quantitative analysis of the molecular alignment in the crystal, improvements are needed both for the stability of the crystal and the *in situ* measurement of the as-grown crystal kept in solutions.

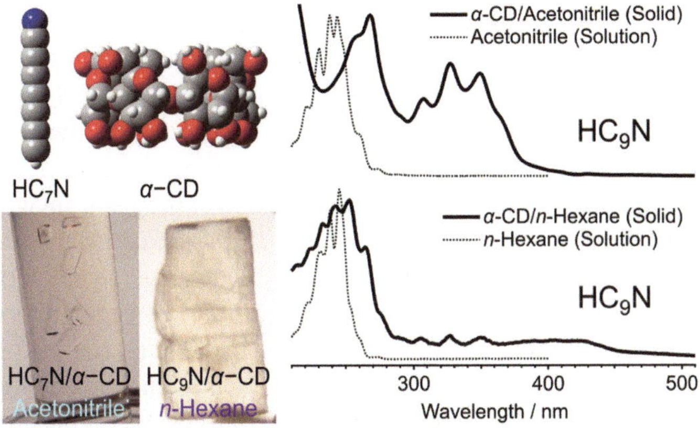

Figure 2.14 The molecular size of HC_7N is compared to that of α–CD (top left). Photos of crystals in millimeter sizes for as-grown inclusion compounds of HC_7N/α–CD/acetonitrile and HC_9N/α–CD/*n*-hexane (bottom left). UV–VIS absorption spectra of HC_9N cyanopolyyne included in solids of α–CD/acetonitrile (top right) and α–CD/*n*-hexane (bottom right). Spectra of HC_9N in relevant solvents are superposed for comparison (dotted line). Crystal fields of cyclodextrin clathrates affect electronic states of HC_9N to redshift the allowed transition in the UV and to intensify the forbidden transition in the near UV [40].

2.4.3 Aligned Polyyne Molecules in PVA Films as UV Polarizer

The transition dipole of the electronic transition, $^1\Sigma_u^+ \leftarrow \tilde{X}\,^1\Sigma_g^+$, relevant to the UV absorption bands of polyyne molecules lies in parallel with the molecular axis and the excitation is stimulated by the electromagnetic wave having the electric-field component oscillating in the direction parallel to the molecular axis. Since the molecular orientation is random in solutions and isotropic in average, no difference is noticeable in absorption intensities between vertical and horizontal polarizations of the incident light used for the measurement. Having a solid material in which polyyne molecules are dispersed and aligned in one direction, the polarization component of electromagnetic waves having the electric vector oscillating in the direction parallel to that of the aligned polyyne molecules is absorbed more strongly than the polarization component oscillating perpendicularly to the direction of the molecular alignment. As a result, when a light source of random polarization is used, transmitted light through the sample is polarized, i.e., the light with a stronger component of linearly polarized light in the direction perpendicular to the molecular alignment. Based on this property, solid solutions with aligned molecules are exploited for a device with which a linearly polarized light is created, or simply a polarizer. Here, preparation of PVA films with aligned polyyne molecules dispersed inside and characterization of their light polarizing properties are introduced [41].

An aqueous solution of PVA is transparent for the electromagnetic wave of wavelengths in between 0.2 and 1.5 μm and is dried to form solid films useful as dispersion media for fixing dispersed molecules inside. After soaking a PVA film in a solution of polyyne molecules in methanol, the softened film is stretched in one direction and kept dry for a few days keeping the length of the stretched film, so as to fix directions of molecular axes of dispersed polyyne molecules as aligned around the direction of stretching of the film. Using boric acid, H_3BO_3 or $B(OH)_3$, solvated in the methanol solution prior to soaking a film, the PVA film after soaking and drying is hardened by cross-linking covalent bonds between polymer chains of PVA assisted by the trivalent bonding of a boron atom with three hydroxyl groups for dehydration.

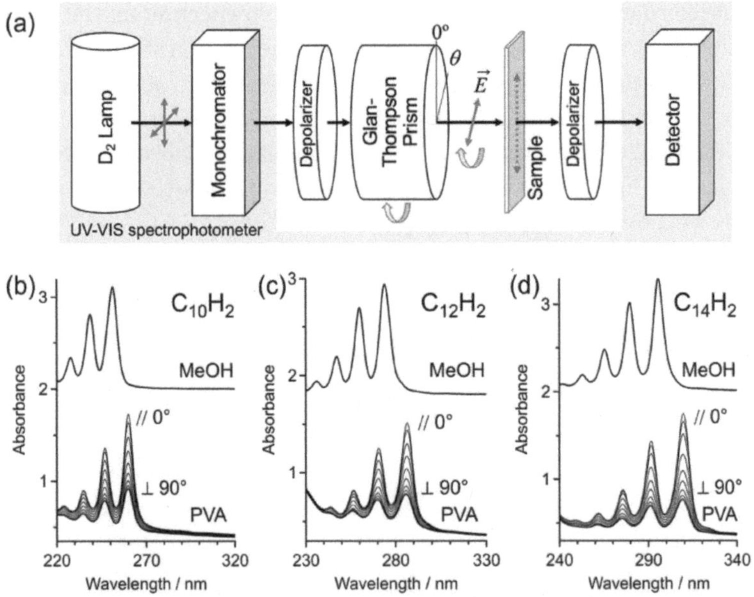

Figure 2.15 UV absorption spectra of polyyne molecules dispersed and aligned in PVA films. (a) Schematic drawing for the measurement system and UV spectra of (b) $C_{10}H_2$, (c) $C_{12}H_2$, and (d) $C_{14}H_2$, for showing the peak shift in solid PVA films (bottom) compared to the methanol solution (top). Absorption intensity is angular dependent on the direction of polarization of incident linearly polarized light with respect to the direction of stretching for the film [41].

The PVA film with aligned polyyne molecules looks colorless and transparent, but the measurement of the UV spectrum reveals the presence of polyynes in the solid solution by their absorption bands. In solid PVA, the peak of UV absorption bands of polyyne molecules redshifts compared to the peak in solutions, but the spectral pattern of vibrational progression is more or less the same as compared in Figs. 2.15(b)–(d). In order to make the incident light linearly polarized, a polarizer, namely a Glan-Thompson prism, is set in the sample room of a commercially available UV-VIS spectrophotometer (JASCO V-670), which is mounted on a rotatable holder for changing the direction of polarization with respect to the direction of stretching of the sample film as illustrated in Fig. 2.15(a). To avoid intrinsic polarization effects by optics used for conducting light from the source to the detector in the spectrophotometer, a pair of depolarizing devices are placed at the entrance and the exit of the sample room. From a series of UV absorption spectra measured by using linearly polarized incident light whose direction of polarization is systematically changed, maximum absorption intensity is observed for the incident light polarized parallel to the direction of stretching of the film, while minimum absorption intensity is observed for the polarized light perpendicular to the direction of stretching of the film as shown in Figs. 2.15(b)–(d).

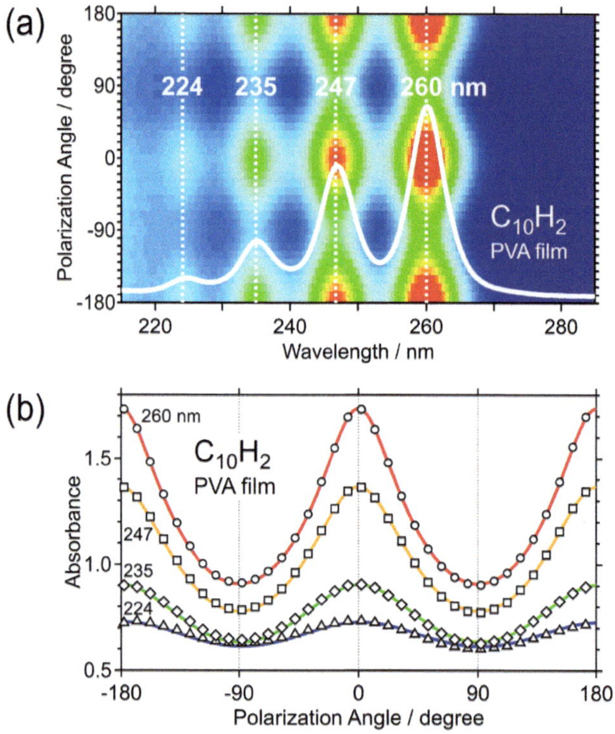

Figure 2.16 Angular dependence of the UV absorption spectra of $C_{10}H_2$ aligned in PVA. (a) 2D-map of the absorption intensity as functions of wavelength and polarization angle. (b) Angular dependence of absorbance at selected wavelengths of 224, 235, 247, and 260 nm [41]. For the fitting curve, see Section 2.4.5.

Four to five bands seen as vibrational structures in the UV absorption spectrum for polyyne molecules belong to a single electronic transition, ${}^1\Sigma_u^+ \leftarrow \tilde{X}\,{}^1\Sigma_g^+$, and the absorption spectrum changes by keeping relative intensity of these bands, thus a single absorption spectrum of linearly polarized light of a specific direction of polarization is insufficient for evaluation of the polarizing effect. The angle dependence of transmitted intensity of the linearly polarized light with respect to the direction of stretching of the film reveals that the transmittance, *T*, is trigonometric (a sine- or a cosine-function) as a function the angle between the direction of polarization of the incident light and the direction of stretching of the film. Therefore, the absorbance, *A*, as a negative logarithm of *T*, draws a characteristic curve in the angular dependence on the plane of polarization as illustrated in Fig. 2.16.

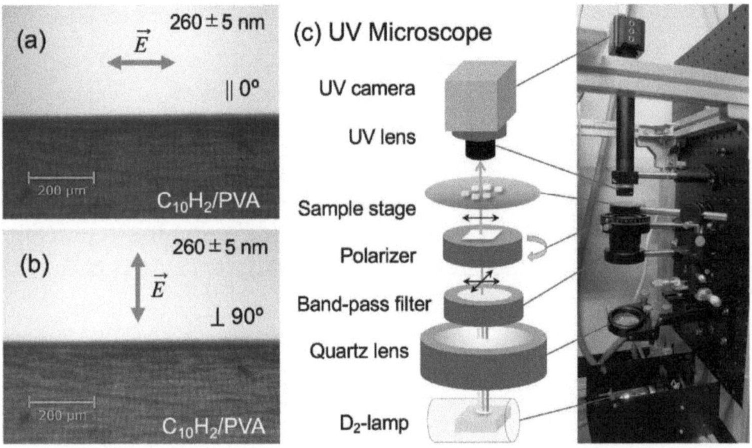

Figure 2.17 Transmission images of a stretched $C_{10}H_2$/PVA film using linearly polarized light at 260 ± 5 nm with the polarization directed (a) parallel and (b) perpendicular to the direction of stretching [41]. (c) UV-polarizing transmission microscope with a rotatable polarizer [15].

Prior to explaining the analysis of polarizing characteristics using the spectrophotometer, a visible example is shown for images taken by using a home-built low-magnitude UV microscope equipped with a source of linearly polarized UV light. The image of a PVA film containing aligned polyyne molecules looks darker when the polarization of the linearly polarized incident light is parallel to the direction of stretching of the film in Fig. 2.17(a), whereas brighter when the polarization of the incident light is perpendicular to the direction of stretching in Fig. 2.17(b). A deuterium lamp is used for the light source to create UV images. A narrow window of wavelengths is selected from the continuous output spectrum of the light source by using a bandpass filter, *e.g.*, 260 ± 5 nm. The diverging beam is paralleled through a quartz lens and conducted to the sample stage of a quartz window. The image of the sample by the transmitted UV light is focused by a camera-lens system and recorded on a CCD camera, both developed for imaging by UV light as in Fig. 2.17(c).

In the following section, the order parameter for molecular orientation is introduced based on the theory of absorption of linearly polarized light. The analysis for the angular dependence of absorption intensity of linearly polarized light is summarized.

2.4.4 Linear Dichroism

Linear dichroism (LD) is defined as the difference in absorbance when the angle of the plane for a linearly polarized incident light is rotated by 90°, i.e., for the measurement of absorbance upon rotation of the polarization angle where $A_{\parallel}$ and $A_{\perp}$ are found maximum and minimum in absorbance, respectively, the difference ΔA is called LD [48].

$$\Delta A = A_{\parallel} - A_{\perp} \equiv LD \qquad \text{(linear dichroism)} \qquad (2.1)$$

In addition to the angle-dependent absorption by aligned molecules, the angle-*independent* absorption by solvent molecules and scattering on the surface of the sample contribute to absorbance. Here, an angle-*independent* component in absorbance, A_0, is introduced, though it is not included in the conventional theory. Then, using the ratio of absorbance, D, for the aligned molecules,

$$D = \frac{A_{\parallel} - A_0}{A_{\perp} - A_0} \qquad \text{(polarization ratio)} \qquad (2.2)$$

the order parameter, S, is defined as follows [48].

$$S = \frac{D-1}{D+2} \qquad \text{(order parameter)} \qquad (2.3)$$

Since, for a sample containing trapped molecules of random orientation, no difference is expected in absorbance between parallel and perpendicular polarization components, i.e., $A_{\parallel} = A_{\perp}$, and since the molecular absorbance for the perpendicular component reaches zero for a sample of complete alignment, $A_\perp \to A_0$, the range of polarization ratio is $1 \le D < \infty$, and hence the order parameter takes $S = 0$ for the random orientation and $S = 1$ for the complete alignment, $0 \le S < 1$.

From the definitions of ΔA in (2.1) and of D in (2.2), relations are deduced between the order parameter, S, defined in (2.3) and the maximum or minimum absorbance, $A_\parallel$ or $A_\perp$, as follows.

$$S = \frac{\Delta A}{3\left(A_\parallel - A_0\right) - 2\Delta A}, \quad S = \frac{\Delta A}{3\left(A_\perp - A_0\right) + \Delta A} \tag{2.4}$$

With these relations, the maximum absorbance, $A_\parallel$, and the minimum absorbance, $A_\perp$, are found to be proportional to their difference, ΔA.

$$A_\parallel = A_0 + \frac{1+2S}{3S}\Delta A, \quad A_\perp = A_0 + \frac{1-S}{3S}\Delta A \tag{2.5}$$

Therefore, after measuring the absorbance spectra, $A_\parallel$ and $A_\perp$, and plotting one of them as a function of their difference, ΔA, the order parameter, S, is deduced from the slope of the linearly fitting line as depicted in Fig. 2.18.

Since, for a sample of random orientation of molecules, $S \to 0$, such as liquid solutions, difference is not discernible between absorbances of $A_\parallel$ and $A_\perp$, thus $\Delta A \to 0$, both expressions in (2.5) for the absorbance, $A_\parallel$ and $A_\perp$, become equal to each other as $A_0 + \Delta A/3S$. The molecular absorbance for this condition, $\Delta A/3S$, is undetermined as 0/0, but the absorbance for the sample of random molecular orientation is defined as usual to be $A = A_0 + \varepsilon d$, using the absorption coefficient, ε. This condition can be employed as a standard of molecular absorption where $d = cl$ is the column density of molecules in the sample, where c denotes concentration and l optical-path length. From comparison of these relations, an important formula is deduced.

$$\varepsilon d = \frac{\Delta A}{3S} \quad \text{or} \quad \Delta A = 3S\varepsilon d \tag{2.6}$$

The difference of absorbance, ΔA, between the maximum, $A_\parallel$, and the minimum, $A_\perp$, absorbances, is proportional both to the order parameter, S, and to the intensity of intrinsic molecular absorption,

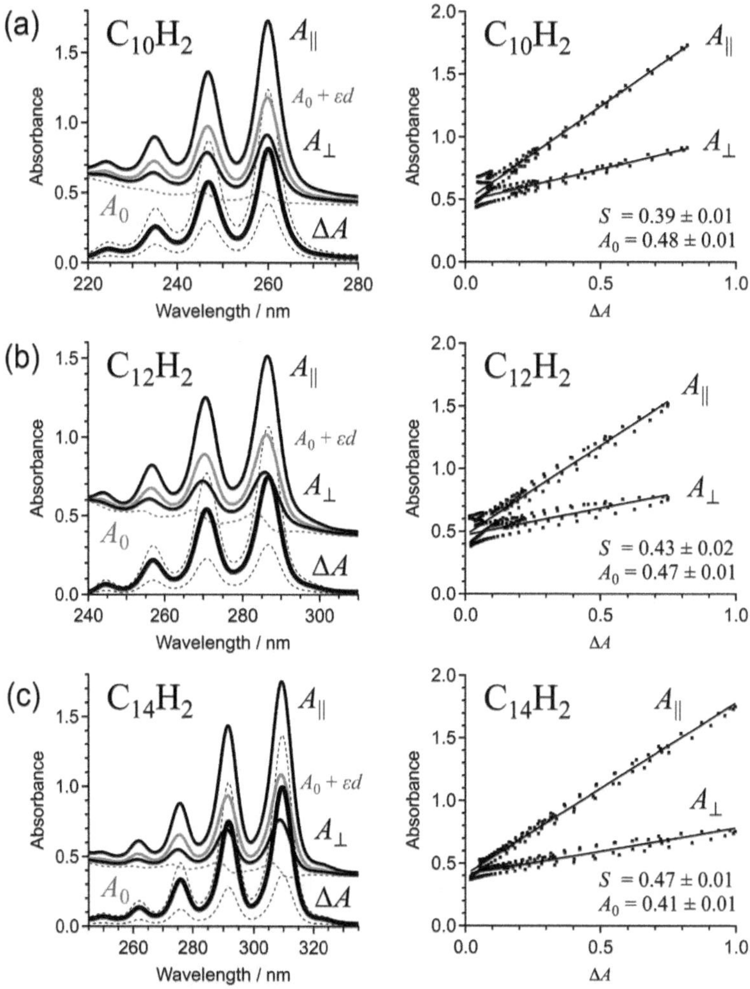

Figure 2.18 Analysis of the order parameter, S, for polyynes (a) $C_{10}H_2$, (b) $C_{12}H_2$, and (c) $C_{14}H_2$ in stretched PVA films. Observed absorbance spectra in the left panel, $A_\parallel$ and $A_\perp$, are plotted as a function of their difference, ΔA, which confirms linearity in dotted plots in the right panel. From the slope and intercept, parameters, S and A_0, are derived, respectively. A major cause for the deviation from linearity is the lack of uniformity in the baseline, A_0. The dashed line in gray in the left panel is a wavelength-dependent baseline, A_0 (λ), for which the slope, S, is fixed for the fitting line in black in the right panel. Dashed lines in the left panel are differences for $A_\parallel$ and $A_\perp$ subtracted by the wavelength-dependent baseline, A_0 (λ) [15].

εd (thus proportional also to the concentration, c, and the thickness of the sample, l, in addition to the absorption coefficient, ε, of the molecule). Using the relation (2.6), equations in (2.5) are expressed simply by the absorption coefficient, ε.

$$A_{\parallel} = A_0 + (1+2S)\varepsilon d, \quad A_{\perp} = A_0 + (1-S)\varepsilon d \tag{2.7}$$

The average weighted by 1:2 of the spectra, $A_{\parallel}$ and $A_{\perp}$, for equations in (2.7) provides a formula without the order parameter, S, as the intrinsic absorbance of the sample, $A_0 + \varepsilon d$.

$$\frac{A_{\parallel} + 2A_{\perp}}{3} = A_0 + \varepsilon d \tag{2.8}$$

This implies that the maximum absorbance spectrum, $A_{\parallel}$, lies by two thirds of ΔA higher than the absorbance spectrum for random orientation, and that the minimum absorbance spectrum, $A_{\perp}$, lies by one third of ΔA lower than the absorbance spectrum for random orientation.

$$A_{\parallel} = A_0 + \varepsilon d + \frac{2}{3}\Delta A, \quad A_{\perp} = A_0 + \varepsilon d - \frac{1}{3}\Delta A \tag{2.9}$$

For the analysis of experimental data, the order parameter, S, is deduced directly from the slope of the plot of $A_{\parallel}$ or $A_{\perp}$ as a function of ΔA using one of the equations in (2.5), which does not depend explicitly on εd. The component in absorbance with no angular dependence, A_0, is deduced from the intercept of the same plot. The intrinsic molecular absorbance, εd, is calculated from Equation (2.6), if necessary. It is also possible to deduce S and A_0 from the slope and intercept, respectively, for the plot of the weighted average of 1:2 using the following equation modified from (2.8).

$$\frac{A_{\parallel} + 2A_{\perp}}{3} = A_0 + \frac{\Delta A}{3S} \tag{2.10}$$

2.4.5 Angular Dependence of Transmittance and Absorbance for Linearly Polarized Light

It is experimentally confirmed that the angular dependence of transmittance is trigonometric as a function of the angle for the direction of linearly polarized incident light with respect to the fixed direction of the sample. For the maximum transmittance, $T_{\perp}$, (or the minimum absorbance, $A_{\perp}$) and the minimum transmittance, $T_{\parallel}$, (or the maximum absorbance, $A_{\parallel}$) successively appear by an increment of 90°, $T_{\perp} > T_{\parallel}$, the angular dependence of transmittance, T, is expressed phenomenologically by the following function as plotted in Fig. 2.19(a).

$$T = \frac{T_{\perp} + T_{\parallel}}{2} - \frac{T_{\perp} - T_{\parallel}}{2}\cos 2\theta \tag{2.11}$$

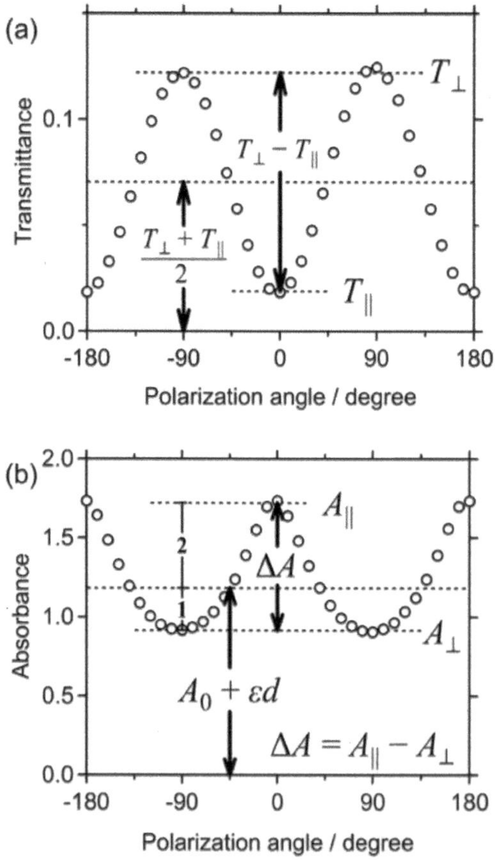

Figure 2.19 Observables and parameters for the polarization-angle dependence of absorption intensity in linear dichroism, $LD = \Delta A$. Angular dependence of (a) transmittance, $T_{\parallel}$ and $T_{\perp}$, and (b) absorbance, $A_{\parallel}$ and $A_{\perp}$. The difference, ΔA, and the absorbance for the random orientation, $A_0 + \varepsilon d$, are depicted in (b). Transmittance shows a trigonometric curve as plotted by circles in (a), while its logarithmic plot as absorbance in (b) shows a narrower peak at the angle for higher absorption [15].

Using the definition of transmittance, T, and absorbance, A, as $T = 10^{-A}$, equations for $A_{\parallel}$ and $A_{\perp}$ in (2.9) are transformed into corresponding transmittance, $T_{\parallel}$ and $T_{\perp}$, respectively, then substitute the variables in (2.11) to transform T in (2.11) into $A = -\log T$, providing the following equation for angular dependence in absorbance.

$$A = A_0 + \varepsilon d + \frac{\Delta A}{6} - \log\left[\cosh\left(\frac{\ln 10}{2}\Delta A\right) - \sinh\left(\frac{\ln 10}{2}\Delta A\right)\cos 2\theta\right] \tag{2.12}$$

When the absorption is strong and the molecular alignment is substantial, the curve of angular dependence for absorbance shows a narrow peak at the absorption maximum and a shallow bottom of valley at the absorption minimum as in Fig. 2.19(b). Numerical simulations of (2.12) with the aid of the relation, $\Delta A = 3S\varepsilon d$, afford visual understanding on the angular dependence in absorbance for the difference of the order parameter, S, associated with the molecular alignment in Fig. 2.20(a) and for the difference of the absorption intensity itself, εd, associated with the molecular concentration in Fig. 2.20(b).

Experimental plots of absorbance measured at various angles are fitted by a curve using Equation (2.12) as a function of the angle, θ. It is noted here that there are only two independent parameters, which can be set $\alpha = A_0 + \varepsilon d$ and $\beta = \Delta A$ to have the following function (2.13) for fitting.

$$A = \alpha + \frac{\beta}{6} - \log\left[\cosh\left(\frac{\ln 10}{2}\beta\right) - \sinh\left(\frac{\ln 10}{2}\beta\right)\cos 2(\theta - \theta_0)\right] \tag{2.13}$$

An offset parameter, θ_0, is included in the fitting function (2.13), which is effective for an improvement in determination of angles for the maximum and minimum absorbances. The fitting parameters, α and β, are deduced to evaluate truly maximum and minimum absorbances, $A_{\parallel}$ and $A_{\perp}$, using the following Equations (2.14).

$$A_{\parallel} = \alpha + \frac{2}{3}\beta \quad \text{and} \quad A_{\perp} = \alpha - \frac{1}{3}\beta \tag{2.14}$$

The parameters, S and A_0, should be determined simultaneously, but it is impossible to determine both of them uniquely from an angle-dependent curve of absorbance measured at a single wavelength alone, because the fitting parameters, α and β, depend on each other through εd, c.f., $\alpha = A_0 + \varepsilon d$ and $\beta = \Delta A = 3S\varepsilon d$. The solution is to use a series of data measured at multiple wavelengths, i.e., pairs of absorption maxima and minima are deduced from the fitting of angle-dependent curves for absorbance measured at multiple wavelengths, λ, then spectra, $A_{\parallel}(\lambda)$ and $A_{\perp}(\lambda)$, are reconstructed to calculate the difference spectrum, $\Delta A(\lambda)$. For the plots of $A_{\parallel}(\lambda)$ and $A_{\perp}(\lambda)$ as a function of $\Delta A(\lambda)$, the linear fitting parameters, namely the slope and the intercept, provide S and A_0, as in the case of Equation (2.5). It is also possible to plot the weighted average of $A_{\parallel}(\lambda)$ and $A_{\perp}(\lambda)$ in (2.10) to determine S and A_0.

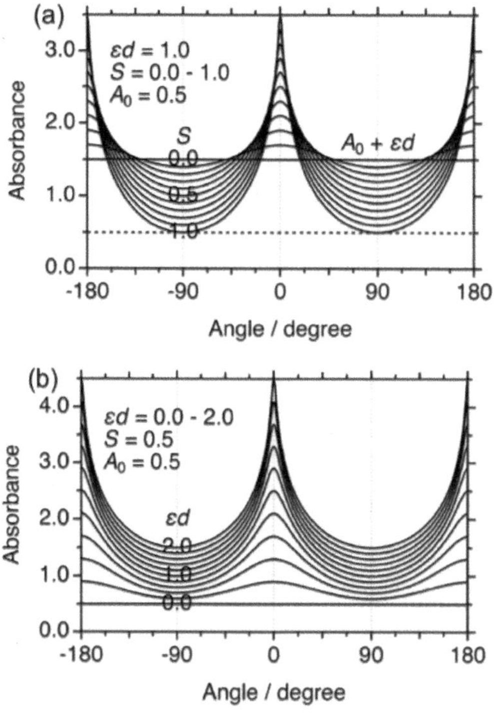

Figure 2.20 Simulations of angular dependence in absorbance for linear dichroism (LD). (a) Dependence on the order parameter, S (0.0–1.0), or the degree of molecular orientation, and (b) dependence on the molecular absorbance, εd (0.0–2.0), or the product of the absorption coefficient and the column density [15].

A gradually increasing baseline toward shorter wavelengths is observed in a typical UV absorption spectrum due to absorption by solvent molecules and scattering on the surface. Instead of attributing a constant absorbance to A_0 throughout the wavelengths in a whole range of the spectrum, improvements are possible for increasing linearity in the plot of $A_{\parallel}$ and $A_{\perp}$ as a function of ΔA by adopting a wavelength-dependent smooth background spectrum, $A_0(\lambda)$, as illustrated in Fig. 2.18. To summarize, for a sample supposed to contain uniformly dispersed and aligned molecules, the order parameter, S, is to be determined uniquely for all the polarization angles throughout the entire wavelength region of the relevant electronic transition.

2.5 Solid Aggregates of Polyynes

Long-term storage for years in a refrigerator at −20°C of a concentrated HPLC fraction containing size-separated polyyne molecules, C_8H_2 and $C_{10}H_2$, in *n*-hexane unexpectedly resulted in precipitation of ingredients as thread-like solid materials of a few millimeters in length. They looked basically white but slowly changing in colors from slightly yellowish to brownish after they were taken out of the solution into the air (Fig. 2.21). The material was soft, not cracking, and easily bent, assuring that it was not a crystal but aggregates of presumably a certain kind of carbonaceous particles from stacking molecules.

Micro-Raman spectroscopy was performed for this material to reveal two major Raman bands appearing in the sp- and sp^2-regions in Fig. 2.21. High-frequency bands peaking at 2180 and 2143 cm^{-1} are attributable to stretching vibrational modes of sp-carbon chains which are comparable to the frequencies of strong Raman signals at 2172 cm^{-1} for C_8H_2 and at 2132 cm^{-1} for $C_{10}H_2$ in solutions of *n*-hexane [28]. The lower frequency band of comparable intensity peaking at 1591 cm^{-1} is associated with the vibrational mode of trivalent sp^2-carbon. A possible mechanism responsible for the spectrum is that a cross-linking bond is formed between adjacent polyyne molecules upon condensation of C_8H_2 and $C_{10}H_2$ molecules onto the surface of a pre-formed aggregate. The cross-linking reaction proceeds by opening the polyynic triple bond, –C≡C–, into the graphitic double bond, >C=C<, introducing a trivalent 2D connection between two polyynic carbon chains. The growth of such sp-sp^2 networks does not necessarily require complete conversion of the sp-carbon moiety into the sp^2-carbon moiety [49], thus the former remains substantially in the solid material.

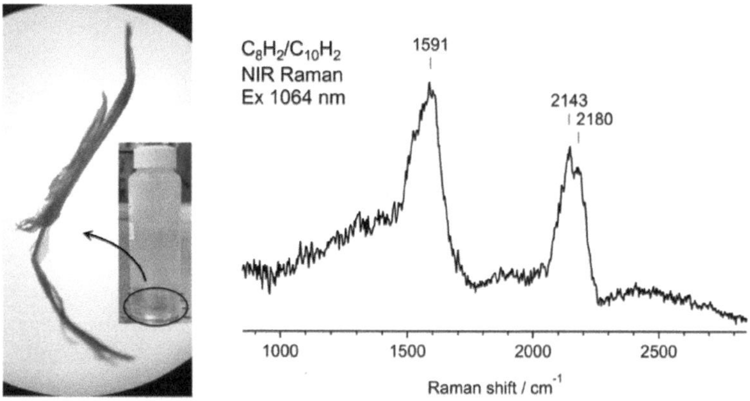

Figure 2.21 Micro-Raman spectrum of thread-like solid aggregates of polyynes, C_8H_2 and $C_{10}H_2$. The Raman band at higher frequencies, 2180 and 2143 cm^{-1}, is attributable to polyynic or acetylenic sp-carbon, while the band peaking at 1591 cm^{-1} to graphitic sp^2-carbon. Measurement performed by D. Strelnikov, S. Lebedkin, and M. M. Kappes at KIT, Germany.

The lack of intensity in-between 1850–2050 cm^{-1} in Fig. 2.21 reveals the absence of cumulenic sp-carbon chains, =C=C= [17, 50, 51]. The relatively narrow band of sp-carbon in Fig. 2.21 implies that the polyynic chains or acetylenic bonds of the precursory C_8H_2 and $C_{10}H_2$ molecules are retained in this material. The single band for sp^2-carbon having a relatively narrow bandwidth of ~150 cm^{-1} indicates a single type of cross-linking bonds. These observations suggest that the solid aggregates of polyyne molecules, C_8H_2 and $C_{10}H_2$, are chemically homogeneous materials in which sp-carbon chains or acetylenic bonds are connected in-between the cross-linking bonds of trivalent sp^2-carbon atoms. Further investigations are needed for crystalline pure condensate of a single polyyne species, desirably free from cross-linking covalent bonds. A key for finding such an ideal material can be unknown handling conditions of the precursor, untested detection techniques in the laboratory, or just searching for it spectroscopically in space. Started from cyanopolyynes in space, science of polyynes is still extending its arms toward a variety of fields, especially as model compounds of truly one-dimensional carbon allotrope of carbyne.

Acknowledgments

Alumni of the laboratory in Kindai University, Koji Okuda, Tatsuya Doi, Tetsushi Murakami, Hiroyuki Nagayama, Yoriko Wada, Mao Saikawa, Asuka Endo, Ryosuke Sata, Kaito Tanaka and all the others are acknowledged for their efforts on experimental studies. Prof. Haruo Shiromaru and Prof. Takeshi Kodama, Tokyo Metropolitan University, and Prof. Joseph Sanderson, University of Waterloo, are acknowledged for the discussion on the formation mechanism of laser ablated polyynes. The author is grateful to Prof. Kenro Hashimoto, The Open University of Japan, Prof. Tohru Sato, Kyoto University, and Prof. Miho Hatanaka, Keio University, for computational chemistry. The work on the polyyne-iodine complex was inspired by Prof. Tatsuhisa Kato, Kyoto University. Prof. Hiromichi Kataura, AIST, kindly provided us with laser ablated carbon nanotubes. Discussions on linear dichroism are acknowledged to Prof. Hal Suzuki, Kindai University. Collaborations on phosphorescence are appreciated with Dr. Urszula Szczepaniak and Prof. Robert Kołos, Polish Academy of Sciences, Prof. Claudine Crépin, CNRS, and Prof. Jean-Claude Guillemin, University of Renes. The micro-Raman measurement was performed in collaboration with Dr. Dmitry Strelnikov, Dr. Sergei Lebedkin, and Prof. Manfred M. Kappes, Karlsruhe Institute of Technology, and Prof. Wolfgang Krätschmer, Max-Planck Institut für Kernphysik.

References

1. Broten, N. W., Oka, T., Avery, L. W., MacLeod, J. M., and Kroto H. W. (1978). The detection of HC_9N in interstellar space, *Astrophys. J.*, **223**, L105–L107.
2. Kroto, H. W., Heath, J. R., O'Brien, S. C., Curl, R. F., and Smalley, R. E. (1985). C_{60}: buckmensterfullerene, *Nature*, **318**, 162–164.
3. Iijima, S. (1991). Herical microtubules of graphitic carbon, *Nature*, **354**, 56–58.
4. Novoselov, K. S., Geim, A. K., Morozov, S. V., Jiang, D., Zhang, Y., Dubonos, S. V., Grigorieva, I. V., and Firsov, A. A. (2004). Electric field effect in atomically thin carbon films, *Science*, **306**, 666–669.

5. Osawa, E. (1970). Superaromaticity, *Kagaku* (*Chemistry*), **25**, 854–863 (in Japanese).

6. Cami, J., Bernard-Salas, J., Peebles, E., and Malek, S. E. (2010). Detection of C_{60} and C_{70} in a young planetary nebula, *Science*, **329**, 1180–1182.

7. Krätschmer, W., Fostiropoulos, K., and Huffman, D. R. (1990). The infrared and ultraviolet absorption spectra of laboratory-produced carbon dust: evidence for the presence of the C_{60} molecule, *Chem. Phys. Lett.*, **170**, 167–170.

8. Eastmond, R., Johnson, T. R., and Walton, D. R. M. (1972). Silylation as a protective method for terminal alkynes in oxidative couplings, a general synthesis of the parent polyynes $H(C{\equiv}C)_nH$ (n = 4–10, 12), *Tetrahedron*, **28**, 4601–4616.

9. Kloster-Jensen, E., Haink, H.-J., and Christen, H. (1974). The electronic spectra of unsubstituted mono- to penta-acetylene in the gas phase and solutions in the range 1100 to 4000 Å, *Helv. Chim. Acta*, **57**, 1731–1744.

10. Diederich, F., Stang, P. J., and Tykwinski, R. R. (2004) *Acetylene Chemistry: Chemistry, Biology, and Material Science,* Wiley-VCH, Weinheim, Germany, DOI: 10.1002/3527605487.

11. Kaiser, K., Scriven, L. M., Schulz, F., Gawel, P., Gross, L., and Anderson, H. L. (2019). An sp-hybridized molecular carbon allotrope, cyclo[18] carbon, *Science*, **365**, 1299–1301.

12. Chalifoux, W. A., and Tykwinski, R. R. (2010). Synthesis of polyynes to model the sp-carbon allotrope carbyne, *Nat. Chem.*, **2**, 967–971.

13. Tsuji, M., Tsuji, T., Kuboyama, S., Yoon, S.-H., Korai, Y., Tsujimoto, T., Kubo, K., Mori, A., and Mochida, I. (2002). Formation of hydrogen-capped polyynes by laser ablation of graphite particles suspended in solution, *Chem. Phys. Lett.*, **355**, 101–108.

14. Cataldo, F. (2005). *Polyynes: Synthesis, Properties, and Applications,* CRC Press, Boca Raton, FL, USA, DOI: 10.1201/978142002.

15. Wakabayashi, T. (2022). Spectroscopic study of polyynes and their composite materials, *TANSO*, **301**, 18–29 (in Japanese); *Carbon*, **192**, 482 (abstract).

16. Wakabayashi, T., Ong, A.-L., Strelnikov, D., and Krätschmer, W. (2004). Flashing carbon on cold surfaces, *J. Phys. Chem. B*, **108**, 3686–3690.

17. Wakabayashi, T., Yamaguchi, Y., and Krätschmer, W. (2005). Formation of small particles by condensation of carbon chain molecules, *Bussei Kenkyu* (*Solid State Physics*), **4**, 490–513 (in Japanese).

18. Endo, H., Taguchi, Y., Matsumoto, J., Achiba, Y., Kodama, T., Shiromaru, H., and Wakabayashi, T. (2015). Anticorrelated formation of fullerenes and polyynes upon laser ablation of graphite under various concentrations of hydrogen sources, *Chem. Phys. Lett.*, **642**, 35–38.

19. Taguchi, Y., Endo, H., Abe, Y., Matsumoto, J., Wakabayashi, T., Kodama, T., Achiba, Y., and Shiromaru, H. (2015). Polyyne formation by graphite laser ablation in argon and propane mixed gases, *Carbon*, **94**, 124–128.

20. Taguchi, Y., Endo, H., Kodama, T., Achiba, Y., Shiromaru, H., Wakabayashi, T., Wales, B., and Sanderson, J. H. (2017). Polyyne formation by ns and fs laser induced breakdown in hydrocarbon gas flow, *Carbon*, **115**, 169–174.

21. Takizawa, N., Kodama, T., Shiromaru, H., Wakabayashi, T., Al-Tuairqi, S., Wang, Qi, Zhang, H., and Sanderson, J. H. (2019). Efficient polyyne formation by ns and fs laser-induced breakdown in ethylene and acetylene gas flow, *Carbon*, **152**, 372–375.

22. Ramadhan, A., Wesolowski, M., Wakabayashi, T., Shiromaru, H., Fujino, T., Kodama, T., Duley, W., and Sanderson, J. (2017). Synthesis of hydrogen- and methyl-capped long-chain polyynes by intense ultrashort laser pulse irradiation of toluene, *Carbon*, **118**, 680–685.

23. Sato, Y., Kodama, T., Shiromaru, H., Sanderson, J. H., Fujino, T., Wada, Y., Wakabayashi, T., and Achiba, Y. (2010). Synthesis of polyyne molecules from hexane by irradiation of intense femtosecond laser pulses, *Carbon*, **48**, 1673–1676.

24. Wakabayashi, T., Saikawa, M., Wada, Y., and Minematsu, T. (2012). Isotope scrambling in the formation of cyanopolyynes by laser ablation of carbon particles in liquid acetonitrile, *Carbon*, **50**, 47–56.

25. Wakabayashi, T., Wada, Y., Iwahara, N., and Sato, T. (2013). Vibronic bands in the HOMO-LUMO excitation of linear polyyne molecules, *J. Phys. Conf. Ser.*, **428**, 012004.

26. Wakabayashi, T. (2008). Optical properties of atomically size-controlled polyyne molecules, *Radiation Chemistry*, **85**, 32–38 (in Japanese).

27. Wakabayashi, T., Tabata, H., Doi, T., Nagayama, H., Okuda, K., Umeda, R., Hisaki, I., Sonoda, M., Tobe, Y., Minematsu, T., Hashimoto, K., and Hayashi, S. (2007). Resonance Raman spectra of polyyne molecules $C_{10}H_2$ and $C_{12}H_2$ in solution, *Chem. Phys. Lett.*, **433**, 296–300.

28. Tabata, H., Fujii, M., Hayashi, S., Doi, T., and Wakabayashi, T. (2006). Raman and surface-enhanced Raman scattering of a series of size-separated polyynes, *Carbon*, **44**, 3168–3176.

29. Tabata, H., Doi, T., Wakabayashi, T., and Hayashi, S. (2009). Interaction of carbon linear chains with silver island films studied by surface-enhanced Raman scattering, *J. Nanoelectron. Optoelectron.*, **4**, 220–223.

30. Hanamura, K., Fujii, M., Wakabayashi, T., and Hayashi, S. (2011). Surface-enhanced Raman scattering of size-separated polyynes (C_8H_2) adsorbed on silver colloidal nanoparticles, *Chem. Phys. Lett.*, **503**, 118–123.

31. Wakabayashi, T., Nagayama, H., Daigoku, K., Kiyooka, Y., and Hashimoto, K. (2007). Laser induced emission spectra of polyyne molecules $C_{2n}H_2$ (n = 5–8), *Chem. Phys. Lett.*, **446**, 65–70.

32. Szczepaniak, U., Ozaki, K., Tanaka, K., Ohnishi, Y., Wada, Y., Guillemin, J.-C., Crépin, C., Kołos, R., Morisawa, Y., Suzuki, H., and Wakabayashi, T. (2020). Phosphorescence excitation mapping and vibrational spectroscopy of HC_9N and $HC_{11}N$ cyanopolyynes in organic solvents, *J. Mol. Struct.*, **1214**, 128201.

33. Wakabayashi, T., Szczepaniak, U., Tanaka, K., Saito, S., Fukumoto, K., Ohnishi, R., Ozaki, K., Yamamoto, T., Suzuki, H., Guillemin, J.-C., Shiromaru, H., Kodama, T., and Hatanaka, M. (2022). Phosphorescence of hydrogen-capped linear polyyne molecules C_8H_2, $C_{10}H_2$ and $C_{12}H_2$ in solid hexane matrices at 20 K, *Photochem.*, **2**, 181–201.

34. Wada, Y., Wakabayashi, T., and Kato, T. (2011). Photoinduced reaction of hydrogen-end-capped polyynes with iodine molecules, *J. Phys. Chem. B*, **115**, 8439–8445.

35. Wada, Y., Morisawa, Y., and Wakabayashi, T. (2012). Spectroscopic characterization of a series of polyyne-iodine molecular complexes $H(C{\equiv}C)_nH(I_6)$ of n = 5–9, *Chem. Phys. Lett.*, **541**, 54–59.

36. Wada, Y., Koma, K., Ohnishi, Y., Sasaki, Y., and Wakabayashi, T. (2012). Photoinduced reaction of methylpolyynes $H(C{\equiv}C)_nCH_3$ (n = 5–7) and polyyne $H(C{\equiv}C)_nH$ with I_2 molecules, *Eur. Phys. J. D*, **66**, 322.

37. Nishide, D., Dohi, H., Wakabayashi, T., Nishibori, E., Aoyagi, S., Ishida, M., Kikuchi, S., Kitaura, R., Sugai, T., Sakata, M., and Shinohara, H. (2006). Single-wall carbon nanotubes encaging linear chain $C_{10}H_2$ polyyne molecules inside, *Chem. Phys. Lett.*, **428**, 356–360.

38. Nishide, D., Wakabayashi, T., Kataura, H., Achiba, Y., and Shinohara, H. (2007). Raman spectroscopy of size-separated linear polyyne molecules $C_{2n}H_2$ (n = 4–6) encapsulated in single-wall carbon nanotubes, *J. Phys. Chem. C*, **111**, 5178–5183.

39. Wakabayashi, T., Murakami, T., Nagayama, H., Nishide, D., Kataura, H., Achiba, Y., Tabata, H., Hayashi, S., and Shinohara, H. (2009). Raman spectral features of longer polyynes $HC_{2n}H$ (n = 4–8) in SWNTs, *Eur. Phys. J. D*, **52**, 79–82.

40. Saikawa, M. (2011). Formation mechanism of laser ablated polyynes in liquids and supramolecular crystallization, *Dissertation for Master's of Science, Graduate School of Science and Engineering, Kindai University* (in Japanese).

41. Sata, R., Suzuki, H., Ueno, N., Morisawa, Y., Hatanaka, M., and Wakabayashi, T. (2019). UV-polarizing linear polyyne molecules aligned in PVA, *Chin. J. Chem. Phys.*, **32**, 175–181.

42. Nakai, H. (2012). Discovery of chemical principles: symmetry rules for degenerate excitations, *J. Comp. Chem. Jpn.*, **11**, 1–16 (in Japanese).

43. Ding, H., Schmidt, T. W., Pino, T., Güthe, F., and Maier, J. P. (2003). Towards bulk behaviour of long hydrogenated carbon chains, *Phys. Chem. Chem. Phys.*, **5**, 4772–4775.

44. Shindo, F., Bénilan, Y., Chaquin, P., Guillemin, J.-C., Jolly, A., and Raulin, F. (2001). IR spectrum of C_8H_2: integrated band intensities and some observational implications, *J. Mol. Spectrosc.*, **210**, 191–195.

45. Pichler, T., Kuzmany, H., Kataura, H., and Achiba, Y. (2001). Metallic polymers of C_{60} inside single-walled carbon nanotubes, *Phys. Rev. Lett.*, **87**, 267401.

46. Malard, L. M., Nishide, D., Dias, L. G., Capaz, R. B., Gomes, A. P., Jorio, A., Achete, C. A., Saito, R., Achiba, Y., Shinohara, H., and Pimenta, M. A. (2007). Resonance Raman study of polyynes encapsulated in single-wall carbon nanotubes, *Phys. Rev. B*, **76**, 233412

47. Aree, T., Jacob, J., Saenger, W., and Hoier, H. (1998). Crystal structure of α-cyclodextrin-acetonitrile-hexahydrate, *Carbohydr. Res.*, **307**, 191–197.

48. Nordén, B., Rodger, A., and Dafforn, T. (2010). *Linear Dichroism and Circular Dichroism*, RSC Publishing, Cambridge, United Kingdom.

49. Yamaguchi, Y., and Wakabayashi, T. (2004). Coagulation of linear carbon molecules into nanoparticles: a molecular dynamics study, *Chem. Phys. Lett.*, **388**, 436–440.

50. Ravagnan, L., Piseri, P., Bruzzi, M., Miglio, S., Bongiorno, G., Baserga, A., Casari, C. S., Li Bassi, A., Lenardi, C., Yamaguchi, Y., Wakabayashi, T.,

Bottani, C. E., and Milani, P. (2007). Influence of cumulenic chains on the vibrational and electronic properties of *sp*-sp^2 amorphous carbon, *Phys. Rev. Lett.*, **98**, 216103.

51. Heimann, R. B., Evsykov, S. E., and Kavan, L. (1999). *Carbyne and Carbynoid Structures,* Springer.

Chapter 3

Growth of Polyynes by Laser Ablation in Solution

Pietro Marabotti and Carlo Spartaco Casari
Department of Energy, Politecnico di Milano,
Via Ponzio 34/3, 20133, Milano, Italy
pietro.marabotti@polimi.it, carlo.casari@polimi.it

3.1 Introduction

Polyynes, a class of carbon-based nanostructures consisting of alternating single and triple bonds, have raised significant attention in the scientific community due to their unique chemical and physical properties including high mechanical strength, exceptional electrical conductivity, and intriguing optical behavior. These characteristics make polyynes appealing for a wide range of applications, including nanoelectronics, photonics, sensors, and energy storage devices.

Polyynes represent carbon atomic wires (CAWs) as one finite realization of the ideal one-dimensional carbon chain, carbyne, the

Emerging Carbyne: Truly One-Dimensional Form of Carbon
Edited by Yahachi Saito

ISBN 978-981-5129-15-1 (Hardcover), 978-1-003-51036-9 (eBook)
www.jennystanford.com

lacking carbon allotrope made by all *sp*-hybridized carbon atoms. The other possible configuration is the cumulene, with all double bonds in an equalized bond structure. CAWs, display properties strongly modulated by the size (i.e., number of carbon atoms) and terminations making them appealing for both fundamental aspects related to conjugation effects and applications.

The synthesis and manipulation of polyynes pose considerable challenges due to their delicate nature and propensity for degradation under conventional processing conditions. Traditional chemical synthetic methods can produce size- and termination-selected structures but often involve harsh reaction conditions. In recent years, pulsed laser ablation (PLA) has emerged as a powerful technique for the controlled synthesis of CAWs in particular in the form of polyynes. PLA is based on high-intensity laser pulses interacting with solid targets, leading to the ejection of ablated species and subsequent nucleation. When applied to polyynes, PLA offers several distinct advantages. Firstly, it provides precise control over the synthesis process, allowing for tailored structures, sizes, and functionalities of the polyynes. Additionally, PLA can be performed in various environments, including liquids, gases, and vacuum, opening a fine-tuning of the growth conditions and properties of the synthesized polyynes. Pulsed laser ablation in liquid (PLAL) has been developed since the 90s. PLAL represents a simple, versatile, and cheap method to synthesize an expanded class of nanomaterials, finding applications in optics, medicine, polymeric nanocomposites, thin films, surface-enhanced Raman scattering (SERS) labeling, antibacterial composites, photonic devices, and additive manufacturing [1–3]. PLAL further allows the generation of metastable phase nanomaterials that cannot be produced otherwise [4–7].

PLAL refers to the process in which nanomaterials are produced by the ablation of a target immersed in a liquid solution using an intense laser beam. The target is usually a solid target, even though dispersed powders or colloidal solutions of nanoparticles can be used too. The employed solvent can essentially vary among all the existing liquid media and can have an active role in the synthesis process. The choice of the laser parameters includes the selection of the wavelength, pulse duration, energy and spot area (i.e., fluence), repetition rate, and the number of pulses or ablation time.

The interaction between the target and the highly intense laser beam (>10^9 W/cm^2) initiates the occurrence of many events starting from the generation of dense plasma from a rapid phase transition of the target material. The interaction between the melted area of the target, the plasma, and the surrounding liquid generates a heavily confined plasma region with a high density of charged species and steep pressure and temperature gradients that induce a fast energy exchange leading to the rapid vaporization of the liquid, the rapid quenching of the ablated material, and the formation and nucleation of nanomaterials. These processes occur in a few hundred nanoseconds [1, 2]. A comprehensive model that explains the role of all the parameters and exactly predicts all the phenomena that are taking place is still missing. Despite this, satisfactory control of the properties of nanomaterials produced by PLAL has been achieved during the last years, and PLAL is indeed a unique method to produce nanomaterials with specific features in many liquid solutions. PLAL meets all the requirements needed to be classified as a "green" synthesis method, including avoiding using chemicals and not necessarily producing waste [1]. By preventing the use of expensive chemicals, PLAL represents a low-cost and easy way to fabricate nanomaterials with a size dispersion below 100 nm on a gram scale per hour [8–10]. However, this significant productivity and the control on nanomaterials size and shape are still far from other synthesis methods [1, 2]. Finally, the absence of universally defined procedures leads to very different results, so each research group produces its specific nanomaterials with its own set of parameters [1].

This chapter aims to provide the recent advancements in pulsed laser ablation in liquids for the synthesis and manipulation of polyynes as a fundamental step to advance their knowledge. To understand the potential and the origin of the actual limitations of PLAL, the physical processes occurring during the formation of nanomaterials and the properties of these products will be described in the following Sections. We will discuss how pulsed laser ablation in liquids (PLAL), offers precise control over the synthesis of polyynes, allowing for tailored structures and properties. By understanding and harnessing the unique characteristics of polyynes, scientists aim to unlock their full potential for advancements in various scientific and technological domains.

3.2 Physical Synthesis Methods: An Overview

The principle of physical methods used to produce CAWs originates from the first attempt of Kroto *et al.* to reproduce the physicochemical conditions to synthesize "long-chain carbon molecules" in carbon stars [11]. Indeed, the *sp*-hybridization represents a metastable phase of carbon and needs strong out-of-equilibrium conditions to be generated. Such conditions can be reproduced in the laboratory as a hot carbon-rich plasma, confined and rapidly quenched by a liquid or gaseous environment (liquid or gas). Figure 3.1 provides the scheme of the main physical techniques employed to synthesize CAWs in the last decades. These methods can be grouped into two macro-classes: pulsed microplasma cluster source and pulsed laser deposition produce nanostructured films with a mixture of *sp* and sp^2 carbon species, while submerged arc discharge and pulsed laser ablation in liquid directly provide isolated carbon atomic wires.

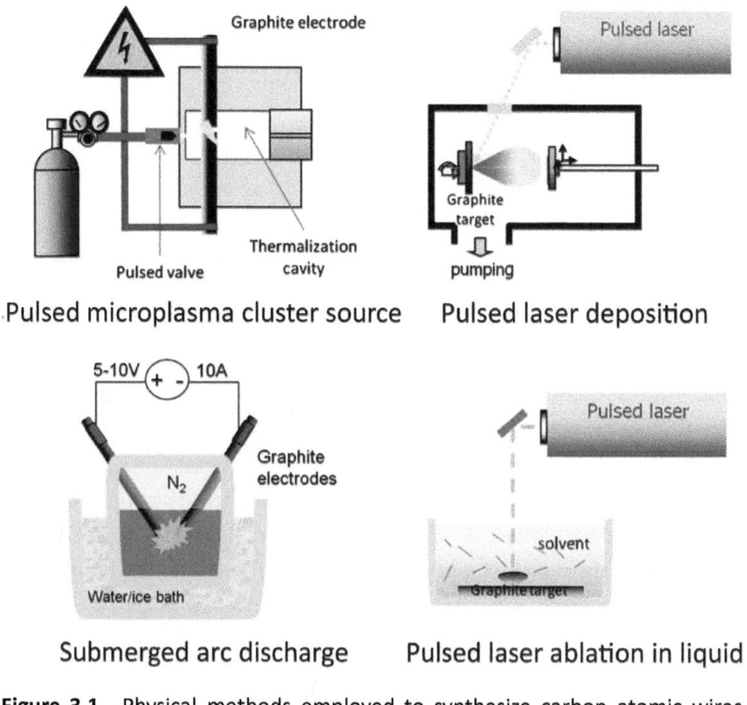

Figure 3.1 Physical methods employed to synthesize carbon atomic wires. Reproduced from Ref. [12] with permission from the Royal Society of Chemistry.

The technique known as pulsed microplasma cluster source (or supersonic cluster beam deposition) was established as a proficient method to synthesize CAWs, as described by Milani *et al.* [13]. This involves the generation of carbon clusters in the gas phase through a pulsed microplasma source. The cluster beam is then collimated and directed towards a substrate to grow nanostructured sp-sp^2 films. *In situ* Raman spectroscopy has revealed the presence of CAWs in these nanomaterials, as reported in previous studies [14–17]. In contrast, pulsed laser deposition relies on creating a plasma region through the ablation by a highly energetic pulsed laser beam of a carbonaceous target in a vacuum or inert gas-filled chamber. The ionized carbon species cool down, nucleate, and condense onto a substrate where sp-sp^2 films are formed [18].

The other class of techniques, able to produce isolated *sp*-carbon chains, includes submerged arc discharge and pulsed laser ablation in liquid. In the first method, an arc discharge is triggered when two conductive electrodes immersed in a dielectric medium are kept in close contact, while an electric potential is applied [19, 20]. The high temperatures and pressure gradients reached due to the confinement of the liquid environment generate the required out-of-equilibrium conditions in which carbon nanostructures clusterize [20]. In general, polydispersed mixtures of polyynes, the alternated form of CAWs, are obtained, together with other carbonaceous byproducts (*e.g.*, amorphous and sp^2carbon moieties, hydrocarbons, macroscopic fragments) [19, 21–28].

Similarly to submerged arc discharge, the plasma in pulsed laser ablation in liquid (PLAL) is ignited by the interaction between a high-power pulsed laser with a solid target or the liquid itself [1–3, 29]. Indeed, in the case of ultrafast laser pulses (pulse duration ≈ *fs*), the laser beam reaches peak intensities high enough to directly dissociate solvent molecules, generating a confined plasma region in the laser focus [30–32]. Otherwise, for longer laser pulses (pulse duration ≥ *ps*), the laser beam is absorbed by a solid target sank (or a powder dispersed) into a liquid medium and generates a plasma region on top of it (or around it) [33–38]. If a carbonaceous target is employed, it represents the main source of carbon for the formation of *sp*-carbon chains. The solvent plays different roles: it confines and quenches the plasma, undergoing pyrolysis and dissociation, and thus providing the small functional groups needed to terminate and stabilize the *sp*-carbon chains. Moreover, if organic solvents are employed, their dissociation may provide new carbon atoms and radicals and act as a secondary carbon source [3, 29]. The main advantage of PLAL is its enhanced flexibility, scalability, and versatility. Indeed, an extensive series of solvents can be employed ranging from water and aqueous solutions, alcohols, polycyclic aromatic hydrocarbons, alkanes, and other organic solvents [29, 32, 36, 39, 40], as well as different target materials, such as carbon-based or metallic solid targets, powders, and colloidal solutions [29, 33, 34, 39, 41]. From the laser side, the choice of pulse duration, laser wavelength, laser energies, and spot sizes, determine the synthesis yield of carbon nanomaterials and the bond hybridization of the products [3, 29, 34, 38].

Among the physical methods, we thoroughly examine pulsed laser ablation in liquid (PLAL) in this Chapter, providing an overview of the physical processes occurring during the ablation, comprising the formation of nanomaterials and, in particular, *sp*-carbon chains, the role of process parameters, solvent, target, and the different characterization techniques employed to evaluate the synthesis yield.

3.3 Physical Processes in Pulsed Laser Ablation in Liquid

In this Section, we will deepen the various phenomena occurring before, during, and after each laser pulse. Specifically, we will focus on ablation involving a laser pulse traveling through a liquid layer and hitting a solid target, and reserve the examination of other conditions (powder target, colloidal solution, or no target) in the next sections.

Before reaching the target, the laser pulse must traverse a liquid layer whose thickness may range from a few millimeters to several centimeters. This phase is crucial to accurately evaluate the deposited fluence because of the many processes that can occur: possible absorption by the liquid, defocusing effects, and, depending on the laser peak power, nonlinear optically-induced phenomena. Solvents that absorb minimally at the desired wavelength are typically chosen for PLAL experiments to maximize the energy transmitted to the target [1, 2, 29]. Most solvents, inorganic or organic, including water, are transparent—or their absorption can be considered negligible—to visible and near-infrared lasers commonly employed in PLAL experiments. Nevertheless, the presence of dissolved powders, solute molecules, nanoparticles, or ablation products may interact with the laser beam, leading to increased absorption and scattering by the liquid layer and reducing the efficiency of ablation [1, 2, 29]. The defocusing effects due to the change of refractive index at the air-liquid interface and for the liquid meniscus must be carefully estimated, especially for solvents with high surface tension and refractive index [42, 43]. In the case of *fs*-laser pulses, in particular, nonlinear optical processes, such as optical breakdown, filamentation, self-focusing, or white light generation, may occur if the electromagnetic field intensity surpasses a threshold, leading to a decrease in energy transmission to the target, and consequently, a reduction in ablation efficacy [1, 2, 44–47].

Upon the laser pulse's arrival at the target, the optical energy is absorbed by the electrons in the material, as depicted in the schematic representation in Fig. 3.2. Nonlinear processes, such as multiphoton absorption and direct photoionization, are activated in the case of ultrashort pulse durations (i.e., $<$ps) with peak intensities $\gg 10^{12}$ W/cm^2, affecting an area equivalent to the spot size [48–51]. In contrast, for longer pulses (i.e., $\geq$ ps), electron and lattice heating occurs simultaneously, as illustrated in Fig. 3.2. The electrons' kinetic energy is transferred to the lattice, leading to thermionic emission, vaporization, boiling, and melting [1, 52]. The target volume involved in these processes exceeds the spot size horizontally due to thermally induced effects and reaches the target's skin depth vertically [1, 52, 53]. The species charged by photoionization and the material heated by the electron-lattice energy transfer separate from the target through coulombic repulsion and thermal processes, respectively [48, 54, 55]. The separated material generates two supersonic shock waves (speed $\geq 10^3$ m/s) propagating toward the target and liquid, as shown in Fig. 3.2 [56, 57].

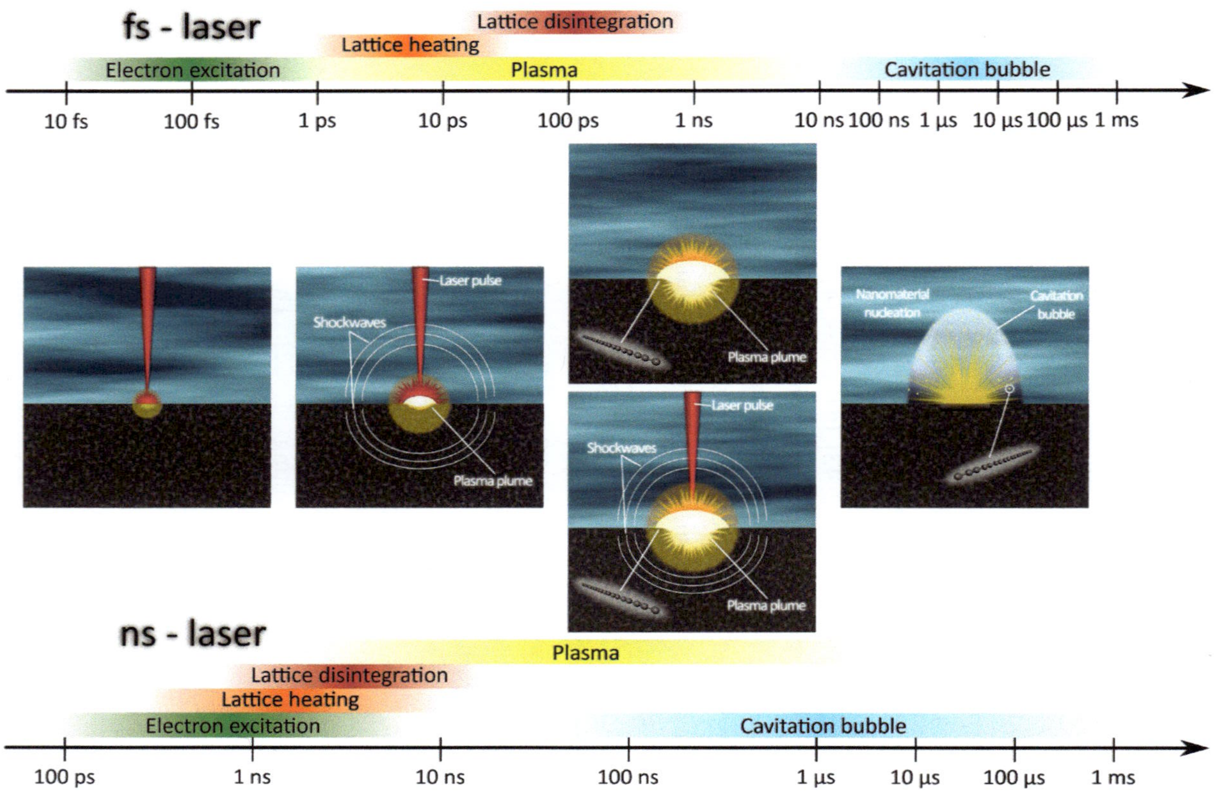

Figure 3.2 Timeline of the physical processes occurring during a PLAL experiment using a *fs*- (upper timeline) or *ns*-laser (lower timeline). Adapted with permission from Ref. [29].

The production of a strong out-of-equilibrium plasma plume results from highly ionized and hot detached materials, heavily confined by the surrounding liquid environment [29, 48, 58–60]. Numerous studies have shown that the confinement induced by the liquid layer during PLAL experiments is better than that experienced in ablations in vacuum or low-gas pressure (*e.g.*, pulsed laser deposition). Consequently, the temperature ($\geq 10^3$ K) and pressure ($\geq 10^{10}$ Pa) reached in PLAL plasmas are significantly higher than in the other cases [1–3, 58, 61, 62]. In the case of ultrashort pulses, the plasma plume lifetime reaches tens of ns due to quenching by the surrounding liquid, as pointed out in Fig. 3.2. The extreme temperature gradients at the plasma-liquid interface induce the dissociation of solvent molecules, generating a high-density region of reactive species such as radicals, ions, and molecular fragments that can chemically react with the ablated material [1, 36, 58, 63–65]. This thermal degradation is assisted by photon- and electron-induced processes due to the impact between solvent molecules and, respectively, energetic UV photons or hot electrons from emitted plasma plume [2]. When the pulse duration is comparable with the plasma lifetime, i.e., for *ns*-lasers, the plasma plume can absorb the pulse energy and extend its duration up to hundreds of *ns* [48, 66–68]. This phenomenon is referred to as "plasma shielding". Although it is still unclear whether nanomaterials start forming during this phase or in subsequent steps, it is well-established that most of the chemistry occurs during the plasma phase.

The rapid evaporation of a solvent, caused by the quenching of the plasma plume, results in the formation of a hot vapor region known as a cavitation bubble. Following ultrashort pulses, this bubble forms after 100 ns, whereas for longer pulses, it occurs after 1 μs, as shown in Fig. 3.2. The nucleation of nanomaterials or nanoparticles has been hypothesized to occur inside or in front of the cavitation bubble, as documented in several works [45, 69–72]. Thus, this phase has a great impact on the size distribution and stoichiometry of nanoparticles/nanomaterials. These products then disperse throughout the liquid layer at supersonic speeds, either alongside or propelled by the shockwaves from the cavitation bubble's collapse upon its quenching by the surrounding liquid [69, 73–75].

To prevent overlap between the next laser pulse and the processes described above, it is crucial to choose an appropriate repetition rate for the pulsed laser used. Typically, a repetition rate of 1–100 Hz is used for *ns*-pulsed lasers, while *fs*- or *ps*-pulsed lasers require a repetition rate of 1–10 kHz. However, as most PLAL setups do not circulate the liquid (see Section 3.7), the influence of previous pulses can still be felt as the resulting nanomaterials become solutes for subsequent pulses. As a result, the amount of scattered or absorbed light may increase during a PLAL experiment, while the ablation yield may decrease.

3.4 Synthesis of Nanomaterial by Pulsed Laser Ablation in Liquid

As anticipated in Section 3.3, the exact instant in the ablation process of nanomaterials' nucleation remains partially unknown. In general, nanomaterials or nanoparticles' formation depends on two players. First, highly energetic species are produced by the interaction between the laser pulse with the target. This serves as the foundation for nanomaterial growth, defining its fundamental chemistry and stoichiometry [2]. Second, reactive species form from the degradation-dissociation of solvent molecules at the plasma-liquid interface, induced by high temperatures, UV photons from the plasma, and hot electrons [2]. These reactive species interact with target atoms and clusters, altering the stoichiometry of the final nanomaterial. Solvent degradation can lead to the formation of hydroxyl radicals and hydrogen peroxide, both of which can oxidize less noble elements than Au, generating oxide layers on nanoparticle surface [2, 76]. When organic solvents are subjected to high temperatures, hot electron impact, or photoinduced reactions, they undergo degradation and release species such as C, N, H_2, and O_2 [2, 29]. These "degradation" products can induce the formation of carbon or N-doped graphitic shells around metal nanoparticles during their clustering phase [77–80]. Finally, the plasma cools rapidly upon contact with the surrounding cold liquid, with cooling rates of up to 10^{12} K/s for *fs*-PLAL and 10^{10} K/s for *ns*-PLAL during the first 200 ns.

The size distribution of nanoparticles is non-uniform, exhibiting a two-peak distribution, as cited in several publications [2, 81–83]. Primary particles with a size of less than 10 nm nucleate during the plasma phase [84–87], while secondary particles, with sizes ranging up to several tens of nm, start forming within the first nanosecond in *fs*- or *ps*-PLAL, following the instability occurring at the interface between the superheated metal layer of the growing nanoparticle and the solvent [88, 89]. Moreover, both types of particles can increase in size within the cavitation bubble, with larger particles and aggregates being favored during this phase [90–92]. Hence, the formation of larger, secondary particles does not occur *via* the coalescence of primary particles but follows a different mechanism.

Among the broad class of nanomaterials fabricated by PLAL, carbon allotropes occupy a privileged place. Indeed, the development of PLAL was driven by the goal of synthesizing exotic carbon nanomaterials such as diamond polytypes like cubic (3C) [93, 94], hexagonal [94–96], body-centered (C_8) [97], and face-centered (*n*) [98, 99] diamonds and diamondlike carbon [100]. Several authors have proposed that the formation of diamonds through PLAL is facilitated by the extreme high-pressure and high-temperature settings in the plasma phase [3, 93]. Typically, these experiments involved using *ns*-lasers and fluences ranging from tens to thousands of J/cm^2, as detailed in Table 3.1, which can generate ideal conditions for creating diamonds instead of graphitic structures.

PLAL has been used to synthesize various carbon allotropes, other than diamonds and similar nanomaterials, as indicated in Table 3.1. Fullerenes such as C60, C70, and other 0D carbon clusters were observed through the ablation of graphite microparticles [119]. Multi-walled carbon nanotubes were formed by ablating graphite in a colloidal solution of Ni or Fe nanoparticles which serves as catalysts like in chemical vapor deposition [108], and graphene exfoliation was achieved by PLAL of a graphite target in liquid nitrogen [118]. Furthermore, as discussed in the following Sections, pulsed laser ablation in liquid has emerged as a powerful technique for producing the finite realization of carbyne and, in particular, of its alternated form, i.e., polyyne [3, 12, 23, 29].

Table 3.1 Experimental conditions and carbon nanostructures synthesized by ablations of carbon targets in different liquid environments

Ref.	Target and Environment	λ [nm]	r [Hz], Duration	τ	f [J/cm^2]	Products
[93]	Pyrolytic graphite in benzene	694	Single pulses	30 ns	20	3C-diamonds of 5–20 μm
[94]	Copper substrate in benzene	308	1–5 pulses	30 ns	1–4	Film of nanodiamond with different polytypes (3C, 2H, and 6H)
[101, 102]	Tungsten foils in cyclohexane and decalin	248	5 Hz, 20 pulses	20 ns	4	Film of particles with 3C and h polytypes of diamond
[103]	Various carbon target in water	532	10 Hz 60 min	7 ns	3.5	3C and h polytypes of diamond
[104]	Single-crystal Si(100) substrate in cyclohexane	248	1 Hz, 2 pulses	23 ns	3.8, 6.6	3C diamond embedded in an amorphous graphitic carbon film
[105]	Graphite in water	532	5 Hz, 30 min	10 ns	100	2H-NP and 3C-NP of 100–300 nm
[106, 107]	Graphite in water or cyclohexane	532	10 Hz, 15–30 min	10–15 ns	28–66	3C-NP embedded in graphite
[108]	Graphite in metal nano-sol (Ni or Fe NP prepared by PLAL in ethanol)	1064	10 Hz	8 ns	36–48	Multi-walled Carbon nanotubes (5–60 nm diameter) and carbon particles (above 100 nm)
[109–111]	Suspension of graphite powder in water	1064	20 Hz, 4–5 h	1.2; 0.4 ms	5000; 1600	3C-NP of 2–13 nm with twinning structure when larger than 4 nm; 3CNP of 3–4 nm

Ref.	Target and Environment	λ [nm]	r [Hz], Duration	τ	f [J/cm^2]	Products
[112]	100–150 nm thick amorphous carbon film on Si wafer in water	532	10 Hz, 20 min	10 ns	10	C_8-NP
[113]	200 nm thick amorphous carbon film on Si wafer in bi-distilled water + ethanol + acetone + salts	532	10 Hz, 15 min	10 ns	100	Micro- and nano-cubes of C8-like structure. Continuous growth using of a vibrator
[114]	Silicon wafer covered by 700 nm thick amorphous carbon films in deionized water	532	1 Hz	10 ns	5.7	3C-NP of 4–7 nm embedded in amorphous carbon
[61]	Graphite in water	355	10 Hz	5 ns	40–1000	2H-NP in a graphitic shell
[73]	Graphite in water under variable pressure	532	10 Hz, 60 min	6 ns	12	DLC NP mixed with graphite like nanometric rods
[4]	Graphite in milli-Q water enriched with Ag NPs	800	10 Hz, 200 min; 100 Hz, 20 min; 1000 Hz, 2 min	100 fs	–	Octahedral-shaped DLC of 1–5 μm; Octahedral-shaped DLC of 1–5 μm; Agglomerates of particles of diamond embedded in DLC having a dimension below 300 nm
[115]	Carbon black suspension (200 nm) in poly(ethylene glycol)	1064	20 Hz, 2–4 h	0.8 ms	80–40000	Carbon NP of 2–6 nm and carbon nanocages of 100–200 nm
[98]	Graphite in water	193; 1064	5 Hz, 10 min	10 ns	5	(193 nm) 3C-NP; (1064 nm) 3C-NP, C_8-NP and n-NP

(Continued)

Table 3.1 (*Continued*)

Ref.	Target and Environment	λ [nm]	r [Hz], Duration	τ	f [J/cm^2]	Products
[99]	Raw detonation 3C-NP in water	532	10 Hz, 5–90 min	10 ns	25.5	n-NP
[116]	Raw detonation 3C-NP in absolute alcohol	532	10 Hz, 5–90 min	10 ns	19	Reversible phase transformations between 3C-NP and carbon onion
[117]	Coal powder in absolute alcohol	532	10 Hz, 120 min	10 ns	25.5	3C-NP (3.1 nm)
[118]	Graphite target in liquid nitrogen	1064	5 Hz, 20 min	10 ns	6	Graphene exfoliation

λ, r, τ, and f represent the laser wavelength, repetition rate, pulse duration, and fluence deposited on the target. The ablation time is also reported. NP stands for nanoparticle.

Adapted with permission from Ref. [3].

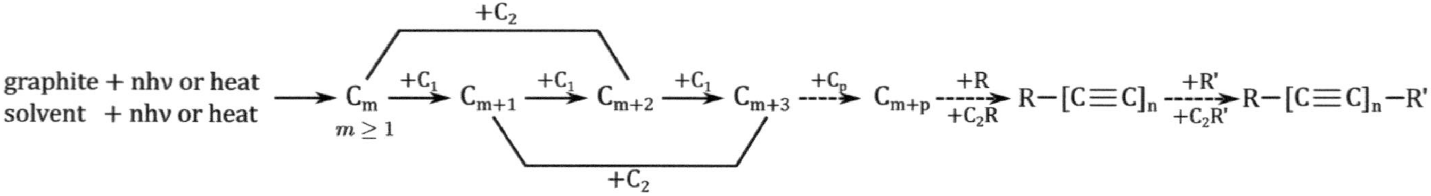

Figure 3.3 Scheme of the formation of polyynes during a PLAL experiment. A carbonaceous element (target or organic solvent) provides the primary carbon cluster (C_m, $m \geq 1$) that grows *via* polymerization reactions (addition of C_p, $p \geq 1$) and is stabilized by the termination process which can involve different small functional groups (R and R′). Reproduced with permission from Ref. [29].

3.5 Formation Mechanisms of Carbon Atomic Wires by PLAL

Among the physical synthesis methods presented in Section 3.2, pulsed laser ablation in liquid is the most used to fabricate carbon atomic wires. PLAL process parameters offer exceptional flexibility and versatility when viewed from a technical standpoint. This includes the ability to vary the target (solid target, powder suspension, colloidal solution, or direct ablation of the liquid), solvent (nearly almost all liquid media and solutions), and laser parameters (pulse duration, energy, wavelength, repetition rate, spot size, and setup geometry), as well as the conditions under which ablation takes place [1–3, 29]. Additionally, PLAL minimizes the number of ablation byproducts and enables crafting polymeric nanocomposites with *sp*-carbon chains in a single step [40].

Due to the ongoing development over the past two decades, we gained a thorough understanding of the mechanism of carbon atomic wires' formation during PLAL, as well as improved control over their production [29]. The main product of PLAL experiments is finite polyyne chains, the alternated form of carbyne, while the synthesis of cumulenic wires has never been reported [29]. For this reason, in the following, we will focus our discussion on polyynes and use terms like "chains" or "wires" referring to polyynes.

Tsuji *et al.* [33] proposed a widely accepted model to explain the growth of hydrogen-capped polyynes (H-capped polyynes, HC_nH) through nanosecond pulsed laser ablation (*ns*-PLAL). The formation of the *sp* skeleton occurs *via* polymerization reaction, in which carbon species such as C_1 atoms, bulkier carbon radicals (C_2), or larger fragments (C_p, $p > 2$) add in a stepwise process to an initial carbon cluster (C_m, $m \geq 1$). Figure 3.3 sketches this mechanism. This elongation process competes with the termination reaction, which aims to stabilize the linear *sp* character by capping the chain with small functional groups [29, 33, 36]. In hydrogen-capped polyynes, this process is referred to as hydrogenation reaction [33] and the termination process occurs with hydrogen atoms or ethynyl radicals (C_2H) provided by the degradation of solvent molecules. Carbon species required for polymerization reactions can originate from the fragmentation of a carbonaceous target or powder, generating the hot confined plasma plume [33, 34, 36, 120]. Instead, the solvent plays a two-fold role: the high temperature at the plasma-liquid interface promotes the degradation of solvent molecules into functional groups, able to terminate and stabilize the growing chain, and, in the case of organic solvents, carbon radicals, acting as a secondary carbon source for the polymerization reaction [33, 34, 36, 120]. Polyynes with small functional groups have been detected so far, as H-capped polyynes or polyynes with other terminations like C...N or CH_3 [33, 36, 121–126]. Despite the use of aromatic compounds, bulky endgroups like those chemically synthesized have never been observed in ablation conditions due to fragmentation and atomization of aromatic molecules caused by high temperatures at the plasma-liquid interface [33, 37, 39, 125, 127, 128]. Consequently, in the scenario that carbon atoms coming from fragmented solvent molecules undergo restructuring into aromatic rings, there is a likelihood that they would seek equilibrium by rearranging into more stable configurations such as sp^2 species [129].

Empirical evidence supporting the proposed model is currently lacking. The rapid timescales of nanomaterials' formation, typically on the order of hundreds of nanoseconds, in conjunction with the intricate plasma plume environment and cavitation bubble phase, hinder the possibility of detecting the growth of carbon atomic wires using existing *in situ* diagnostic techniques. Additionally, simulating by computational methods the PLAL process is a complex task, necessitating simultaneous modeling of the confined plasma dynamics, fast chemical reactions within the plasma and at the plasma-liquid interface, and the nucleation of nanomaterials under extreme out-of-equilibrium conditions [29]. These complexities complicate the prediction of which form of nanomaterials will manifest during a PLAL experiment. As discussed in Section 3.4 and documented in Table 3.1, each ablation could yield multiple varieties of nanomaterials, such as carbon quantum dots (0D) and nanodiamonds (quasi-0D), carbon atomic wires (1D), carbon nanotubes (quasi-1D), graphene sheets (2D), and macroscopic carbonaceous fragments (3D) [3, 130]. However, the optimal selection of PLAL parameters may promote the synthesis of one class of nanomaterials over the others. The solvent and target materials play crucial roles in the growth of *sp*-carbon chains by PLAL, as detailed in Sections 3.8 and 3.9, respectively. Additionally, numerous process variables impact the relationship between polymerization and termination reactions, including the laser wavelength and fluence, pulse duration, and setup geometries which will be clarified further in Section 3.7.

3.6 How to Characterize Carbon Atomic Wires' Mixtures Produced by PLAL

The successful production of polyynes through PLAL was confirmed by various spectroscopic analyses. These included UV-Vis absorption [131, 132] that was usually coupled with HPLC [33, 34, 36, 133], IR absorption [122, 132, 134, 135], laser-induced emission [136], laser-induced phosphorescence [135, 137], Raman spectroscopy [138, 139], surface-enhanced Raman scattering [36, 120, 139–143], and resonance Raman spectroscopy [30, 129, 144]. However, it has never been reported that cumulenic CAWs were synthesized by PLAL [29]. In this Section, we will particularly describe the most used and powerful approaches to characterize polyynes as UV-Vis spectroscopy, high-performance liquid chromatography, and Raman-based techniques.

3.6.1 UV-Vis Absorption Spectroscopy

UV-Vis absorption spectroscopy is a fundamental tool for examining molecules and nanomaterials' optical and electronic properties. The vibronic absorption patterns of polyynes are characterized by sharp and well-defined peaks in the absorption spectra that range from visible to low UV (below 200 nm). These peaks correspond to HOMO-LUMO transitions between specific vibrational levels of the ground and excited electronic states. Hydrogen-capped polyynes, for instance, exhibit red-shifted vibronic patterns with increasing chain length, as demonstrated in Fig. 3.4a [36, 128], due to the enhanced π-electron conjugation and the subsequent decrease of the HOMO-LUMO energy gap [145, 146]. The chain length can be assessed by comparing the UV-Vis absorption spectra of polyynes with the same termination. However, different endgroups modulate π-electron conjugation and lead to distinct absorption spectra, which can be recognized from the manifold of polyynes produced by PLAL [36, 121, 123, 125, 126]. For instance, cyano-capped polyynes demonstrate

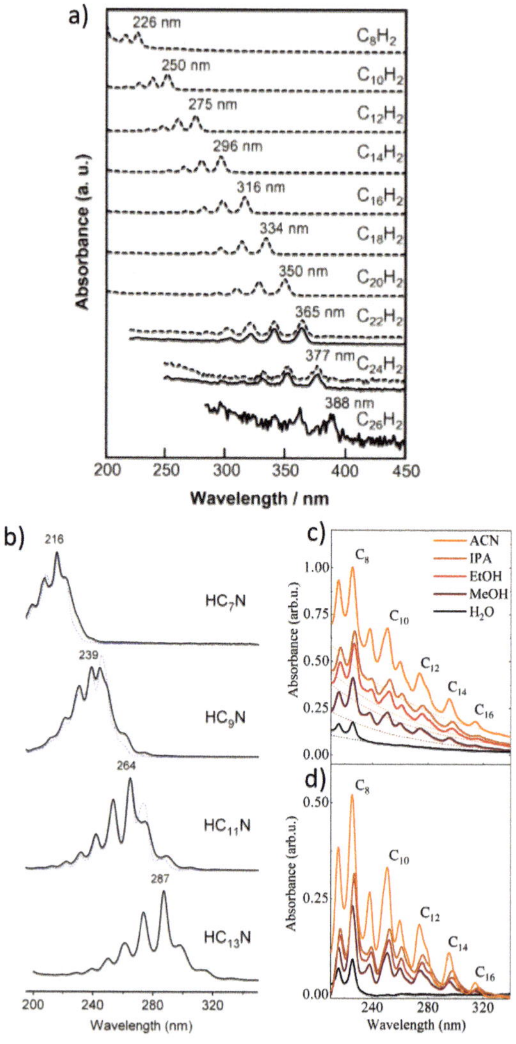

Figure 3.4 UV-Vis absorption of (a) hydrogen-capped polyynes (adapted with permission from Ref. [128]) and (b) cyano-capped polyynes (adapted with permission from Ref. [121]) as a function of their chain lengths. The spectra are extracted from HPLC analyses. UV-Vis spectra of polyynes mixtures obtained from PLAL in various solvents such as acetonitrile (ACN), isopropanol (IPA), methanol (MeOH), ethanol (EtOH), and water (H_2O) before (c) and after (d) the background was subtracted (represented by dotted lines in panel c). The most intense vibronic transition of hydrogen polyynes with 8 (C_8), 10 (C_{10}), 12 (C_{12}), 14 (C_{14}), and 16 (C_{16}) carbon atoms are labeled. Adapted with permission from Ref. [36] with the permission from the PCCP Owner Societies.

bathochromic effects in their absorption peaks (Fig. 3.4b) [36, 121]. In the case of hetero-terminations, each "bright" excited state accessible through dipole-allowed electron transitions can result in a vibronic pattern, leading to a more complex absorption spectrum [36, 121, 126].

UV-Vis spectroscopy allows for fast and direct detection of the successful fabrication of polyynes after an ablation experiment. UV-Vis spectra of polyyne mixtures produced by PLAL show a series of peaks above an absorption background, as reported in Fig. 3.4c [36]. The sharp peaks are assigned to hydrogen-capped polyynes' vibronic series that are the main products of PLAL experiments in almost all ablation solvents. The background absorbs more at shorter wavelengths, then decreases approximately as an exponential function. This feature represents the absorption of byproducts synthesized along with polyynes during the PLAL process. The byproducts range from hydrocarbons to graphene fragments, carbon nanoparticles, fullerenes, and amorphous carbon [29, 35, 36, 133, 147, 148]. By comparing the signal of polyynes *versus* that of byproducts, it is possible to define an *index of purity,* defined for the first time in this framework by Taguchi and coworkers [149]. In this way, it is possible to understand the synthesis efficiency of each solvent and set of process parameters and optimize the ablation condition to improve polyynes' synthesis yield.

3.6.2 High-Performance Liquid Chromatography

High-performance liquid chromatography (HPLC) is a widely used technique for purifying and separating solutions containing impurities or different compounds. HPLC exploits the specific interactions of size- and termination-selected polyynes with a mobile and a stationary phase (or column) to separate and thus analyze them. The sample is injected into the HPLC system and driven towards the stationary phase by the mobile phase. The time required for an analyte to leave the column, known as retention time, is influenced by factors such as the chain length, chemical affinity with the stationary phase, chemical interaction with the mobile phase, and pressure from the mobile phase flow.

Two chromatography methods have been exploited to analyze polyynes: normal-phase and reversed-phase chromatography.

Normal-phase chromatography involves the use of a polar stationary phase and non-polar solvents such as n-hexane as the mobile phase [33, 34, 37, 126, 128, 133]. On the other hand, reversed-phase chromatography utilizes hydrophobic-functionalized silica micro-particles, such as alkyl ligand chains, along with an aqueous solution containing an organic solvent like acetonitrile or methanol [36, 121, 150]. In both approaches, longer polyynes exhibit longer retention times. The termination of the chains plays a crucial role in determining their polarity and subsequent elution from the column, depending on the stationary phase used [36]. Thus, HPLC allows for the precise separation and collection of polyynes based on their chain length and termination by utilizing their specific retention times. This HPLC capability enables the isolation of individual chain species for further characterization, making it a crucial tool in the study of polyynes produced by PLAL. Moreover, by employing an approach based on batch loading, we established a method aimed at concentrating, separating, and collecting size-selected polyynes by PLAL in water [150]. This technique restores the appeal of ablation in water since, as we will discuss in the following Sections, it is one of the worst ablation environments for polyynes.

Several detection methods may be coupled with the HPLC system. Concerning polyynes, the most common ones are mass spectroscopy or UV-Vis spectroscopy. In particular, UV-Vis spectroscopy is commonly used as the detecting technique due to polyynes' distinctive absorption spectrum and its correlation with their π-electron conjugation [33, 34, 36, 37, 121, 126, 128, 137, 150]. The UV-Vis spectra presented in Figs. 3.4a and 3.4b were extracted from the UV-Vis detectors integrated into different HPLC setups [121, 128]. In HPLC, the analytical output is a chromatogram representing the detected signal intensity as a function of retention time. In the case of UV-Vis detectors, the chromatogram illustrates the absorbance at a specific wavelength versus the retention time. Each peak in the chromatogram corresponds to a compound that absorbs light at that particular wavelength and elutes from the column at a specific retention time. The concentration of the analyte can be determined by integrating the area under its chromatographic peak and comparing it to a calibration curve.

3.6.3 Raman Spectroscopy

Raman spectroscopy plays a significant role in investigating the properties of *sp*-carbon chains due to the strong connection between their vibrational and electronic properties as a result of their strong electron-phonon coupling [12, 151]. Polyynes exhibit a distinct Raman-active collective oscillation known as the ECC mode [152] or α mode [133], with a frequency range from 1800 to 2300 cm^{-1}, as reported in Fig. 3.5a. In polyynes, this mode corresponds to an in-phase C≡C stretching [12, 151]. Modifications in the chain length or terminations influence the BLA of polyynes with a direct effect on the position and shape of the ECC mode [133, 151, 153, 154]. Additionally, polyynes may exhibit other Raman-active collective modes due to their finite length or coupling with terminations [12, 151]. For example, hydrogen-capped polyynes display a weak vibration called the β mode, whose frequency and intensity vary with the chain length (see Fig. 3.5a) [133]. Beyond the α mode, polyynes may exhibit vibrational modes and marker bands specific to the functional groups present in their terminations, aiding in structural characterization [155–159].

However, the efficiency of the Raman process is poor and a large quantity of material is needed to collect a good spectrum. This requirement is usually not satisfied by polyynes' mixtures produced by PLAL: indeed, the final concentration of polyynes in solution ranges from 10^{-4} to 10^{-8} mol/L which is far from the detection limit of conventional Raman spectrometers (10^{-2} mol/L). Moreover, byproducts usually efficiently absorb Raman lasers, reducing the scattered light available at the detector. Despite these limitations, the significant Raman cross section of polyynes allowed Tabata *et al.* [133] to record non-resonance Raman spectra of size-selected hydrogen-capped polyynes (HC_nH, $n = 8$–16) separated *via* HPLC and collected in n-hexane. This result was achieved by concentrating ≈1.5 L of polyynes' mixtures synthesized by PLAL in n-hexane through several steps of separation and purifications to delete byproducts and employing a high-power Raman setup (600 mW).

3.6.4 Surface-Enhanced Raman Spectroscopy

Given the low concentration of polyynes obtained from PLAL experiments, a Raman signal enhancer is needed to perform vibrational spectroscopy. In this framework, surface-enhanced Raman scattering and resonance Raman spectroscopy emerge as the best methods to investigate low-concentrated solutions of polyynes.

Surface-enhanced Raman spectroscopy (SERS) employs metallic nanoparticles or nanostructured metal surfaces to enhance the Raman signal of a sample [160, 161]. The enhancement can reach factors of 10^{10} or 10^{11}, enabling the measurement of Raman spectra from single molecules [162]. SERS enhancement is modeled through the electromagnetic and chemical effects [160]. In the electromagnetic theory, the Raman laser is greatly enhanced when it matches the surface plasmon of the metal nanostructure, amplifying the scattered light like an antenna. Surface roughness or curvature is necessary to excite a surface plasmon. The Raman enhancement factor depends on the size of the particle and its distance from the surface. Chemical enhancement, on the other hand, involves a charge transfer mechanism between the adsorbate and metal nanoparticles. It results in the generation of new electronic states, facilitating matching resonance Raman conditions. The chemical effect enhances the Raman signal by up to a factor of 10^2 and typically occurs simultaneously with the electromagnetic SERS effect.

SERS spectroscopy is commonly used to analyze polyynes synthesized through PLAL, given the low concentrations of these samples. When a colloidal solution of metal nanoparticles (NPs) is added to polyynes, they interact chemically with the NPs, forming polyynes-NPs aggregates [141, 163]. This interaction is confirmed by a higher wavelength band in the UV-Vis absorption spectrum of the polyynes-NPs solution compared to the pristine colloidal solution. SERS spectra of polyynes exhibit distinct differences compared to their corresponding Raman spectra. An example of SERS spectra of hydrogen-capped polyynes is displayed in Fig. 3.5b. First, the ECC mode down-shifts in frequency and broadens [139–141, 164]. ECC peak's modifications are attributed to charge transfer between the nanowires and metal nanoparticles, resulting in a modified geometry. Additionally, a new band emerges at low frequencies below 2000 cm^{-1}. The origin of the low-frequency band remains a subject of debate. Some propose that interaction and charge transfer induced by metal nanoparticles lead to cumulenic-like structures with lower ECC mode frequencies [140, 141]. Alternatively, metal nanoparticle aggregates may facilitate the formation of longer wires and the subsequent downshift of their ECC mode [140, 141, 165]. Another perspective suggests that the two SERS bands arise from the broadening and downshifting of the α and β modes of polyynes due to their interaction with metal nanoparticles [139, 166]. Resonance with surface plasmons of nanostructures or nanoparticles of varying dimensions could explain the changes in the shape of the SERS bands depending on the Raman excitation employed [151]. Despite numerous studies, the interpretation of SERS spectra of polyynes remains unresolved due to the complex physicochemical environment and interactions during the SERS enhancement.

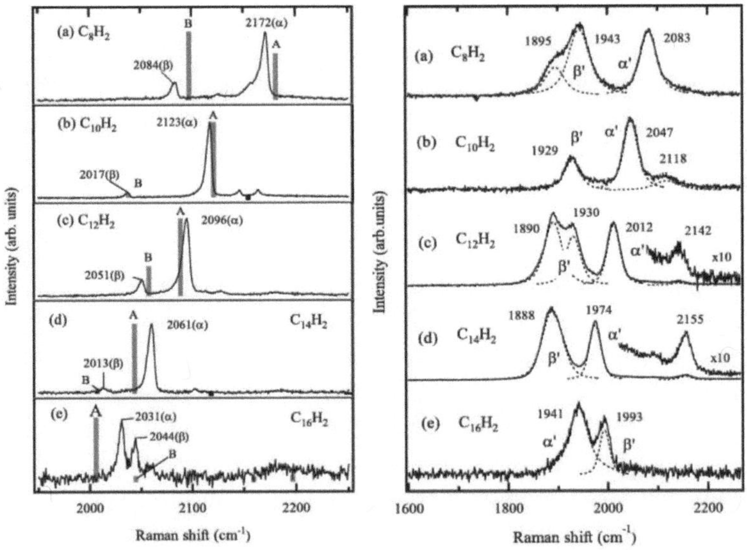

Figure 3.5 Left panel: Non-resonant Raman spectra of (a) HC_8H, (b) $HC_{10}H$, (c) $HC_{12}H$, (d) $HC_{14}H$, and (e) $HC_{16}H$ in n-hexane. Right panel: SERS spectra of (a) HC_8H, (b) $HC_{10}H$, (c) $HC_{12}H$, (d) $HC_{14}H$, and (e) $HC_{16}H$ in n-hexane with a Ag nanostructured film. Adapted with permission from Ref. [133].

The sensitivity of SERS enables us to develop an *in situ* probe to monitor the formation of polyynes by PLAL [43]. We used an innovative surface-enhanced Raman probe: a polymeric pellet functionalized with silver nanoparticles (AgNPs) put within the ablation solution during PLAL experiments. In parallel, we continuously collected SERS spectra from the pellet during the ablation. This solution does not alter the ablation environment since AgNPs remain stuck on the pellet's surface while allowing polyynes to interact with AgNPs and thus be detected during their formation. We observed an intriguing evolution of the SERS signal of polyynes and byproducts that we attributed to the simultaneous fabrication and degradation of polyynes caused by crosslink reactions during synthesis. Ex situ measurement confirmed that what we observed on the pellet effectively simulate what happens in the ablation solution. We developed a computational method based on size-selected SERS spectra of hydrogen-capped polyynes for analyzing the dynamics of long *versus* short chains. We found that shorter polyynes showed less degradation probability, explaining the final size distribution of polyynes biased toward shorter chains. By combining *in operando* analysis and *in situ* SERS spectroscopy, we gained valuable insights into the behavior and characteristics of polyynes at different stages of their synthesis, contributing to a deeper understanding of the ablation process.

3.6.5 Resonance Raman Spectroscopy

Resonance Raman spectroscopy is another method to significantly increase Raman intensity. It requires that the excitation wavelength matches an electronic absorption band of the analyte, resulting in a remarkable magnification of the scattered Raman beam [167, 168]. This technique selectively enhances vibrations with notable changes in bond length or force constant associated with the electronic transition [167]. Moreover, resonance Raman spectroscopy allows for observing selected vibrations and high-order scattering processes [168, 169]. It is particularly useful for analyzing low-concentration analytes in diluted solutions.

Resonance Raman spectroscopy has been rarely used in the examination of polyynes. This can be attributed to a technical constraint: the unavailability of finely tunable UV lasers, particularly below 350 nm, where most polyynes exhibit electronic transitions. Only a small number of studies have reported the pre-resonance or resonance analysis of short polyynes. Wakabayashi and coworkers

[136] made use of pre-resonance Raman spectroscopy to study the vibrational fingerprint of hydrogen-capped polyynes ($HC_{10}H$ and $HC_{12}H$) and the evolution of the α and β modes depending on the Raman laser wavelength (still in pre-resonance conditions). Sato *et al.* [30] collected a resonance Raman spectrum of HC_8H in n-hexane, separated from a mixture obtained by *fs*-PLAL in hexane. Resonance Raman was employed to validate the assumption of the formation of polyynes from this kind of experiment.

However, the capability of resonance Raman spectroscopy goes well beyond these examples. Indeed, we employed the resonance condition to characterize a series of low-concentrated, size-selected hydrogen-capped polyynes produced by PLAL [129]. Resonance Raman not only allowed us to detect and characterize the fundamental vibrational properties of these molecules but also to observe high-order scattering processes like overtones of the α mode up to the fifth order. These results allowed us to precisely determine the electronic and vibrational structure of the ground and excited state and discover polyynes' size-dependent electron-phonon coupling. In addition, the sensitivity and selectivity of resonance Raman scattering would be ideal for establishing an *in situ* probe to monitor the formation and degradation mechanisms of polyynes during their synthesis by PLAL.

3.7 Understanding the Importance of PLAL Parameters for Carbon Atomic Wires Fabrication

3.7.1 The Role of the Laser Wavelength and Fluence

In PLAL experiments, the fluence threshold of the target and the synthesis efficiency of nanomaterials are influenced by the laser wavelength [1–3, 130]. Carbon nanomaterials' fabrication appears to be facilitated by longer wavelengths in the near-infrared region, whereas those of shorter wavelengths in the ultraviolet range promote the creation of allotropes based on sp^3 hybridization or oxidized forms of carbon nanostructures (*e.g.*, graphene oxide) [59, 170–172].

When studying the role of laser wavelength in polyyne formation, it is important to consider two distinct cases. First, in the case of solid targets, longer wavelengths were found to maximize polyyne yield, as pointed out in Fig. 3.6a [38, 41]. Shorter wavelengths were observed to scatter more readily due to impurities and byproducts formed during the ablation, resulting in reduced synthesis efficiency [41]. Additionally, given that the absorption of polyynes shifts toward the visible by increasing the chain length, "long" chains like H-capped polyynes with more than 20 carbon atoms begin to absorb shorter laser wavelengths as the third and fourth harmonics of the Nd:YAG laser (355 and 266 nm, respectively) [37, 41, 43]. Conversely, if the target is a suspension of micro- or nanoparticles, as in the case of graphite and fullerene powders, it was noted that the production of polyynes increased with decreasing laser wavelength, as indicated in Fig. 3.6b [33, 34, 173, 174]. This can be attributed to the fact that carbon powders are more efficient at absorbing short wavelengths than bulk targets. In micro-plasma regions around carbon particles, polyynes can grow. Diffusion processes lead to the reduction of carbon radical density around dispersed nanoparticles, which biases size distribution towards shorter chains in PLAL experiments with powder targets [173, 174]. Concentration optimization of carbonaceous nanoparticles results in better yields of polyynes than ablation in similar process conditions using corresponding bulk targets [173, 174].

Another critical process parameter in the synthesis of nanomaterials by PLAL is laser fluence. The selected fluence is intimately linked to the properties of the produced nanomaterials, including their yield and dimensionality. For carbon nanomaterials synthesized by PLAL, it has been observed that fluence values equal to or less than 1 J/cm^2 promote the formation of layered structures but reduce productivity [130]. Conversely, high fluences greater than 10 J/cm^2 favor the synthesis of multilayered systems with more irregular geometries and induce the phase transition towards sp^3-like carbon nanostructures [130]. In the case of *sp*-carbon chains, the impact of laser fluence is not yet fully understood. Unclear or incomplete descriptions of setup geometries and ablation conditions prevent the reliability of PLAL results. Therefore, we published a systematic study aimed to elucidate the significance of fluence in the production of polyynes through PLAL [29]. Specifically, our goal is to distinguish the contribution of laser energy and spot size from that of fluence.

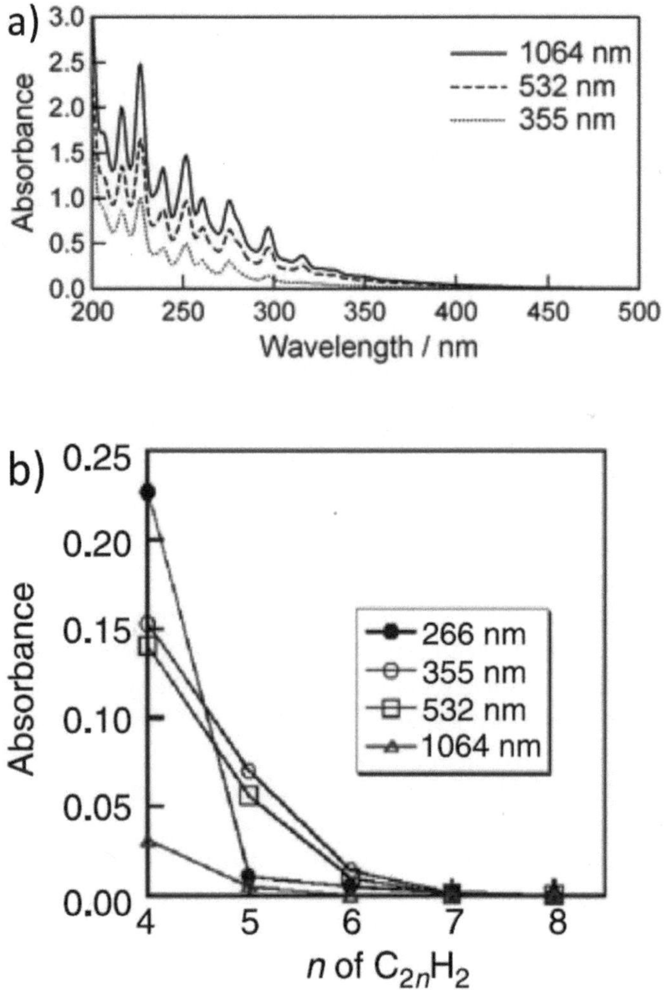

Figure 3.6 Impact of varying laser wavelengths on polyyne synthesis yield when ablating (a) a graphite pellet [41] or (b) a dispersed graphite powder [173] in n-hexane. Adapted with permission from Refs. [41, 173].

This analysis is based on HPLC analyses of several ablation experiments in the same solvent (acetonitrile) and setup conditions (geometry and parameters), and by carefully varying the laser energy and spot size. We noted that the fluence alone is not an adequate parameter of merit when evaluating the synthesis yield of polyynes by PLAL [29, 36]. As reported in Fig. 3.7a, the synthesis efficiency of a hydrogen-capped polyyne chosen as a benchmark, i.e., HC_8H, does not follow any trend with the fluence but shows different results for very similar fluence values. Instead, we can obtain a more coherent picture if we decouple the effect of laser energy and spot size from the fluence. Figure 3.7b presents this comparison: the experimental data for the HC_8H concentration shown in Fig. 3.7a is overlaid with the predicted concentrations of HC_8H, obtained through an *ad hoc* function based on a double-sigmoid shape, using experimental conditions that were not tested [29]. Figure 3.7b shows that the production yield of HC_8H increases as the overall fluence increases. However, for sizable fluences featuring small spot radii, a decrease in the synthesis yield is observed in both the experimental and fitted data. The synthesis yield also varies with the length of the polyynes, as evidenced by the concentration of $HC_{16}H$ in Fig. 3.7c, which shows a maximum at 86.5 mJ/pulse and 1.50 mm of spot radius (fluence of 1.22 J/cm^2) and a limited region in which its formation is favored.

These results suggest that high laser energies are beneficial if the spot radius is larger than approximately 1.50 mm for most chains. Indeed, in ablation experiments, the spot size of the laser beam plays a simple role in the synthesis yield: an increase in spot size results in an increment of ablated material and, thus, in the synthesis yield of polyynes. However, the effect of laser energy is more intriguing, as an increase in this parameter does not always coincide with a growth in the synthesis yield. Longer hydrogen-capped polyynes are produced more efficiently for fluences less than 1 J/cm^2, while shorter chains are promoted for increasing fluence [37]. For higher fluences, other carbon nanostructures may be favored by PLAL synthesis instead of long polyynes [3, 147]. An optimized concentration ratio between ionized carbon atoms, carbon radicals, and terminating groups, such as atomic hydrogen, is necessary to promote the formation of long polyynes compared to shorter ones or more stable allotropes of carbon.

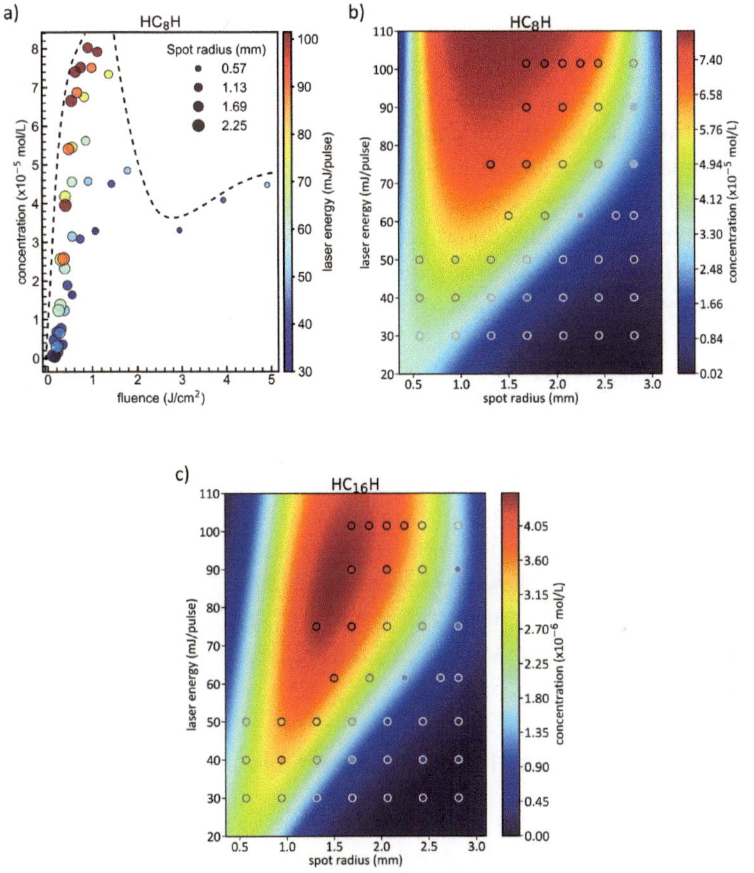

Figure 3.7 (a) Experimental concentrations of HC_8H determined using HPLC analysis of polyyne mixtures synthesized at varying laser fluences. The laser energy is represented by a color scale, while the spot radius is indicated by the size of the circles. The concentration of HC_8H as a function of both the spot radius and laser energy is shown in panel (b). Similarly, panel (c) shows the concentration of $HC_{16}H$. The 2D surfaces representing each polyyne are obtained by fitting the experimental data points. Adapted with permission from Ref. [29].

3.7.2 The Impact of the Pulse Duration: *ns*-PLAL vs *fs*-PLAL

The duration of the laser pulse has a significant impact on the synthesis and ablation yield of *sp*-carbon chains through PLAL.

"Long" pulses (>μs) are not effective in synthesizing polyynes and instead are better suited for producing nanodiamonds [110, 130]. As a result, short pulses ranging from fs to ns are typically used for synthesizing polyynes. While there is no evidence to suggest that the mechanism for polyynes' growth *via* polymerization and termination reactions (see Section 3.5) is different between *ns*- and *fs*-PLAL, the pathway bringing to the fabrication of carbon radicals and terminating functional groups differs [29]. Ultrashort pulses in *fs*-PLAL with power densities ≥ 10^{15} W/cm^2 experiments can directly dissociate solvent molecules into atomic, molecular, and parent ions [31, 32, 124, 126, 175, 176]. This explains why bulk or powder targets are not commonly used for *fs*-PLAL as discussed in Section 3.9. Organic solvents' C–C and C–H bonds can be broken directly by photo-induced processes like photoionization, generating short polyynes and cumulenes such as HC_2H, $H_2C_3H_2$, and HC_4H in a single step without any polymerization. Detecting short polyynes during an *fs*-PLAL experiment through *in situ* time-of-flight mass spectroscopy validates this observation [32, 175]. Long polyynes, on the other hand, typically require secondary (i.e., polymerization) reactions with carbon ions (C^+, C_2^+, and C_3^+) similar to what happens in *ns*-PLAL [31, 32, 175, 176].

3.7.3 The Effect of Setup Geometries

In this framework, distinct PLAL setup geometries have been designed to synthesize CAWs depending on the laser pulse's duration and chosen target, as reported in Fig. 3.8. In the case of *fs*-lasers, the most common geometry, sketched in Fig. 3.8a implies the use of a focal lens to focus the laser inside the liquid layer through the side walls of a quartz vial or at the liquid meniscus from the top. As anticipated before, the peak powers reached in the laser focus (≥10^{15} W/cm^2) are high enough to quickly induce the direct dissociation of solvent molecules whose ions and electrons generate a bright plasma filament in liquid itself [30–32, 125, 126, 175, 176]. The advantage of this geometry is the scarce production of other-than-polyynic products, as reported by Refs. [31, 32, 125, 126, 175], which makes *fs*-PLAL and this geometry particularly efficient for the fabrication of carbon atomic wires. However, amorphous carbon nanoparticles [176] and other byproducts [30] have been detected in some experiments.

Figure 3.8b presents a slightly different technique than PLAL since the confinement environment is a gas instead of a solvent. Indeed, this technique is called laser-induced breakdown (LIB) and implies focusing a high-power pulsed laser into a chamber with a continuous flux of hydrocarbons like octane, hexane, propane, ethylene, and acetylene [177, 178]. Both *ns* and *fs* lasers were employed in these experiments. The high-power density reached in the focal spot induces the generation of dense plasma spots or filaments from solvent dissociation reactions and allows for the fabrication of carbon atomic wires. The products of this process are collected through a nozzle to a tank of hexane, cooled at −80 or −40 °C [177, 178]. Despite the production of carbon soot and several byproducts, the authors pointed out the efficient production of polyynes and obtained solutions (in hexane) with a high purity if compared to *ns*-PLAL in n-hexane [177, 178]. In particular, *fs*-LIB experiments resulted in the purest polyynes' mixture, especially when ethylene and acetylene gases were employed [178].

Going back to more conventional *ns*-PLAL setups, Fig. 3.8c shows the case in which carbonaceous powders are dissolved into a solvent, while no bulk target is employed [33, 34, 121, 122, 133, 139, 144, 174, 180, 182–184]. Tsuji *et al.* employed this system and graphite particles to first demonstrate the feasibility of synthesizing polyynes by PLAL [33]. Usually, a focal lens is used to control the beam size inside the liquid layer. However, unlike in the case of direct ablation of the liquid (Fig. 3.8a), the focal point is outside the vial where a beamstop interrupts the laser path. Solvents are often stirred to keep in constant motion the particles and grant a uniform ablation. Similar to bulk targets, plasma regions are formed around powder particles that efficiently absorb the laser pulse. These micro-plasmas provide carbon atoms, ions, and radicals needed for polymerization reactions, while the solvent quenches and confines the plasma and supplies, by pyrolysis or dissociation, the functional groups to stabilize the *sp*-chains [29, 33, 34]. Even though polyynes

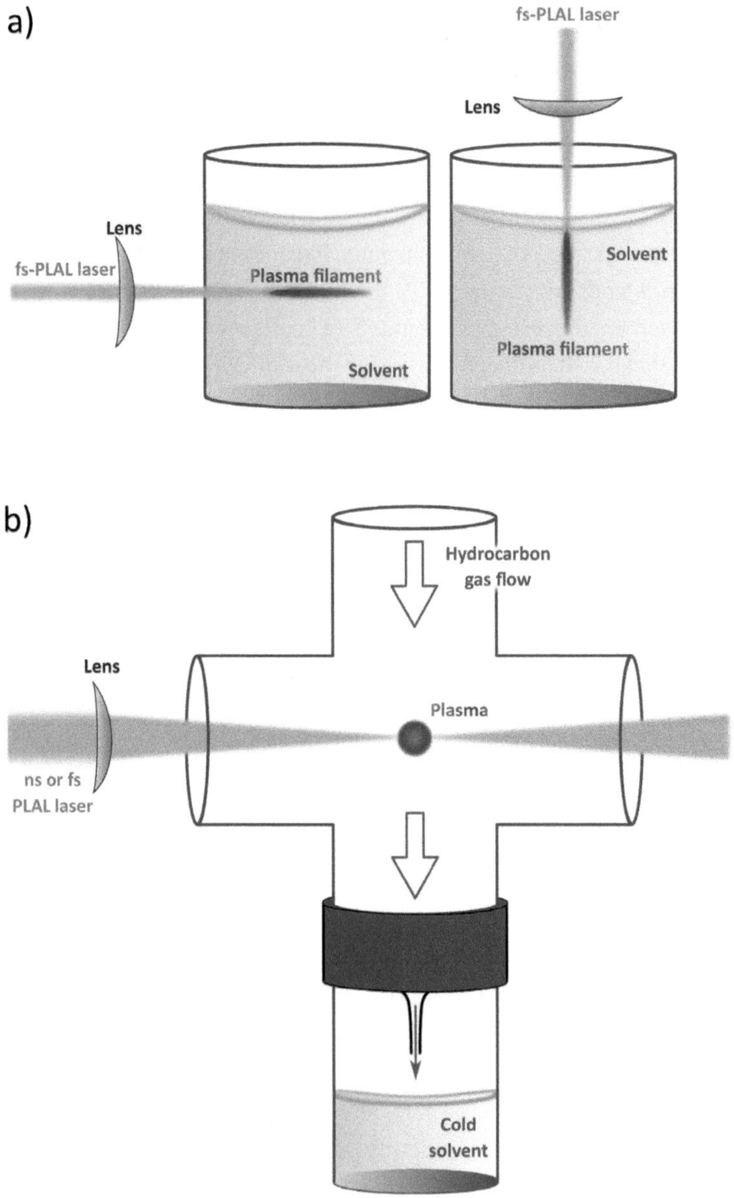
a)
fs-PLAL laser
Lens
Lens
fs-PLAL laser
Plasma filament
Solvent
Solvent
Plasma filament
b)
Hydrocarbon
gas flow
Lens
Plasma
ns or fs
PLAL laser
Cold
solvent

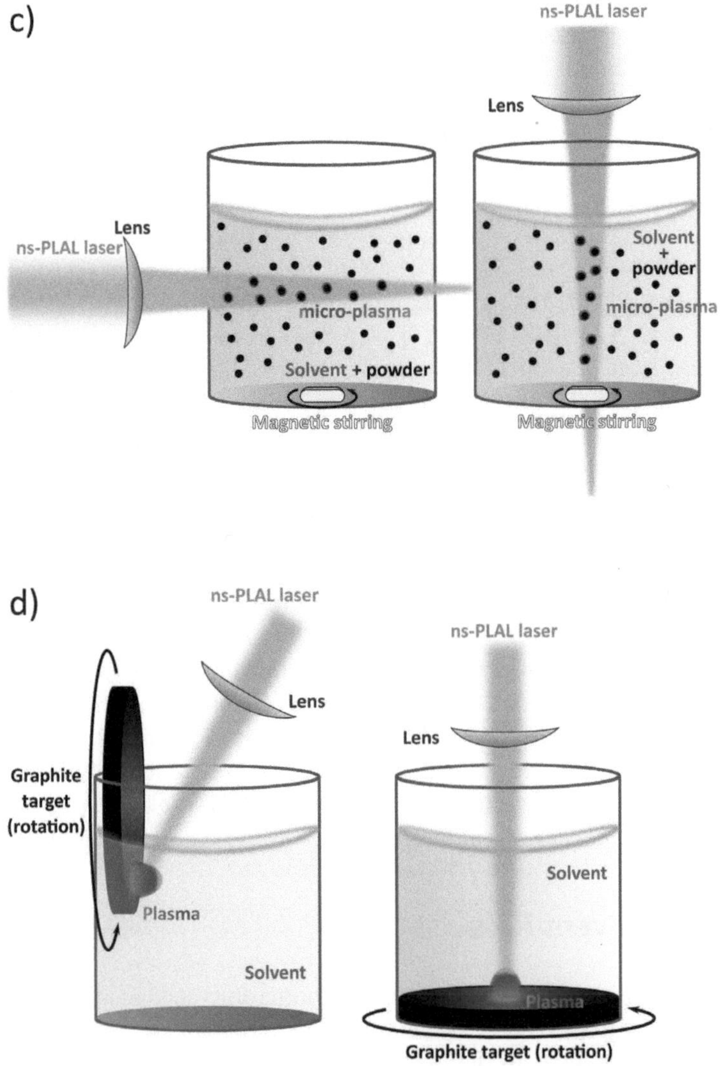

Figure 3.8 (a) Femtosecond PLAL setup direct ablating the solvent in its bulk or focused at its meniscus [126, 179]. (b) Ablation of a flux of hydrocarbons in a chamber by focusing on a *ns*- or *fs*-laser. The products are thus collected in the solvent kept at liquid nitrogen temperature [177, 178]. (c) Nanosecond PLAL setup for the ablation of powders [33, 180]. The solution is kept in continuous motion by magnetic stirring. (d) Nanosecond PLAL setup for ablation of solid targets [36, 174, 181]. The target is continuously rotated to provide uniform ablation.

with no longer than 20 carbon atoms were detected from ablations with this geometry, the synthesis efficiency in terms of polyynes' concentration is much higher compared to the use of solid targets [133, 174]. Only Arutyunyan *et al.* [179] employed a *fs*-laser in this specific geometry obtaining an efficient synthesis of polyyne in hexane up to $HC_{20}H$.

The other setup frequently used in *ns*-PLAL is illustrated in Fig. 3.8d. In this case, the target is a bulk disk, pellet, or rod made of graphite or some other carbon-based compound and is partially immersed [127, 181] or sunk into the solvent [35–39, 41, 120, 123, 128, 143, 174, 182, 185–187]. The laser beam is focused onto the target from the outside with a lens, while the distance between the lens and the target controls the spot size. Usually, the laser is defocused, i.e., the focal point is beyond the target surface, to prevent heavy damage and the generation of holes which may change the geometry and, thus, the ablation process. For the same reason, the target is kept rotating by directly operating on target itself [127, 181] or moving the stage that holds the vial in a circular or spiral path [36]. Even though the amount of byproducts produced is higher with this geometry as measured in Ref. [177, 178], this setup is the simplest and most versatile one and allowed for the production of the longest isolated chain synthesized by physical methods, i.e., $HC_{30}H$ [37]. This result was achieved because of the higher number of carbon precursors generated above the target, confined by viscous solvents like decalin, and that can develop a denser plasma region in which long polyynes can nucleate.

3.8 Solvent Role

The importance of the solvent in the termination reaction of polyynes lies in its ability to furnish the functional groups necessary for this process. Moreover, in the case of organic solvents, it also serves as a (primary or) supplementary carbon source. To demonstrate the impact of the surrounding liquid environment, Table 3.2 showcases the experimental conditions and resulting polyynic species obtained from the ablation of a graphite pellet immersed in various liquids, including aqueous solutions and organic solvents. This Section highlights the significance of the solvent in polyynic synthesis and the potential variation in the resultant products based on the chosen liquid environment.

Table 3.2 Polyynes derived from the ablation of graphite pellets under various experimental conditions in different solvents

Solvent	λ [nm]	r [Hz], duration	τ	f, ε, d	Polyynes
			Water		
Water [35, 120]	532	10 Hz, 20 min	5 ns	0.4 J/cm^2	C_nH_2 (n = 6,8)
Water [182]	1064	10 Hz, 5 min	7 ns	5.1 J/cm^2, 160 mJ, 2 mm	C_nH_2 (n = 8, 10)
Water [143]	532	10 Hz, 20 min	5 ns	0.4 J/cm^2	C_nH_2 (n = 6–10)
Water at different pH [181]	1064	10 Hz, 20 min	5 ns	40 mJ, 2 mm	C_nH_2 (n = 6–10)
Water+D_2O [127]	1064	10 Hz, 20 min	5 ns	40 mJ, 2 mm	C_nH_2 (n = 6–10)
Water [123]	532	10 Hz, 20 min	5 ns	0.5 J/cm^2	C_nH_2 (n = 6, 8)
Water [38]	1064; 532; 355; 266	10 Hz, up to 60 min	5 ns	1.3 J/cm^2, 20 mJ–40 mJ	C_nH_2 (n = 6–10); C_nH_2 (n = 6–10); C_nH_2 (n = 6–10); C_nH_2 (n = 6, 8)
Water [36]	532	10 Hz, 15 min	6 ns	2.8 J/cm^2	C_nH_2 (n = 6–10), HC_8CH_3
			Colloidal Solutions		
Pt colloidal dispersion [188]	1064	10 Hz, 20 min	5 ns	0.4 J/cm^2	Pt NPs encapsulated in carbon atomic wires shell
Au colloidal dispersion [189]	1064	10 Hz, 20 min	5 ns	0.98 J/cm^2	Au NPs and carbon atomic wires nanosystems
Ag colloidal dispersion [190]	1064	10 Hz, 20 min	5 ns	0.98 J/cm^2	Ag NPs and carbon atomic wires nanosystems

(*Continued*)

Table 3.2 (*Continued*)

Solvent	λ [nm]	r [Hz], duration	τ	f, ε, d	Polyynes
			Organic Solvents		
Acetonitrile [35, 120]	532	10 Hz, 20 min	5 ns	0.4 J/cm^2	C_nH_2 (n = 6–16)
n-hexane [174]	532	10 Hz, 5 min	5 ns	0.31 J/cm^2	C_nH_2 (n = 6–20)
n-hexane [186]	532	10 Hz, 180 min	5 ns	0.31 J/cm^2	C_nH_2 (n = 6–22)
decalin [128]	532	10 Hz, 15–80 min	5 ns	0.31 J/cm^2	C_nH_2 (n = 6–26)
methanol;	532	10 Hz, 2.5–5 min	5 ns	22 mJ	C_nH_2 (n = 6–16);
ethanol;					C_nH_2 (n = 6–16);
1-propanol;					C_nH_2 (n = 6–16);
1-butanol;					C_nH_2 (n = 6–16);
t-butyl alcohol;					C_nH_2 (n = 6–18);
n-hexane;					C_nH_2 (n = 6–18);
decalin [39]		10 Hz, 52.5 min			C_nH_2 (n = 6–28)
n-hexane [41]	1064;	10 Hz	5 ns	40 mJ	C_nH_2 (n = 6–22);
	532;				C_nH_2 (n = 6–22);
	355				C_nH_2 (n = 6–20)

Solvent	λ [nm]	r [Hz], duration	τ	f, ε, d	Polyynes
decalin [37]	1064	10 Hz; 180 min	5 ns	0.57 J/cm^2	C_nH_2 (n = 6–30)
n-hexane;	1064	10 Hz; 20 min	5 ns	20 mJ	C_nH_2 (n = 6–14);
c-hexane;					C_nH_2 (n = 6–16);
n-heptane;					C_nH_2 (n = 6–14);
n-octane[187]					C_nH_2 (n = 6–16)
acetonitrile;	1064	10 Hz, 20 min	5 ns	40 mJ, 2 mm	C_nH_2 (n = 6–16);
n-hexane;					C_nH_2 (n = 6–16);
c-hexane [127]					C_nH_2 (n = 6–16)
acetonitrile;	532	10 Hz, 20 min	5 ns	0.5 J/cm^2	C_nH_2 (n = 6–14), HC_nN (n = 7–11), C_nN_2 (n = 6, 8);
methanol;					C_nH_2 (n = 6–12);
c-hexane [123]					C_nH_2 (n = 6–12)
ethanol [185]	1064	1 Hz	9 ns	20–200 mJ	C_nH_2 (n = 6–18)
liquid N2 [124]	800	1000, 20 min	35 fs	25 J/cm^2	C_8N_2
TEOS [191]	1064	10 Hz, 5 min	240 µs	1 J, 0.055 mm	C_nH_2 (n = 8–16), Si–H doped carbon nanoribbon

(*Continued*)

Table 3.2 *(Continued)*

Solvent	λ [nm]	r [Hz], duration	τ	f, ε, d	Polyynes
acetonitrile;	532	10 Hz, 15 min	6 ns	2.8 J/cm^2	C_nH_2 (n = 6–22), HC_nCH_3 (n = 6–18), HC_nN (n = 7–13);
methanol;					C_nH_2 (n = 6–22), HC_nCH_3 (n = 6–18)
ethanol;					C_nH_2 (n = 6–22), HCnCH$_3$ (n = 6–18)
isopropanol [36]					C_nH_2 (n = 6–22), HC_nCH_3 (n = 6–18)
			Gaseous Environments		
Mix of argon and propane [149]	532	10 Hz, 60 s	6 ns	0.5 J/cm^2, 150 mJ, 30 mm^2	HC_nH (n = 6–16), HC_nCH_3 (n = 12–16)

Parameters such as laser wavelength (λ), repetition rate (r), pulse duration (τ), fluence (f), energy (ε), and spot size (d) are also included if provided.

Adapted with permission from Ref. [29].

3.8.1 Water

Water is a low-cost, safe, and environmentally friendly solvent for the synthesis of polyynes by PLAL [35, 36, 38, 120, 143, 182]. Ablations in water have the advantage of avoiding the presence of any carbon source apart from that coming from the ablated target, minimizing the formation of contaminant carbon materials, including carbon-based byproducts, compared to organic solvents [143, 182]. Furthermore, it makes it easier to study the synthesis yield of a specific target and other process parameters. Ablation experiments in pure water preferentially produce short hydrogen-capped polyynes, with a maximum of 10 carbon atoms (i.e., $HC_{10}H$), as observed in several studies [35, 36, 120, 143, 182]. However, "longer" polyynes (i.e., $C_{10}H_2$) can be obtained by employing lower laser intensity and longer wavelength [38]. Adding HCl to water can increase the yield of shorter chains in acidic conditions, as stated by Shin *et al.* [181].

Notably, we have observed a methyl-capped polyyne, C_8CH_3, from PLAL in water [36]. Despite its lower concentration than the hydrogen-capped polyynes, this finding is significant as it suggests that terminating polyynes with a methyl group is feasible through the binding of three hydrogen atoms at one edge of the growing chain.

Although water has several advantages as a solvent for PLAL, there are also some limitations. For example, the hydrogen generation rate in water is the lowest among all the solvents, hindering the termination reaction and letting carbon atoms reorganize in other structures [34, 36]. Furthermore, water is an oxygenated solvent, which can lead to possible oxidative reactions of the *sp*-chains, resulting in the loss of the *sp*-character [3, 36, 130]. Finally, water is a polar solvent, while polyynes are non-polar molecules, making them hardly soluble in it and reducing their stability. Despite these limitations, water remains a popular choice for PLAL experiments due to its low cost, environmental friendliness, and ease of use.

3.8.2 Colloidal Solutions

Instead of pure water, colloidal dispersion of metal nanoparticles may serve as liquid environments for ablation experiments [188–190].

Ablations in colloidal solution of metal nanoparticles are motivated by the observation of enhanced stability of polyynes when mixed in these solutions or dried onto metal surfaces [120, 141, 151, 192]. These PLAL experiments brought to hybrid nanosystems composed of small metal nanoparticles of Pt, Au, or Ag encapsulated in a network of carbon atomic wires [188–190]. While metal nanoparticles stabilize carbon chains, polyynes prevent nanoparticle aggregation and coalescence in larger agglomerates. Moreover, carbon atomic wires with Ag or Au nanoparticle complexes demonstrated potential wound-healing [190] and anti-inflammatory [188] properties, respectively.

3.8.3 Organic Solvents

Several organic solvents have been used for polyynes' synthesis by PLAL. The main advantage of organic solvents compared to water is their contribution as secondary carbon sources for polymerization reactions and the possibility to provide different terminating groups.

Acetonitrile has low polarity and a low amount of oxygen dissolved, making it a good environment for the growth of polyynes. PLAL of a graphite target in acetonitrile has been shown to form hydrogen-capped polyynes (HC_nH) with lengths up to 22 carbon atoms [35, 36,120, 123, 187]. Besides hydrogen-capped polyynes, chains terminated with methyl (HC_nCH_3) and cyano (HC_nCN) groups at one side were detected, up to, respectively, 18 and 12 carbon atoms [36, 123]. Moreover, Forte *et al.* observed the synthesis of dicyano-capped chains (CNC_nCN or $NC_{n+2}N$) up to 8 *sp*-carbon atoms [123]. The effective formation of polyynes in acetonitrile results from its tendency to degrade during laser ablation, with a high tendency toward carbonization [36]. The dissociation of acetonitrile may provide functional groups like methyl and cyano radicals that can terminate the chain, as testified by the wide range and high concentration of hetero-capped wires recognized in Refs. [36, 123]. The coking reaction of acetonitrile promotes the polymerization of polyynes by providing carbon species in the plasma region. However, this also leads to a higher concentration of carbon-based byproducts, resulting in a higher UV-Vis background and a lower purity than, for example, alcohols or n-hexane [36, 149].

According to the work of Russo *et al.* [124], exposing graphite to *fs*-ablation in liquid nitrogen leads to the creation of dicyano-capped polyynes comprising 8 *sp*-carbon atoms. This outcome was observed only under particular fluence conditions, specifically within the range from 20 to 30 J/cm^2, where N_2 molecules become ionized and can bind with polyynes to yield cyanopolyynes.

Concerning alcohols as the solvent for graphite ablation experiments, when using methanol resulted in fewer byproducts versus other solvents like acetonitrile, ethanol, and isopropanol and allowed for the synthesis of hydrogen-capped polyynes up to 22 carbons [36, 39, 123]. Using polar solvents like methanol induces a positive charge onto hydrogen atoms due to the increased negative charge along the carbon chain [123]. As the chain length increases, the net charge becomes neutral in the central region. The high energy density in the focus region promotes the formation of carbon chain anions that are neutralized by the acidity of the polar solvent [123]. We also produced and detected methyl-capped polyynes up to 18 carbon atoms [36]. Ethanol was another solvent used for PLAL, which led to synthesizing hydrogen-capped polyynes up to 16 carbon atoms in previous studies [39], while we were able to reach a length of 22 carbon atoms [36]. Other alcohols like 1-propanol, 1-butanol, and tert-butyl alcohol were also studied, and hydrogen-capped polyynes up to 16 and 18 carbon atoms were synthesized for the first two solvents and t-butyl alcohol, respectively [39]. However, the effect of solvent polarity on polyynes formation was not observed in these studies. We found that the yield of polyynes increased as the solvent polarity decreased, which suggests that solvent properties like oxygen dissolved, carbonization tendency, and C/H ratio play a crucial role in polyynes synthesis [29, 36]. In this framework, we extensively investigated the impact of the solvent's C/H ratio on polyynes yield [29], which revealed a reduction in the formation of short polyynes and an enhancement in the concentration of longer ones as the C/H ratio of the solvent increased, ranging from 0 for water to 0.49 for isopropanol, with intermediate values of 0.33 for methanol and 0.4 for ethanol.

Various studies have highlighted the crucial role of alkanes in the production of polyynes through PLAL [39, 41, 123, 127, 174, 186, 187]. Cyclohexane has been found to promote the formation of carbon chain anions, producing hydrogen-capped polyynes up to 16 carbon atoms in Refs. [123, 127, 187]. Similarly, n-hexane has been widely employed as a solvent for PLAL to obtain hydrogen-capped polyynes with longer chain lengths of up to 22 carbon atoms [39, 41, 127, 174, 186, 187]. The concentration of polyynes synthesized in n-hexane is higher than in water due to the ease of fragmentation of carbon-carbon bonds of n-hexane. The fragmented species then interact with the graphitic plasma to produce longer polyynes [187]. By comparing PLAL in alkanes, polyynes' yield seems strongly affected by the C/H ratio of the solvent molecules rather than other properties such as bond dissociation energy, thermal conductivity, and total mass of hydrogen atoms per volume of solvent [187].

In a study by Wu *et al.,* the authors investigated the production of polyynes using tetraethyl orthosilicate (TEOS) as a solvent and a graphite pellet as a target [191]. They were able to obtain hydrogen-capped polyynes with up to 16 carbon atoms. The ablation process in TEOS involves its decomposition, generation of C_2 radicals from both TEOS and the graphite target, growth of polyynes, and finally their hydrogenation. In addition to polyynes, carbon nanoplates/ribbons composed of graphene-based lamellae with Si–H dopant were also generated through this process.

The production of long hydrogen-capped polyynes by PLAL depends on the choice of solvent. Decalin looks like an effective solvent for this task due to its high viscosity (3.042 mPa s at 25 °C [193]) and high C/H ratio (0.56), which promote the growth of sp-chains [29, 37, 39]. Additionally, decalin's low polarity (dielectric constant 2.1 at 20 °C [194]) contributes to the stabilization of polyynes, resulting in their higher concentration in decalin compared to other solvents. Employing tetralin as a solvent for PLAL led to a decrease in short polyynes and an increase in longer ones [195]. This may be attributed to tetralin's higher C/H ratio, which inhibits the termination reaction on both ends of the *sp*-carbon chains, resulting in longer polyynes. Similar to decalin, tetralin's high viscosity (2.015 mPa-s at 25 °C [193]) and low polarity (dielectric constant 2.7 at 20 °C [196]) may also contribute to an increase in the synthesis yield of longer polyynes. Overall, decalin and tetralin are potentially useful solvents for promoting chain growth and, indeed, Matsutani *et al.* produced the longest hydrogen-capped polyyne ($HC_{30}H$) by PLAL in decalin [37].

3.8.4 Gaseous Environments

The PLAL experiment by Taguchi *et al.* [149] was designed similarly to the experimental setup shown in Fig. 3.8b. In this experiment, a graphite target with many small holes (1 mm) was ablated using a flux of Ar/propane mixed gas, and the resulting products were collected in n-hexane maintained at –80 °C. Although this approach cannot be considered as ablation in liquid, the flow of propane gas provides a low-pressure hydrocarbon environment, leading to the growth of polyynes with the help of carbon, hydrogen, and methyl radicals coming from the dissociation of propane in the plasma plume. The authors observed hydrogen- and methyl-capped chains of up to 16 carbon atoms from HPLC analysis of the mixture in n-hexane. The yield of longer polyynes increased with a higher percentage of propane in the gas mix, indicating that this environment is more favorable for froming and stabilizing polyynes. Furthermore, the final mixture of polyynes synthesized in this experiment exhibited a higher purity and selectivity than ablations in liquid solvents, such as n-hexane, demonstrating the superiority of this method for polyynes synthesis.

3.8.5 Summary

The properties of an efficient solvent for the synthesis of polyynes by PLAL can be determined by understanding why water is not suitable for this purpose. Indeed, organic solvents have a higher concentration of polyynes than water by at least one order of magnitude, with the maximum chain length achieved being more than twice that from ablations in water. The presence of carbon atoms in organic solvents acts as a secondary carbon source for growing polyyne chains, leading to the formation of a bigger number of chains and the fabrication of much longer chains [23, 29, 35–37, 39, 127, 128, 173, 186]. Moreover, the solvent must provide the terminating groups needed to stabilize the *sp*-carbon chains and avoid any reorganization into more stable other-than-*sp* structures [36]. The balance between carbon and hydrogen atoms, known as the C/H ratio, is a critical solvent parameter that determines the best ablation environment [26, 29, 36, 37, 149]. Organic solvents have a higher hydrogen generation rate than water, and high values of the C/H ratio improve the synthesis yield of polyynes and bias the size distribution towards longer chains. However, if the hydrogen surrounding the plasma plume is depleted too much, the stabilization by termination reactions of the growing chains is reduced. Organic solvents can provide terminations other than hydrogen, like cyano (CN) and methyl (CH_3) groups, producing cyano-, dicyano-, and methyl-capped polyynes [36, 121, 123–126]. The solvent's carbonization tendency, polarity, oxygen solubility, and viscosity also impact polyynes' synthesis yield and stability. Solvents that undergo carbonization rapidly at high temperatures produce more byproducts, reducing the stability of polyynes due to crosslinking reactions of *sp*-carbon chains with byproducts [26, 36]. Solvent polarity defines the solubility of polyynes and their stability since higher solubility means chains are more dispersed in the liquid volume and less liable to interact [36, 39, 123, 149]. Polyynes become less stable in the presence of oxygen, which can be found in solvent molecules such as alcohols or dissolved in gaseous form in the solvent [29, 36]. This is due to an increased likelihood of oxidative reactions. The synthesis of longer polyynes appears to depend also on viscosity, as observed in the production of the longest polyyne ever synthesized by PLAL ($HC_{30}H$), achieved by using decalin as the ablation solvent, which has a high viscosity of 3.042 mPa s at 300 K [37, 128]. The specific contribution of each solvent property needs further investigation due to the complexity of the phenomenon involved during polyynes synthesis by PLAL.

3.9 Target Role

Choosing the appropriate target is crucial for determining the final products of ablation because it governs the primary interaction with the laser pulse, as well as subsequent optical, electronic, and thermal processes [1–3]. The various targets used in pulsed laser ablation for carbon allotrope synthesis, including solid targets, powder targets, direct ablation of the solvent, and metal targets, are classified in Table 3.3 [29].

Table 3.3 Experimental parameters and resulting polyynes obtained from ablation experiments of various targets in different liquids

Target	Solvent	λ [nm]	***r*** **[Hz], duration**	τ	f, ε, d	Polyynes
			Solid Targets			
shungite [197]	water	1064	50 Hz	2 ms	1–3 J	SERS *sp*–band
PTCDA pellet [174]	n-hexane	532	10 Hz, 5 min	5 ns	0.31 J/cm^2	C_nH_2 (n = 6–18)
PTCDA pellet [39]	methanol; ethanol; 1-propanol; 1-butanol; t-butyl alcohol; n-hexane	532	10 Hz; 30–45 min	5 ns	22 mJ	C_nH_2 (n = 6–10); C_nH_2 (n = 6–10); C_nH_2 (n = 6–10); C_nH_2 (n = 6–16); C_nH_2 (n = 6–18); C_nH_2 (n = 6–18)
fullerene pellet [41]	n-hexane	1064; 532; 355	10 Hz	5 ns	40 mJ	C_nH_2 (n = 6–22);C_nH_2 (n = 6–22); C_nH_2 (n = 6–20)
fullerene pellet [41]	methanol	1064; 532; 355	10 Hz	5 ns	40 mJ	C_nH_2 (n = 6–20)
			Powder Targets			
graphite particles [179]	water; n-hexane	800	1000 Hz	160 fs	(0.5, 1) mJ	C_nH_2 (n = 4–12); C_nH_2 (n = 4–20)

(*Continued*)

Table 3.3 (*Continued*)

Target	Solvent	λ [nm]	*r* [Hz], duration	τ	*f*, ε, *d*	Polyynes
graphite particles [182]	water	1064	10 Hz, 5 min	7 ns	5.1 J/cm^2, 160 mJ, 2 mm	–
graphite powder; PTCDA powder [174]	n-hexane	532	10 Hz, 5 min	5 ns	0.31	C_nH_2 (n = 6–20); C_nH_2 (n = 6–18)
graphite particles [144, 183]	n-hexane	532	10 Hz, 60 min	5 ns	100 mJ, 100 mm	C_nH_2 (n = 8–12)
diamond nanoparticles [180]	ethanol	532	20 Hz, 0–120 min	7 ns	1.3 J/cm^2, 2 mm	C_nH_2 (n = 8–16)
diamond nanoparticles [133]	ethanol	532	20 Hz, 0–160 min	7 ns	1.3 J/cm^2	C_nH_2 (n = 8–20)
diamond nanoparticles; carbon onions; graphite powder [184]	ethanol	532	20 Hz, 5–10 min	7 ns	1.3 J/cm^2, 2 mm	C_nH_2 (n = 8–16); C_nH_2 (n = 8–14); C_nH_2 (n = 8–14)
graphite powder [139]	ethanol	532	10 Hz, 180 min	5 ns	0.21 mJ, 9 mm	C_nH_2 (n = 8–16)
graphite powder [33]	benzene	355	10 Hz	5–9 ns	200 J/cm^2, 40 mJ	C_nH_2 (n = 10–16)
graphite powder [33]	toluene	1064; 532; 355	10 Hz	1 µs; 5–9 ns; 5–9 ns	200 J/cm^2, 40 mJ	C_nH_2 (n = 10–14); C_nH_2 (n = 10–16); C_nH_2 (n = 10–16)

Target	Solvent	λ [nm]	*r* [Hz], duration	τ	*f*, ε, *d*	Polyynes
graphite powder [33]	n-hexane	355	10 Hz	5–9 ns	200 J/cm^2, 40 mJ	C_nH_2 (n = 8–14)
fullerene particles [34]	n-hexane	1064; 532; 355; 266	10 Hz, 40–120 min	5–9 ns	200 J/cm^2, 20–80 mJ	C_8H_2; C_nH_2 (n = 8,10); C_nH_2 (n = 8–12); C_8H_2
fullerene particles [34]	methanol	1064; 532; 355; 266	10 Hz, 40–60 min	5–9 ns	200 J/cm^2, 20–80 mJ	–; –; C_nH_2 (n = 8,10); C_8H_2
graphite particles [122]	n-hexane +iodine; CCl_4 + iodine	1064	10 Hz	5 ns	800 mJ	$C_nH_2I_6$ (n = 10–18); $CC_nH_2I_6$ (n = 10–14)
graphite powder [121]	acetonitrile	532	10 Hz, 300 min	5 ns	0.4 J	C_nH_2 (n = 8–14), HC_nN (n = 7–13)

(*Continued*)

Table 3.3 (*Continued*)

Target	Solvent	λ [nm]	*r* [Hz], duration	τ	*f*, ε, *d*	Polyynes
			Direct ablation of the Solvent			
no target [31]	acetone	800	1000 Hz	90 fs	300 μJ, 10 μm	C_6H_2
no target [125, 126]	toluene	800	1000 Hz	35 fs	300 μJ	C_nH_2 (n = 12–18); HC_nCH_3 (n = 12, 14)
no target [30]	n-hexane; decane	800	1000 Hz, 30–300 min	100 fs	0.9 μJ	C_nH_2 (n = 8–12); C_nH_2 (n = 8–10);
no target [176]	benzene	800	1000 Hz, 90 min	120 fs	300 μJ	C_nH_2 (n = 8,14)
no target [32, 175]	n-octane	800	100 Hz, 120 min	100 fs	300 μJ	C_nH_2 (n = 8–14)
			Gaseous Environemnt			
no target [177]	Ar/propane gas Ar/hexane gas Ar/octane gas Ar/benzene gas Ar/toluene gas	532; 1064	30 Hz, 60 s	– ns	150 mJ; 460 mJ	C_nH_2 (n = 6–10); C_nH_2 (n = 6–10); C_nH_2 (n = 6–12); C_nH_2 (n = 6–12); C_nH_2 (n = 6–10)
no target [177]	Ar/hexane gas Ar/toluene gas	800	1 kHz, 5 min; 1 kHz, 10 min	35 fs	1.5 mJ	C_nH_2 (n = 6–12);

Target	**Solvent**	**λ [nm]**	***r* [Hz], duration**	**τ**	**f, ε, d**	**Polyynes**
no target [178]	Ethylene gas Acetylene gas	532	– ns	150 mJ	C_nH_2 (n = 6–12); C_nH_2 (n = 6–12)	–
no target [178]	Ethylene gas Acetylene gas	800	35 fs	2 mJ	C_nH_2 (n = 6–12); C_nH_2 (n = 6–12)	–
			Metal Targets			
Gold target [198]	ethanol	532	5 Hz	10 ns	500 mJ, <1 mm	carbyne crystals (finite length)
Gold target [199]	ethanol	532	20 Hz	9 ns	400 mJ, <1 mm	Au–pseudocarbyne
Gold target [200]	ethanol	532	20 Hz	9 ns	400 mJ, <1 mm	Au–pseudocarbyne

λ, r, τ, f, ε, and d denote the laser wavelength, repetition rate, pulse duration, fluence, energy, and spot size, respectively.
Adapted with permission from Ref. [29].

3.9.1 Solid Targets

To some extent, the performance of non-graphitic solid targets is similar to that of graphite. When a perylene-tetracarboxylic dianhydride (PTCDA) pellet was used, slightly shorter chains were obtained due to the presence of carboxylic dianhydride groups [39, 174]. These groups led to a decrease in carbon radical production because they transfer part of the energy of the laser to the solvent, lowering the ablation efficiency. Polar solvents, in particular, were affected by this phenomenon. In contrast, fullerene pellets are more efficient at producing long polyynes than PTCDA pellets, and they can produce chains up to 22 carbon atoms in n-hexane [41]. Similar to bulk graphite targets, fullerene pellets have a lower production yield of polyynes with shorter wavelengths of the laser pulse. In this case, the absorption of visible laser pulses by fullerene particles detached from the pellet and suspended in the liquid reduced the amount of energy reaching the bulk target, resulting in a reduction of the synthesis yield [41].

3.9.2 Powder Targets

When using dispersed particles or powders in ablation experiments, the physics of the process is noticeably different compared to when using a solid target. The hybridization of carbon atoms in these materials affects the formation mechanism and final size distribution of polyynes. For example, when using graphite or carbon onion particles (sp^2-carbon), the formation of carbon radicals is similar to that of bulk targets, with laser pulse absorption leading to the break of carbon bonds and the subsequent formation of capped *sp*-chains with hydrogen atoms from the solvent [144, 179, 182, 183]. The size distribution of the resulting polyynes is independent of the particle size [133, 184]. In contrast, when using dispersed nanodiamonds (sp^3-carbon), the size distribution of the polyynes depends on the particle size and ablation times, with longer polyynes produced from smaller particles and shorter ablation times [133, 180, 184]. This can be explained by a two-step mechanism of radical carbon formation, where the laser pulse first induces a phase transition from nanodiamonds to graphitic nanoparticles, followed by the direct vaporization of the latter to produce carbon radicals. Furthermore, using powders instead of bulk targets changes the effect of laser wavelength on ablation efficiency. This effect has been discussed in detail in Section 3.7. It is important to note that this variation in ablation efficiency due to the use of dispersed particles or powders has implications for the scalability and reproducibility of PLAL-based polyynes synthesis.

3.9.3 Direct Ablation of the Solvent

The use of *fs*-pulsed lasers with high power densities (i.e., $\geq 10^{15}$ W/cm^2) in *fs*-PLAL for polyynes' synthesis eliminates the need for a target, as outlined in Table 3.3. As discussed in Section 3.3, the mechanism of polyyne formation using *fs*-laser involves the direct ionization and dissociation of solvent molecules. Short polyynes with up to HC_6H can be synthesized from parent ions of hydrocarbons and alkanes, as several studies suggest [30–32, 126, 175, 176]. Longer polyynes with up to 18 carbon atoms follow a different synthesis path that involves several secondary reactions with molecular fragments, carbon ions, and radicals [30, 175, 201]. Interestingly, the ablation of toluene results in the formation of methyl-capped polyynes, demonstrating the successful dissociation and termination with CH_3 ions [126]. The ability to terminate polyynes with CH_3 ions opens up new possibilities for synthesizing functionalized polyynes with various end groups.

3.9.4 Gaseous Environments

A subsequent experiment based on the work of Taguchi *et al.* [149] involved the use of *ns*- and *fs*-lasers for the direct ablation of hydrocarbon gases, referred to as *ns*- and *fs*-laser-induced breakdown (LIB) [177, 178]. The researchers examined various hydrocarbons such as hexane, octane, benzene, toluene, ethylene, and acetylene, the first four mixed with argon in different proportions [177, 178]. The ns-LIB synthesis produced polyynes with carbon chains up to 12 atoms, while *fs*-LIB in toluene did not yield polyynes, likely due to laser filamentation, which reduces the peak intensity of the plasma below the threshold required for polyynes' growth [177]. Of the hydrocarbons tested, ethylene and acetylene were the most effective in generating polyynes [178]. Moreover, these hydrocarbons produced the purest polyynes and the highest number of them compared to other hydrocarbons, although Takizawa *et al.* observed the visible production of soot collected in liquid hexane after the ablation. The increased polyyne production could be attributed to the laser ablation of soot, which provided additional carbon radicals for polyynes' polymerization [178].

3.9.5 Metal Targets

The ablation of metal targets under specific experimental conditions has resulted in the synthesis of diverse *sp* compounds, whose structure and composition are still under debate. For instance, Pan *et al.* claimed the formation of a compound they called "carbyne nanocrystals", featuring a kinked structure composed of 8 *sp*-carbon atoms, from the ablation of a gold target in ethanol [198]. In this process, the gold atoms produced in the plasma phase are believed to participate in the dehydrogenation of ethanol and form Au–H complexes, facilitating the formation of carbon-carbon triple bonds from the ethanol skeleton [198]. Various spectroscopic characterizations have provided evidence in support of the unique structure of this compound [198, 202, 203].

However, in similar ablation experiments under the same conditions, Kim *et al.* reported the presence of a different class of nanostructures[199, 200]. According to their findings, the products are *sp*-carbon chains, ranging up to 16 carbon atoms, bonded to sub-nanometer gold nanoparticles, referred to as Au-pseudocarbynes [199, 200, 204]. The spectroscopic characterization, including Raman, Fourier-transform IR, and UV-Vis absorption spectroscopy, of these compounds closely resemble that of the carbyne nanocrystals reported by Pan *et al.* [198]. Nevertheless, the correct attribution of the structure of these compounds is still an open issue.

3.10 Encapsulation of Polyynes by PLAL as a Route to Stabilization

In this section, we describe possible stabilization strategies that include the encapsulation of polyynes produced by PLAL in polymeric matrices or single- and double-walled carbon nanotubes (SWCNTs and DWCNTs). These methods allow for extending polyynes' lifetime by avoiding crosslinking and oxidative reactions that are the main degradation pathways for polyynes [12].

3.10.1 Polymeric Matrices

The majority of the studies on nanocomposites used poly(vinyl alcohol) (PVA) as the matrix due to its low cost, chemical stability, solubility in water, and good film formation properties, as reported in Table 3.4 [209, 210]. Okada *et al.* [205] were the first to use PVA in a nanocomposite film to prevent the transformation of polyynes into sp^2 structures. They synthesized polyynes by ablating a graphite powder suspension dispersed in methanol, followed by adding Ag colloidal solution and PVA granules to fabricate the nanocomposite film. The solution is dried in air for 48 h at 30 °C and the encapsulation was verified by SERS and UV-Vis spectroscopy. SERS spectra showed that the sp-carbon chains were stable for at least one month. UV-Vis spectra of PVA/Ag colloid showed a narrow absorption band at 425 nm, while the nanocomposite film had a broadened and shifted plasmonic absorption band, indicating the aggregation of Ag nanoparticles when mixed with polyyne solutions, as already observed by D'Urso *et al.* [163].

Recently, Sata *et al.* [206] studied the angular dependence of absorption intensity of linearly polarized UV light in solid PVA films embedding aligned polyynes. They fabricated a PVA film and thus dipped it in a polyyne/boric acid solution for 3 h. Polyynes were produced by PLAL in methanol of carbon particles and separated by HPLC to obtain size-selected solutions of chains from 10 to 14 *sp*-carbon atoms. The polymer film was stretched in one direction to properly align the polyynes encapsulated within it and then left to dry to restore its hardness. This alignment was verified through polarization-dependent UV-Vis spectroscopy, which showed that the absorption of polyynes is maximized when the light is parallel to the stretching direction of the film.

In our recent study, Peggiani *et al.* [40] successfully synthesized polyynes using laser ablation in a solution in which PVA was already dissolved, allowing for the direct encapsulation of polyynes in the polymer. This method was employed since adding the polymer after the synthesis process was found to negatively affect the *sp*-carbon chains' stability due to the heating processes needed to dissolve PVA [15] and prevent complete blending between polyynes, metal nanoparticles, and polymer. To prepare the polyynes-based nanocomposite, we performed the ablation of a graphite pellet in

Table 3.4 Polyynes by PLAL stabilized following different strategies

Matrix	Solvent	Target	λ [nm]	r [Hz], duration	τ	f, ε, d	Polyynes
Polymeric matrices							
PVA [205]	methanol	graphite powder	532 nm	20 Hz	7 ns	1.2 J/cm^2	–
PVA [206]	methanol[a]	carbon particles	–	–	–	–	C_nH_2 (n = 10–14)
PVA [40]	water	graphite pellet	532 nm	10 Hz, 15–30 min	5–7 ns	5.2 J/cm^2, 1.4 mm	C_nH_2 (n = 6–10)
PMMA [25]	acetone	Ag pellet + graphite pellet	532 nm	10 Hz, 5 min + 15 min	5–7 ns	2 J/cm^2	–
Nanotubes							
SWCNTs [183]	n-hexane	graphite particles (0.5 g/mL)	532 nm	10 Hz, 60 min	5–9 ns	100 mJ, 10 mm	$HC_{10}H$[b]
SWCNTs, DWCNTs [207, 208]	n-hexane	graphite particles	–	–	–	–	HC_nH (n = 10,12)[b]

Parameters such as laser wavelength (λ), repetition rate (r), pulse duration (τ), fluence (f), energy (ε), and spot size (d) are also included if provided.
Adapted with permission from Ref. [29].

[a]Boric acid is mixed with the size-selected polyynes and then PVA is added to the solution.

[b]Separated by HPLC after the synthesis.

an aqueous solution of PVA. Ag nanoparticles were added to the PVA/polyynes solution to prepare the nanocomposite film, which was drop-casted and allowed to dry for 24 hours. SERS analyses were performed on the liquid and solid samples, revealing that the nanocomposite film demonstrated enhanced polyynes' stability for up to 11 months, thanks to the immobilization of the *sp*-carbon chains and inhibition of the reactions between them.

Our study, carried out by Peggiani *et al.* [25], discusses the synthesis of polyynes through pulsed laser ablation in a PMMA solution, as indicated in Table 3.4. PMMA was chosen due to its transparency, compatibility with silver nanoparticles, and water resistance, which is fundamental for waterproof applications. A two-step ablation process was employed: the first step implies the ablation (5 min) of an Ag solid target in a PMMA/acetone solution to produce Ag nanoparticles; thus, the mixture of Ag nanoparticles, PMMA, and acetone serves as the solvent for the second step which involves the ablation (15 min) of a graphite pellet to synthesize polyynes. Despite the low concentration of Ag nanoparticles and polyynes, the SERS signal of *sp*-carbon chains was visible in the nanocomposite film. The film enhances polyynes' stability to at least 5 months, demonstrating longer stability than non-encapsulated hydrogen-capped polyynes.

3.10.2 Nanotubes

Carbon nanotubes represent the perfect protective environment for polyynes: indeed, they can impede crosslinking reactions among chains, oxidation reactions, and other chemical-physical degradation processes. Nishide *et al.* [183] were the first who developed the method to encapsulate *sp*-carbon chains produced by PLAL in single-walled carbon nanotubes (SWCNTs). After synthesizing polyynes by PLAL in n-hexane and selecting the hydrogen-capped polyynes with 10 *sp*-carbon atoms ($HC_{10}H$), they mixed the $HC_{10}H$ solution with the SWCNTs and kept at 80 °C for two days. The resulting solid was washed with n-hexane and dried at 80 °C for 24 hours. Finally, Raman spectroscopy and X-ray diffraction confirmed the realization of this hybrid material and its stability well above 670 K. Similarly, Malard *et al.* [207] and Moura *et al.* [208] encapsulated polyynes into SWCTNs and DWCNTs using different methods. To encapsulate polyynes inside SWCNTs, a piece of nanotube film was dipped inside a separated polyyne dispersion in n-hexane and kept at 80 °C for 24 h under vacuum. The resulting samples were carefully washed with n-hexane to completely remove any polyynes remained attached outside. Instead, DWNTs were heated at 823 K in dry air for 30 min to remove their endcaps before a small piece of open-ended DWNTs was dipped in a methanol solution of polyyne and kept at 353 K for 24 h.

3.11 Conclusions

In this chapter we have discussed the state of the art of research on the synthesis of polyynes by PLAL. PLAL is a versatile technique able to efficiently produce different types of polyynes with various terminations and lengths. The use of laser ablation enables the adoption of different targets and the *contactless* process based on laser-matter interaction allows the use of *in situ* characterization techniques aimed at understanding the formation mechanisms. Through PLAL, researchers have achieved remarkable progress in the synthesis of polyynes with controlled sizes and properties. The liquid environment in PLAL plays a crucial role in enhancing control over the growth process, minimizing thermal damage, and enabling the incorporation of functional additives. By carefully selecting the liquid medium and optimizing the laser parameters, it is possible to achieve high-quality polyynes with desirable properties. Furthermore, the ability to synthesize polyynes with specific structures and sizes using PLAL opens up new avenues for exploring their unique chemical and physical properties. The availability of well-defined polyynes facilitates the investigation of their optical, electronic, and mechanical characteristics. This deeper understanding of polyynes synthesized *via* PLAL enables us to explore their potential applications in various fields. The possibility to produce different chains with tailored terminations and the realization of solid materials with long-term stability can open the exploitation of the peculiar properties of *sp*-carbon chains in advanced applications. For instance, the peculiar nonlinear optical properties are appealing in optoelectronics while applications in biosensing and organic electronics have been recently reported [159, 211, 212]. The current research on polyynes synthesized through

PLAL highlights their promising potential in nanotechnology, materials science, biomedicine, and energy storage. Polyynes exhibit exceptional properties such as high mechanical strength, excellent electrical conductivity, and intriguing optical behavior. By harnessing these properties, polyynes can be utilized in the development of advanced nanomaterials, functional devices, sensors, and energy storage systems. Looking ahead, future research should focus on further optimizing the PLAL process parameters to achieve even greater control over the properties of the synthesized polyynes. Exploring novel liquid environments, investigating the influence of laser parameters, and integrating additional functional additives during PLAL are areas that hold great potential for advancing the synthesis of polyynes with enhanced properties and functionalities.

Acknowledgements

The authors acknowledge funding from the European Research Council (ERC) under the European Union's Horizon 2020 research and innovation program ERC-Consolidator Grant (ERC CoG 2016 EspLORE grant agreement no. 724610, website: www.esplore.polimi.it).

References

1. V. Amendola, and M. Meneghetti. What controls the composition and the structure of nanomaterials generated by laser ablation in liquid solution? *Physical Chemistry Chemical Physics,* 15(9):3027–3046, **2013**. doi:10.1039/C2CP42895D.
2. A. Kanitz, M. Kalus, E. Gurevich, A. Ostendorf, S. Barcikowski, and D. Amans. Review on experimental and theoretical investigations of the early stage, femtoseconds to microseconds processes during laser ablation in liquid-phase for the synthesis of colloidal nanoparticles. *Plasma Sources Science and Technology,* 28(10):103001, **2019**. doi:10.1088/1361-6595/ab3dbe.
3. D. Amans, M. Diouf, J. Lam, G. Ledoux, and C. Dujardin. Origin of the nano-carbon allotropes in pulsed laser ablation in liquids synthesis. *Journal of Colloid and Inter/Ace Science,* 489:114–125, **2017**. doi:10.1016/j.jcis.2016. 08.017.

4. A. Santagata, A. De Bonis, A. De Giacomo, M. Dell'Aglio, A. Laurita, G. Senesi, R. Gaudiuso, S. Orlando, R. Teghil, and G. Parisi. Carbon-based nanostructures obtained in water by ultrashort laser pulses. *The Journal of Physical Chemistry C,* 115(12):5160–5164, **2011**. doi:10.1021/jp1094239.
5. P. Wagener, J. Jakobi, C. Rehbock, V. S. K. Chakravadhanula, C. Thede, U. Wiedwald, M. Bartsch, L. Kienle, and S. Barcikowski. Solvent-surface interactions control the phase structure in laser-generated iron-gold core-shell nanoparticles. *Scientific Reports,* 6(1):1–12, **2016**. doi:10.1038/srep23352.
6. A. Tymoczko, M. Kamp, O. Prymak, C. Rehbock, J. Jakobi, U. Schurmann, L. Kienle, and S. Barcikowski. How the crystal structure and phase segregation of Au-Fe alloy nanoparticles are ruled by the molar fraction and size. *Nanoscale,* 10(35):16434–16437, **2018**. doi:10.1039/C8NR03962C.
7. Z. Swiatkowska-Warkocka, A. Pyatenko, F. Krok, B. R. Jany, and M. Marszalek. Synthesis of new metastable nanoalloys of immiscible metals with a pulse laser technique. *Scientific Reports,* 5(1):1–6, **2015**. doi:10.1038/srep09849.
8. C. L. Sajti, R. Sattari, B. N. Chichkov, and S. Barcikowski. Gram scale synthesis of pure ceramic nanoparticles by laser ablation in liquid. *The Journal of Physical Chemistry C,* 114(6):2421–2427, **2010**. doi:10.1021/jp906960g.
9. S. Jendrzej, B. Gokce, M. Epple, and S. Barcikowski. How size determines the value of gold: economic aspects of wet chemical and laser-based metal colloid synthesis. *ChemPhysChem,* 18(9):1012–1019, **2017**. doi:10.1002/cphc.201601139.
10. R. Streubel, S. Barcikowski, and B. Gokce. Continuous multigram nanoparticle synthesis by high-power, high-repetition-rate ultrafast laser ablation in liquids. *Optics Letters,* 41(7):1486–1489, **2016**. doi:10.1364/OL.41.001486.
11. H. W. Kroto, J. R. Heath, S. C. O'Brien, R. F. Curl, and R. E. Smalley. C60: buckminsterfullerene. *Nature,* 318(6042):162–163, **1985**. doi:10.1038/318162a0.
12. C. S. Casari, M. Tommasini, R. R. Tykwinski, and A. Milani. Carbon-atom wires: 1-D systems with tunable properties. *Nanoscale,* 8(8):4414–4435, **2016**. doi:10.1039/C5NR06175J.
13. P. Milani, and S. Iannotta. *Cluster Beam Synthesis of Nanostructured Materials.* 1st edn, Springer Series in Cluster Physics, Springer Berlin, Heidelberg, **2012**, ISBN 978-3-642-59899-9.

14. L. Ravagnan, F. Siviero, C. Lenardi, P. Piseri, E. Barborini, P. Milani, C. S. Casari, A. L. Bassi, and C. E. Bot-tani. Cluster-beam deposition and in situ characterization of carbyne-rich carbon films. *Physical Review Letters,* 89(28):285506, **2002**. doi:10.1103/PhysRevLett.89.285506.

15. C. S. Casari, A. L. Bassi, L. Ravagnan, F. Siviero, C. Lenardi, P. Piseri, G. Bongiorno, C. E. Bottani, and P. Milani. Chemical and thermal stability of carbyne-like structures in cluster-assembled carbon films. *Physical Review B,* 69(7):075422, **2004**. doi:10.1103/PhysRevB.69.075422.

16. L. Ravagnan, P. Piseri, M. Bruzzi, S. Miglio, G. Bongiorno, A. Baserga, C. S. Casari, A. L. Bassi, C. Lenardi, Y. Ya-maguchi *et al.* Influence of cumulenic chains on the vibrational and electronic properties of sp-sp^2 amorphous carbon. *Physical Review Letters,* 98(21):216103, **2007**. doi:10.1103/PhysRevLett.98.216103.

17. M. Bruzzi, P. Piseri, S. Miglio, G. Bongiorno, E. Barborini, C. Ducati, J. Robertson, and P. Milani. Electrical conduction in nanostructured carbon and carbon-metal films grown by supersonic cluster beam deposition. *The European Physical Journal B-Condensed Matter and Complex Systems,* 36(1):3–13, **2003**. doi:10.1140/epjb/e2003-00311-4.

18. C. S. Casari, C. S. Giannuzzi, and V. Russo. Carbon-atom wires produced by nanosecond pulsed laser deposition in a background gas. *Carbon,* 104:190–195, **2016**. doi:10.1016/j.carbon.2016.03.056.

19. F. Cataldo. Simple generation and detection of polyynes in an arc discharge between graphite electrodes submerged in various solvents. *Carbon (New York, NY),* 41(13):2671–2674, 2003. doi:10.1016/S0008-6223(03)00345-2.

20. H. Lange, M. Sioda, A. Huczko, Y. Zhu, H. Kroto, and D. Walton. Nanocarbon production by arc discharge in water. *Carbon,* 41(8):1617–1623, **2003**. doi:10.1016/S0008-6223(03)00111-8.

21. F. Cataldo. Synthesis of polyynes in a submerged electric arc in organic solvents. *Carbon,* 42(1):129–142, **2004**. doi:10.1016/j.carbon.2003.10.016.

22. F. Cataldo. Submerged electric arc between graphite electrodes: a one-pot tool for the synthesis of long-chain polyynes in solution. *Tetrahedron Letters,* 45(1):141–144, **2004**. doi:10.1016/j.tetlet.2003.10.100.

23. F. Cataldo. *Polyynes: Synthesis, Properties, and Applications,* CRC Press, 2005. ISBN 9781574445121.

24. S. Peggiani, A. Senis, A. Facibeni, A. Milani, P. Serafini, G. Cerrato, A. Lucotti, M. Tommasini, D. Fazzi, C. Castiglioni *et al.* Size-selected

polyynes synthesised by submerged arc discharge in water. *Chemical Physics Letters*, 740:137054, **2020**. doi:10.1016/j.cplett.2019.137054.

25. S. Peggiani. *Fabrication and Characterization of Carbon-Atom Wires and Wire-Based Nanocomposites.* Ph.D. Thesis, Energy and Nuclear Science and Technology, **2021**.
26. F. Cataldo. Polyynes and cyanopolyynes synthesis from the submerged electric arc: about the role played by the electrodes and solvents in polyynes formation. *Tetrahedron,* 60(19):4265–4274, **2004**. doi:10.1016/j.tet.2004.03.033.
27. F. Cataldo. Polyynes: a new class of carbon allotropes. About the formation of dicyanopolyynes from an electric arc between graphite electrodes in liquid nitrogen. *Polyhedron,* 23(11):1889–1896, **2004**. doi:10.1016/j.poly.2004.04.024.
28. Y. Wu, Y. Zhang, T. Zhu, H. Li, Y. Liu, and X. Zhao. Effects of precursor molecules on polyyne formation by arc discharge between two copper electrodes. *Chemical Physics Letters,* 730:64–69, **2019**. doi:10.1016/j.cplett.2019.05.040.
29. P. Marabotti, S. Peggiani, A. Vidale, and C. S. Casari. Pulsed laser ablation in liquid of sp-carbon chains: status and recent advances. *Chinese Physics B,* 31(12):125202, **2022**. doi:10.1088/1674-1056/ac81b2.
30. Y. Sato, T. Kodama, H. Shiromaru, J. Sanderson, T. Fujino, Y. Wada, T. Wakabayashi, and Y. Achiba. Synthesis of polyyne molecules from hexane by irradiation of intense femtosecond laser pulses. *Carbon,* 48(5):1673–1676, **2010**. doi:10.1016/j.carbon.2009.12.036.
31. A. Hu, J. Sanderson, A. Zaidi, C. Wang, T. Zhang, Y. Zhou, and W. Duley. Direct synthesis of polyyne molecules in acetone by dissociation using femtosecond laser irradiation. *Carbon,* 46(13):1823–1825, **2008**. doi:10.1016/j.carbon.2008.07.036.
32. A. Zaidi, A. Hu, D. Henneke, and W. Duley. Femtosecond laser irradiation of liquid alkanes: mechanism of polyyne formation. *Chemical Physics Letters,* 723:151–154, **2019**. doi:10.1016/j.cplett.2019.01.036.
33. M. Tsuji, T. Tsuji, S. Kuboyama, S.-H. Yoon, Y. Korai, T. Tsujimoto, K. Kubo, A. Mori, and I. Mochida. Formation of hydrogen-capped polyynes by laser ablation of graphite particles suspended in solution. *Chemical Physics Letters,* 355(1–2):101–108, **2002**. doi:10.1016/S0009-2614(02)00192-6.
34. M. Tsuji, S. Kuboyama, T. Matsuzaki, and T. Tsuji. Formation of hydrogen-capped polyynes by laser ablation of C60 particles suspended in solution. *Carbon,* 41(11):2141–2148, **2003**. doi:10.1016/S0008-6223(03)00241-0.

35. G. Compagnini, V. Mita, R. S. Cataliotti, L. D'Urso, and O. Puglisi. Short polyyne chains produced by pulsed laser ablation of graphite in water. *Carbon,* 45(12):2456–2458, **2007**. doi:10.1016/j.carbon.2007.07.002.

36. S. Peggiani, P. Marabotti, R. A. Lotti, A. Facibeni, P. Serafini, A. Milani, V. Russo, A. L. Bassi, and C. S. Casari. Solvent-dependent termination, size and stability in polyynes synthesized via laser ablation in liquids. *Physical Chemistry Chemical Physics,* 22(45):26312–26321, **2020**. doi:10.1039/D0CP04132G.

37. R. Matsutani, K. Inoue, T. Sanada, N. Wada, and K. Kojima. Preparation of long-chain polyynes of $C_{28}H_2$ and $C_{30}H_2$ by liquid-phase laser ablation. *Journal of Photochemistry and Photobiology A: Chemistry,* 240:1–4, **2012**. doi:10.1016/j.jphotochem.2012.05.004.

38. Y. E. Park, S. K. Shin, and S. M. Park. The physical effects on the formation of polyynes by laser ablation. *Chemical Physics Letters,* 568:112–116, **2013**. doi:10.1016/j.cplett.2013.03.016.

39. R. Matsutani, T. Kakimoto, H. Tanaka, and K. Kojima. Preparation of polyynes by liquid-phase laser ablation using different irradiation target materials and solvents. *Carbon,* 49(1):77–81, **2011**. doi:10.1016/j.carbon.2010.08.044.

40. S. Peggiani, A. Facibeni, A. Milani, C. Castiglioni, V. Russo, A. L. Bassi, and C. S. Casari. In situ synthesis of polyynes in a polymer matrix via pulsed laser ablation in a liquid. *Materials Advances,* 1(8):2729–2736, **2020**. doi:10.1039/D0MA00545B.

41. R. Matsutani, K. Inoue, N. Wada, and K. Kojima. Wavelength dependence of polyyne preparation by liquid-phase laser ablation using pellet targets. *Chemical Communications,* 47(20):5840–5842, **2011**. doi:10.1039/c1cc00102g.

42. A. Menendez-Manjon, P. Wagener, and S. Barcikowski. Transfer-matrix method for efficient ablation by pulsed laser ablation and nanoparticle generation in liquids. *The Journal of Physical Chemistry C,* 115(12):5108–5114, **2011**. doi:10.1021/jp109370q.

43. P. Marabotti, S. Peggiani, A. Facibeni, P. Serafini, A. Milani, V. Russo, A. L. Bassi, and C. S. Casari. In situ surface-enhanced Raman spectroscopy to investigate polyyne formation during pulsed laser ablation in liquid. *Carbon,* 189:219–229, **2022**. doi:10.1016/j.carbon.2021.12.060.

44. L. Berthe, R. Fabbro, P. Peyre, and E. Bartnicki. Wavelength dependent of laser shock-wave generation in the water-confinement regime. *Journal of Applied Physics,* 85(11):7552–7555, **1999**. doi:10.1063/1.370553.

45. W. T. Nichols, T. Sasaki, and N. Koshizaki. Laser ablation of a platinum target in water. I. Ablation mechanisms. *Journal of Applied Physics,* 100(11):114912, **2006**. doi:10.1063/1.2390640.

46. A. V. Kabashin, and M. Meunier. Synthesis of colloidal nanoparticles during femtosecond laser ablation of gold in water. *Journal of Applied Physics,* 94(12):7941–7943, **2003**. doi:10.1063/1.1626793.

47. J. S. Hoppius, S. Maragkaki, A. Kanitz, P. Gregorcic, and E. L. Gurevich. Optimization of femtosecond laser processing in liquids. *Applied Surface Science,* 467:255–260, **2019**. doi:10.1016/j.apsusc.2018.10.121.

48. C. Momma, B. N. Chichkov, S. Nolte, F. von Alvensleben, A. Tunnermann, H. Welling, and B. Wellegehausen. Short-pulse laser ablation of solid targets. *Optics Communications,* 129(1–2):134–142, **1996**. doi:10.1016/0030-4018(96)00250-7.

49. A. Kaiser, B. Rethfeld, M. Vicanek, and G. Simon. Microscopic processes in dielectrics under irradiation by subpicosecond laser pulses. *Physical Review B,* 61(17):11437, **2000**. doi:10.1103/PhysRevB.61.11437.

50. V. Gruzdev. Photoionization rate in wide band-gap crystals. *Physical Review B,* 75(20):205106, **2007**. doi:10.1103/PhysRevB.75.205106.

51. V. Temnov, K. Sokolowski-Tinten, P. Zhou, A. El-Khamhawy, and D. Von Der Linde. Multiphoton ionization in dielectrics: comparison of circular and linear polarization. *Physical Review Letters,* 97(23):237403, **2006**. doi:10.1103/PhysRevLett.97.237403.

52. L. V. Zhigilei, P. B. Kodali, and B. J. Garrison. A microscopic view of laser ablation. *The Journal of Physical Chemistry B,* 102:2845–2853, **1998**. doi:10.1021/jp9733781.

53. L. V. Zhigilei, Z. Lin, and D. S. Ivanov. Atomistic modeling of short pulse laser ablation of metals: connections between melting, spallation, and phase explosion. *The Journal of Physical Chemistry C,* 113(27):11892–11906, **2009**. doi:10.1021/jp902294m.

54. D. Perez, and L. J. Lewis. Molecular-dynamics study of ablation of solids under femtosecond laser pulses. *Physical Review B,* 67(18):184102, **2003**. doi:10.1103/PhysRevB.67.184102.

55. D. Perez, L. K. Beland, D. Deryng, L. J. Lewis, and M. Meunier. Numerical study of the thermal ablation of wet solids by ultrashort laser pulses. *Physical Review B,* 77(1):014108, **2008**. doi:10.1103/PhysRevB.77.014108.

56. Y. Sano, N. Mukai, K. Okazaki, and M. Obata. Residual stress improvement in metal surface by underwater laser irradiation. *Nuclear Instruments and Methods in Physics Research Section B: Beam Interactions with*

Materials and Atoms, 121(1–4):432–436, **1997**. doi:10.1016/S0168-583X(96)00551-4.

57. L. Berthe, R. Fabbro, P. Peyre, L. Tollier, and E. Bartnicki. Shock waves from a water-confined laser-generated plasma. *Journal of Applied Physics,* 82(6):2826–2832, **1997**. doi:10.1063/1.366113.

58. T. Sakka, S. Iwanaga, Y. H. Ogata, A. Matsunawa, and T. Takemoto. Laser ablation at solid-liquid interfaces: an approach from optical emission spectra. *The Journal of Chemical Physics,* 112(19):8645–8653, **2000**. doi:10.1063/1.481465.

59. H. W. Kang, H. Lee, and A. J. Welch. Laser ablation in a liquid-confined environment using a nanosecond laser pulse. *Journal of Applied Physics,* 103(8):083101, **2008**. doi:10.1063/1.2905314.

60. B. Kumar, and R. K. Thareja. Synthesis of nanoparticles in laser ablation of aluminum in liquid. *Journal of Applied Physics,* 108(6):064906, **2010**. doi:10.1063/1.3486517.

61. D. Amans, A.-C. Chenus, G. Ledoux, C. Dujardin, C. Reynaud, O. Sublemontier, K. Masenelli-Varlot, and O. Guillois. Nanodiamond synthesis by pulsed laser ablation in liquids. *Diamond and Related Materials,* 18(2–3):177–180, **2009**. doi:10.1016/j.diamond.2008.10.035.

62. M. Dell'Aglio, R. Gaudiuso, O. De Pascale, and A. De Giacomo. Mechanisms and processes of pulsed laser ablation in liquids during nanoparticle production. *Applied Surface Science,* 348:4–9, **2015**. doi:10.1016/j.apsusc.2015.01.082.

63. T. Sakka, K. Takatani, Y. H. Ogata, and M. Mabuchi. Laser ablation at the solid-liquid interface: transient absorption of continuous spectral emission by ablated aluminium atoms. *Journal of Physics D: Applied Physics,* 35(1):65, **2001**. doi:10.1088/0022-3727/35/1/312.

64. P. Boyer, and M. Meunier. Modeling solvent influence on growth mechanism of nanoparticles (Au, Co) synthesized by surfactant free laser processes. *The Journal of Physical Chemistry C,* 116(14):8014–8019, **2012**. doi:10.1021/jp2092994.

65. T. Begildayeva, S. J. Lee, Y. Yu, J. Park, T. H. Kim, J. Theerthagiri, A. Ahn, H. J. Jung, and M. Y. Choi. Production of copper nanoparticles exhibiting various morphologies via pulsed laser ablation in different solvents and their catalytic activity for reduction of toxic nitroaromatic compounds. *Journal of Hazardous Materials,* 409:124412, **2021**. doi:10.1016/j.jhazmat.2020.124412.

66. J. Yoo, S. Jeong, X. Mao, R. Greif, and R. Russo. Evidence for phase-explosion and generation of large particles during high power nanosecond laser ablation of silicon. *Applied Physics Letters,* 76(6):783–785, **2000**. doi:10.1063/1.125894.

67. Q. Lu, S. S. Mao, X. Mao, and R. E. Russo. Delayed phase explosion during high-power nanosecond laser ablation of silicon. *Applied Physics Letters,* 80(17):3072–3074, **2002**. doi:10.1063/1.1473862.

68. T. Sakka, S. Masai, K. Fukami, and Y. H. Ogata. Spectral profile of atomic emission lines and effects of pulse duration on laser ablation in liquid. *Spectrochimica Acta Part B: Atomic Spectroscopy,* 64(10):981–985, **2009**. doi:10.1016/j.sab.2009.07.018.

69. T. Tsuji, Y. Okazaki, Y. Tsuboi, and M. Tsuji. Nanosecond time-resolved observations of laser ablation of silver in water. *Japanese Journal of Applied Physics,* 46(4R):1533, **2007**. doi:10.1143/JJAP.46.1533.

70. D. G. Shchukin, and H. Mohwald. Sonochemical nanosynthesis at the engineered interface of a cavitation microbubble. *Physical Chemistry Chemical Physics,* 8(30):3496–3506, **2006**. doi:10.1039/B606104D.

71. T. E. Itina. On nanoparticle formation by laser ablation in liquids. *The Journal of Physical Chemistry C,* 115(12):5044–5048, **2011**. doi:10.1021/jp1090944.

72. T. Tsuji, Y. Tsuboi, N. Kitamura, and M. Tsuji. Microsecond-resolved imaging of laser ablation at solid-liquid interface: investigation of formation process of nano-size metal colloids. *Applied Surface Science,* 229(1–4):365–371, **2004**. doi:10.1016/j.apsusc.2004.02.013.

73. A. De Giacomo, A. De Bonis, M. Dell'Aglio, O. De Pascale, R. Gaudiuso, S. Orlando, A. Santagata, G. Senesi, F. Taccogna, and R. Teghil. Laser ablation of graphite in water in a range of pressure from 1 to 146 atm using single and double pulse techniques for the production of carbon nanostructures. *The Journal of Physical Chemistry* C, 115(12):5123–5130, **2011**. doi:10.1021/jp109389c.

74. R. Tanabe, T. T. Nguyen, T. Sugiura, and Y. Ito. Bubble dynamics in metal nanoparticle formation by laser ablation in liquid studied through high-speed laser stroboscopic videography. *Applied Surface Science,* 351:327–331, **2015**. doi:10.1016/j.apsusc.2015.05.030.

75. A. De Bonis, M. Sansone, L. D'Alessio, A. Galasso, A. Santagata, and R. Teghil. Dynamics of laser-induced bubble and nanoparticles generation during ultra-short laser ablation of Pd in liquid. *Journal of Physics D: Applied Physics,* 46(44):445301, **2013**. doi:10.1088/0022-3727/46/44/445301.

76. V. K. Meader, M. G. John, L. M. Frias Batista, S. Ahsan, and K. M. Tibbetts. Radical chemistry in a femtosecond laser plasma: photochemical reduction of Ag+ in liquid ammonia solution. *Molecules,* 23(3):532, **2018**. doi:10.3390/molecules23030532.

77. V. Amendola, S. Polizzi, and M. Meneghetti. Laser ablation synthesis of silver nanoparticles embedded in graphitic carbon matrix. *Science of Advanced Materials,* 4(3–4):497–500, **2012**. doi:10.1166/sam.2012.1308.

78. V. Amendola, S. Polizzi, and M. Meneghetti. Laser ablation synthesis of gold nanoparticles in organic solvents. *The Journal of Physical Chemistry B,* 110(14):7232–7237, **2006**. doi:10.1021/jp0605092.

79. V. Amendola, P. Riello, and M. Meneghetti. Magnetic nanoparticles of iron carbide, iron oxide, iron@ iron oxide, and metal iron synthesized by laser ablation in organic solvents. *The Journal of Physical Chemistry C,* 115(12):5140–5146, **2011**. doi:10.1021/jp109371m.

80. H. J. Jung, and M. Y. Choi. One-pot synthesis of graphitic and nitrogen-doped graphitic layers on nickel nanoparticles produced by pulsed laser ablation in liquid: solvent as the carbon and nitrogen source. *Applied Surface Science,* 457:1050–1056, **2018**. doi:10.1016/j.apsusc.2018.07.036.

81. J.-P. Sylvestre, A. V. Kabashin, E. Sacher, and M. Meunier. Femtosecond laser ablation of gold in water: influence of the laser-produced plasma on the nanoparticle size distribution. *Applied Physics A,* 80(4):753–758, **2005**. doi:10.1007/s00339-004-3081-4.

82. G. Marzun, J. Nakamura, X. Zhang, S. Barcikowski, and P. Wagener. Size control and supporting of palladium nanoparticles made by laser ablation in saline solution as a facile route to heterogeneous catalysts. *Applied Surface Science,* 348:75–84, **2015**. doi:10.1016/j.apsusc.2015.01.108.

83. G. Cristoforetti, E. Pitzalis, R. Spiniello, R. Ishak, F. Giammanco, M. Muniz-Miranda, and S. Caporali. Physico-chemical properties of Pd nanoparticles produced by Pulsed Laser Ablation in different organic solvents. *Applied Surface Science,* 258(7):3289–3297, **2012**. doi:10.1016/j.apsusc.2011.11.084.

84. B. Kumar, D. Yadav, and R. K. Thareja. Growth dynamics of nanoparticles in laser produced plasma in liquid ambient. *Journal of Applied Physics,* 110(7):074903, **2011**. doi:10.1063/1.3642968.

85. M. Dell'Aglio, A. Santagata, G. Valenza, A. De Stradis, and A. De Giacomo. Study of the effect of water pressure on plasma and cavitation bubble induced by pulsed laser ablation in liquid of silver and missed

variations of observable nanoparticle features. *ChemPhysChem,* 18(9):1165–1174, **2017**. doi:10.1002/cphc.201601231.

86. S. Ibrahimkutty, P. Wagener, T. D. S. Rolo, D. Karpov, A. Menzel, T. Baumbach, S. Barcikowski, and A. Plech. A hierarchical view on material formation during pulsed-laser synthesis of nanoparticles in liquid. *Scientific Reports,* 5(1):1–11, **2015**. doi:10.1038/srep16313.
87. S. Reich, A. Letzel, A. Menzel, N. Kretzschmar, B. Gokce, S. Barcikowski, and A. Plech. Early appearance of crystalline nanoparticles in pulsed laser ablation in liquids dynamics. *Nanoscale,* 11(14):6962–6969, **2019**. doi:10.1039/C9NR01203F.
88. C.-Y. Shih, C. Wu, M. V. Shugaev, and L. V. Zhigilei. Atomistic modeling of nanoparticle generation in short pulse laser ablation of thin metal films in water. *Journal of Colloid and Interface Science,* 489:3–17, **2017**. doi:10.1016/j.jcis.2016.10.029.
89. C.-Y. Shih, R. Streubel, J. Heberle, A. Letzel, M. V. Shugaev, C. Wu, M. Schmidt, B. Gokce, S. Barcikowski, and L. V. Zhigilei. Two mechanisms of nanoparticle generation in picosecond laser ablation in liquids: the origin of the bimodalsize distribution. *Nanoscale,* 10(15):6900–6910, **2018**. doi:10.1039/C7NR08614H.
90. S. Ibrahimkutty, P. Wagener, A. Menzel, A. Plech, and S. Barcikowski. Nanoparticle formation in a cavitation bubble after pulsed laser ablation in liquid studied with high time resolution small angle x-ray scattering. *Applied Physics Letters,* 101(10):103104, **2012**. doi:10.1063/1.4750250.
91. P. Wagener, S. Ibrahimkutty, A. Menzel, A. Plech, and S. Barcikowski. Dynamics of silver nanoparticle formation and agglomeration inside the cavitation bubble after pulsed laser ablation in liquid. *Physical Chemistry Chemical Physics,* 15(9):3068–3074, **2013**. doi:10.1039/C2CP42592K.
92. A. Letzel, B. Gokce, P. Wagener, S. Ibrahimkutty, A. Menzel, A. Plech, and S. Barcikowski. Size quenching during laser synthesis of colloids happens already in the vapor phase of the cavitation bubble. *The Journal of Physical Chemistry C,* 121(9):5356–5365, **2017**. doi:10.1021/acs.jpcc.6b12554.
93. S. Ogale, A. Malshe, S. Kanetkar, and S. Kshirsagar. Formation of diamond particulates by pulsed ruby laser irradiation of graphite immersed in benzene. *Solid State Communications,* 84(4):371–373, **1992**. doi:10.1016/0038-1098(92)90479-S.
94. J. Singh, M. Vellaikal, and J. Narayan. Laser-enhanced synthesis and processing of diamond films from liquid hydrocarbons. *Journal of Applied Physics,* 73(9):4351–4356, **1993**. doi:10.1063/1.352819.

95. C. Frondel, and U. B. Marvin. Lonsdaleite, a hexagonal polymorph of diamond. *Nature,* 214(5088):587–589, **1967**. doi:10.1038/214587a0.

96. P. Nemeth, L. A. Garvie, T. Aoki, N. Dubrovinskaia, L. Dubrovinsky, and P. R. Buseck. Lonsdaleite is faulted and twinned cubic diamond and does not exist as a discrete material. *Nature Communications,* 5(1):1–5, **2014**. doi:10.1038/ncomms6447.

97. R. L. Johnston, and R. Hoffmann. Superdense carbon, C8: supercubane or analog of y-Si? *Journal of the American Chemical Society,* 111(3):810–819, **1989**. doi:10.1021/ja00185a004.

98. S. Z. Mortazavi, P. Parvin, A. Reyhani, S. Mirershadi, and R. Sadighi-Bonabi. Generation of various carbon nanostructures in water using IR/UV laser ablation. *Journal of Physics D: Applied Physics,* 46(16):165303, **2013**. doi:10.1088/0022-3727/46/16/165303.

99. J. Xiao, J. Li, P. Liu, and G. Yang. A new phase transformation path from nanodiamond to new-diamond via an intermediate carbon onion. *Nanoscale,* 6(24):15098–15106, **2014**. doi:10.1039/C4NR05246C.

100. Y. Wu, F. Jelezko, M. B. Plenio, and T. Weil. Diamond quantum devices in biology. *Angewandte Chemie International Edition,* 55(23):6586–6598, **2016**. doi:10.1002/anie.201506556.

101. A. Sharma, R. Vispute, D. Joag, S. Ogale, S. D. Joag, P. Ayyub, M. Multani, G. Dey, and S. Banerjee. Diamond nucleation at the organic liquid-metal interface by laser-induced reactive quenching. *Materials Letters,* 17(1–2):42–48, **1993**. doi:10.1016/0167-577X(93)90145-N.

102. A. Sharma, H. Salunke, G. Das, P. Ayyub, and M. Multani. Electronic structure of the 4H polytype of diamond. *Journal of Physics: Condensed Matter,* 8(31):5801, **1996**. doi:10.1088/0953-8984/8/31/013.

103. Y. Wang, R. Yu, Z. Liu, R. Huang, L. Zheng, X. Zhang, G. Hu, and Z. Zhang. Production of diamond nanospherulite at carbon-water interface by laser ablation and its characterization by TEM. *Chemical Journal of Chinese Universities-Chinese,* 18(1):124–126, **1997.**

104. Y. Lu, S. Huang, X. Wang, and Z. Shen. Laser-assisted growth of diamond particulates on a silicon surface from a cyclohexane liquid. *Applied Physics A: Materials Science & Processing,* 66(5), **1998**. doi:10.1007/s003390050710.

105. G.-W. Yang, J.-B. Wang, and Q.-X. Liu. Preparation of nano-crystalline diamonds using pulsed laser induced reactive quenching. *Journal of Physics: Condensed Matter,* 10(35):7923, **1998**. doi:10.1088/0953-8984/10/35/024.

106. S. Pearce, S. Henley, F. Claeyssens, P. May, K. Hallam, J. Smith, and K. Rosser. Production of nanocrystalline diamond by laser ablation at the solid/liquid interface. *Diamond and Related Materials,* 13(4–8):661–665, **2004**. doi:10.1016/j.diamond.2003.08.027.

107. L. Yang, P. May, L. Yin, J. Smith, and K. Rosser. Growth of diamond nanocrystals by pulsed laser ablation of graphite in liquid. *Diamond and Related Materials,* 16(4–7):725–729, **2007**. doi:10.1016/j.diamond.2006.11.010.

108. C. Chen, W. Chen, and Y. Zhang. Synthesis of carbon nano-tubes by pulsed laser ablation at normal pressure in metal nano-sol. *Physica E: Low-Dimensional Systems and Nanostructures,* 28(2):121–127, **2005**. doi:10.1016/j. physe.2005.02.006.

109. P. Bai, S. Hu, T. Zhang, J. Sun, and S. Cao. Effect of laser pulse parameters on the size and fluorescence of nanodiamonds formed upon pulsed-laser irradiation. *Materials Research Bulletin,* 45(7):826–829, **2010**. doi:10.1016/j.materresbull.2010.03.009.

110. J. Sun, S.-L. Hu, X.-W. Du, Y.-W. Lei, and L. Jiang. Ultrafine diamond synthesized by long-pulse-width laser. *Applied Physics Letters,* 89(18):183115, **2006**. doi:10.1063/1.2385210.

111. S. Hu, J. Sun, X. Du, F. Tian, and L. Jiang. The formation of multiply twinning structure and photoluminescence of well-dispersed nanodiamonds produced by pulsed-laser irradiation. *Diamond and Related Materials,* 17(2):142–146, **2008**. doi:10.1016/j.diamond.2007.11.009.

112. P. Liu, H. Cui, and G. Yang. Synthesis of body-centered cubic carbon nanocrystals. *Crystal Growth and Design,* 8(2):581–586, **2008**. doi:10.1021/cg7006777.

113. P. Liu, Y. Cao, C. Wang, X. Chen, and G. Yang. Micro- and nanocubes of carbon with C8-like and blue luminescence. *Nano Letters,* 8(8):2570–2575, **2008**. doi:10.1021/nl801392v.

114. P. Liu, C. Wang, J. Chen, N. Xu, G. Yang, N. Ke, and J. Xu. Localized nanodiamond crystallization and field emission performance improvement of amorphous carbon upon laser irradiation in liquid. *The Journal of Physical Chemistry* C, 113(28):12154–12161, **2009**. doi:10.1021/jp901359b.

115. S. Hu, Y. Dong, J. Yang, J. Liu, and S. Cao. Simultaneous synthesis of luminescent carbon nanoparticles and carbon nanocages by laser ablation of carbon black suspension and their optical limiting properties. *Journal of Materials Chemistry,* 22(5):1957–1961, **2012**. doi:10.1039/C1JM14510J.

116. J. Xiao, G. Ouyang, P. Liu, C. Wang, and G. Yang. Reversible nanodiamond-carbon onion phase transformations. *Nano Letters,* 14(6):3645–3652, **2014**. doi:10.1021/nl5014234.

117. J. Xiao, P. Liu, and G. Yang. Nanodiamonds from coal under ambient conditions. *Nanoscale,* 7(14):6114–6125, **2015**. doi:10.1039/C4NR06186A.

118. S. Z. Mortazavi, P. Parvin, and A. Reyhani. Fabrication of graphene based on Q-switched Nd: YAG laser ablation of graphite target in liquid nitrogen. *Laser Physics Letters,* 9(7):547, **2012**. doi:10.7452/lapl.201210033.

119. A. Fojtik, and A. Henglein. Laser ablation of films and suspended particles in a solvent: formation of cluster and colloid solutions. *Berichte-Bunsengesellschaftfur Physikalische Chemie,* 97:252–252, **1993.**

120. G. Compagnini, V. Mita, L. D'Urso, R. S. Cataliotti, and O. Puglisi. Spectroscopic study of polyynes obtained by laser ablation in liquids. *Journal of Raman Spectroscopy: An International Journal for Original Work in all Aspects of Raman Spectroscopy, Including Higher Order Processes, and also Brillouin and Rayleigh Scattering,* 39(2):177–181, **2008**. doi:10.1002/jrs.1837.

121. T. Wakabayashi, M. Saikawa, Y. Wada, and T. Minematsu. Isotope scrambling in the formation of cyanopolyynes by laser ablation of carbon particles in liquid acetonitrile. *Carbon,* 50(1):47–56, **2012**. doi:10.1016/j.carbon.2011.07.053.

122. Y. Wada, K. Koma, Y. Ohnishi, Y. Sasaki, and T. Wakabayashi. Photoinduced reaction of methylpolyynes $H(C{\equiv}C)_nCH_3$ (n = 5–7) and polyyne $H(C{\equiv}C)_5H$ with I_2 molecules. *The European Physical Journal D,* 66(12):1–6, **2012**. doi:10.1140/epjd/e2012-30545-x.

123. G. Forte, L. D'Urso, E. Fazio, S. Patane, F. Neri, O. Puglisi, and G. Compagnini. The effects of liquid environments on the optical properties of linear carbon chains prepared by laser ablation generated plasmas. *Applied Surface Science,* 272:76–81, **2013**. doi:10.1016/j.apsusc.2012.03.156.

124. P. Russo, A. Hu, G. Compagnini, W. W. Duley, and N. Y. Zhou. Femtosecond laser ablation of highly oriented pyrolytic graphite: a green route for large-scale production of porous graphene and graphene quantum dots. *Nanoscale,* 6(4):2381–2389, **2014**. doi:10.1039/C3NR05572H.

125. A. Ramadhan, M. Wesolowski, T. Wakabayashi, H. Shiromaru, T. Fujino, T. Kodama, W. Duley, and J. Sanderson. Generation of polyyne and methylpolyyne molecules from toluene by intense femtosecond laser

pulse irradiation. *Journal of Physics: Conference Series,* 635:112125, **2015**. doi:10.1088/1742-6596/635/11/112125.

126. A. Ramadhan, M. Wesolowski, T. Wakabayashi, H. Shiromaru, T. Fujino, T. Kodama, W. Duley, and J. Sanderson. Synthesis of hydrogen-and methyl-capped long-chain polyynes by intense ultrashort laser pulse irradiation of toluene. *Carbon,* 118:680–685, **2017**. doi:10.1016/j.carbon.2017.03.096.

127. S.-K. Shin, and S.-M. Park. Preparation of polyynes by the laser ablation of graphite in water and organic solvents. *Bulletin of the Korean Chemical Society,* 33(2):597–600, **2012**. doi:10.5012/bkcs.2012.33.2.597.

128. K. Inoue, R. Matsutani, T. Sanada, and K. Kojima. Preparation of long-chain polyynes of $C_{24}H_2$ and $C_{26}H_2$ by liquid-phase laser ablation in decalin. *Carbon,* 48(14):4209–4211, **2010**. doi:10.1016/j.carbon.2010.07.020.

129. P. Marabotti, M. Tommasini, C. Castiglioni, P. Serafini, S. Peggiani, M. Tortora, B. Rossi, A. Li Bassi, V. Russo, and C. S. Casari. Electron-phonon coupling and vibrational properties of size-selected linear carbon chains by resonance Raman scattering. *Nature Communications,* 13(1):5052, **2022**. doi:10.1038/s41467-022-32801-3.

130. G. K. Yogesh, S. Shukla, D. Sastikumar, and P. Koinkar. Progress in pulsed laser ablation in liquid (PLAL) technique for the synthesis of carbon nanomaterials: a review. *Applied Physics A,* 127(11):1–40, **2021**. doi: 10.1007/s00339-021-04951-6.

131. R. Eastmond, T. Johnson, and D. Walton. Silylation as a protective method for terminal alkynes in oxidative couplings: a general synthesis of the parent polyynes $H(C{=}C)_nH$ (n = 4–10, 12). *Tetrahedron,* 28(17):4601–4616, **1972**. doi:10.1016/0040-4020(72)80041-3.

132. F. Shindo, Y. Benilan, J.-C. Guillemin, P. Chaquin, A. Jolly, and F. Raulin. Ultraviolet and infrared spectrum of C_6H_2 revisited and vapor pressure curve in Titan's atmosphere. *Planetary and Space Science,* 51(1):9–17, **2003**. doi:10.1016/S0032-0633(02)00151-4.

133. H. Tabata, *et al.* Laser ablation of diamond particles suspended in ethanol: Effective formation of long polyynes. *Carbon,* 44:522–529, **2006**. doi:10.1016/j.carbon.2005.08.007.

134. F. Shindo, Y. Benilan, P. Chaquin, J.-C. Guillemin, A. Jolly, and F. Raulin. IR spectrum of C_8H_2: integrated band intensities and some observational implications. *Journal of Molecular Spectroscopy,* 210(2):191–195, **2001**. doi:10.1006/jmsp.2001.8459.

135. U. Szczepaniak, K. Ozaki, K. Tanaka, Y. Ohnishi, Y. Wada, J.-C. Guillemin, C. Crepin, R. Kotos, Y. Mori-sawa, H. Suzuki, *et al.* Phosphorescence excitation mapping and vibrational spectroscopy of HC_9N and $HC_{11}N$ cyanopolyynes in organic solvents. *Journal of Molecular Structure,* 1214:128201, **2020**. doi:10.1016/j.molstruc. 2020.128201.

136. T. Wakabayashi, H. Nagayama, K. Daigoku, Y. Kiyooka, and K. Hashimoto. Laser induced emission spectra of polyyne molecules $C_{2n}H_2$ (n = 5–8). *Chemical Physics Letters,* 446(1–3):65–70, **2007**. doi:10.1016/j.cplett.2007.08.057.

137. T. Wakabayashi, U. Szczepaniak, K. Tanaka, S. Saito, K. Fukumoto, R. Ohnishi, K. Ozaki, T. Yamamoto, H. Suzuki, J.-C. Guillemin, *et al.* Phosphorescence of hydrogen-capped linear polyyne molecules C_8H_2, $C_{10}H_2$ and $C_{12}H_2$ in solid hexane matrices at 20 K. *Photochem,* 2(1):181–201, **2022**. doi:10.3390/photochem2010014.

138. M. Turowski, C. Crepin, S. Douin, M. Gronowski, I. Couturier-Tamburelli, N. Pietri, A. Wasiak, and R. Kotos. Low temperature Raman spectra of cyanobutadiyne (HC_5N). *Vibrational Spectroscopy,* 62:268–272, **2012**. doi:10.1016/j.vibspec.2012.05.005.

139. H. Tabata, M. Fujii, S. Hayashi, T. Doi, and T. Wakabayashi. Raman and surface-enhanced Raman scattering of a series of size-separated polyynes. *Carbon,* 44(15):3168–3176, **2006**. doi:10.1016/j.carbon.2006.07.004.

140. G. Compagnini, G. Patane, L. D'Urso, O. Puglisi, R. S. Cataliotti, and B. Pignataro. On the interaction of carbon nanowires with noble metals through a study of their surface-enhanced Raman spectra. *The Journal of Physical Chemistry C,* 112(51):20301–20306, **2008**. doi:10.1021/jp807969c.

141. A. Lucotti, C. S. Casari, M. Tommasini, A. L. Bassi, D. Fazzi, V. Russo, M. Del Zoppo, C. Castiglioni, F. Cataldo, C. E. Bottani, *et al.* sp Carbon chain interaction with silver nanoparticles probed by Surface Enhanced Raman Scattering. *Chemical Physics Letters,* 478(1–3):45–50, **2009**. doi:10.1016/j.cplett.2009.06.030.

142. F. Cataldo. The role of Raman spectroscopy in the research on sp-hybridized carbon chains: carbynoid structures polyynes and metal polyynides. *Journal of Raman Spectroscopy,* 39(2):169–176, **2008**. doi:10.1002/jrs.1830.

143. G. Grasso, L. D'Urso, E. Messina, F. Cataldo, O. Puglisi, G. Spoto, and G. Compagnini. A mass spectrometry and surface enhanced Raman spectroscopy study of the interaction between linear carbon chains and noble metals. *Carbon,* 47(11):2611–2619, **2009**. doi:10.1016/j.carbon.2009.05.013.

144. T. Wakabayashi, H. Tabata, T. Doi, H. Nagayama, K. Okuda, R. Umeda, I. Hisaki, M. Sonoda, Y. Tobe, T. Minematsu, *et al.* Resonance Raman spectra of polyyne molecules $C_{10}H_2$ and $C_{12}H_2$ in solution. *Chemical Physics Letters,* 433(4–6):296–300, **2007**. doi:10.1016/j.cplett.2006.11.077.

145. A. Milani, M. Tommasini, M. Del Zoppo, C. Castiglioni, and G. Zerbi. Carbon nanowires: phonon and π-electron confinement. *Physical Review B,* 74(15):153418, **2006**. doi:10.1103/PhysRevB.74.153418.

146. A. Milani, M. Tommasini, and G. Zerbi. Connection among Raman wavenumbers, bond length alternation and energy gap in polyynes. *Journal of Raman Spectroscopy,* 40(12):1931–1934, **2009**. doi:10.1002/jrs.2342.

147. J. Hobley, T. Nakamori, S. Kajimoto, M. Kasuya, K. Hatanaka, H. Fukumura, and S. Nishio. Formation of 3,4,9,10-perylenetetracarboxylic dianhydride nanoparticles with perylene and polyyne byproducts by 355 nm nanosecond pulsed laser ablation of microcrystal suspensions. *Journal of Photochemistry and Photobiology A: Chemistry,* 189(1):105–113, **2007**. doi:10.1016/j.jphotochem.2007.01.016.

148. G. K. Yogesh, E. Shuaib, A. K. Priya, P. Rohini, S. V. Anandhan, U. M. Krishnan, V. Kalyanavalli, S. Shukla, and D. Sastikumar. Synthesis of water-soluble fluorescent carbon nanoparticles (CNPs) from nanosecond pulsed laser ablation in ethanol. *Optics & Laser Technology,* 135:106717, **2021**. doi:10.1016/j.optlastec.2020.106717.

149. Y. Taguchi, H. Endo, Y. Abe, J. Matsumoto, T. Wakabayashi, T. Kodama, Y. Achiba, and H. Shiromaru. Polyyne formation by graphite laser ablation in argon and propane mixed gases. *Carbon,* 94:124–128, **2015**. doi:10.1016/j.carbon.2015.06.058.

150. S. Peggiani, A. Facibeni, P. Marabotti, A. Vidale, S. Scotti, and C. S. Casari. A single liquid chromatography procedure to concentrate, separate and collect size-selected polyynes produced by pulsed laser ablation in water. *Fullerenes, Nanotubes and Carbon Nanostructures,* 1–7, **2022**. doi:10.1080/1536383X.2022.2137498.

151. A. Milani, M. Tommasini, V. Russo, A. L. Bassi, A. Lucotti, F. Cataldo, and C. S. Casari. Raman spectroscopy as a tool to investigate the structure and electronic properties of carbon-atom wires. *Beilstein Journal of Nanotechnology,* 6(1):480–491, **2015**. doi:10.3762/bjnano.6.49.

152. C. Castiglioni, M. Tommasini, and G. Zerbi. Raman spectroscopy of polyconjugated molecules and materials: confinement effect in one and two dimensions. *Philosophical Transactions of the Royal Society of London. Series A: Mathematical, Physical and Engineering Sciences,* 362(1824):2425–2459, **2004**. doi:10.1098/rsta.2004.1448.

153. M. Tommasini, A. Milani, D. Fazzi, A. Lucotti, C. Castiglioni, J. A. Januszewski, D. Wendinger, and R. R. Tykwinski. π-conjugation and end group effects in long cumulenes: Raman spectroscopy and DFT calculations. *The Journal of Physical Chemistry C,* 118(45):26415–26425, **2014**. doi:10.1021/jp509724d.

154. A. Milani, V. Barbieri, A. Facibeni, V. Russo, A. Li Bassi, A. Lucotti, M. Tommasini, M. D. Tzirakis, F. Diederich, and C. S. Casari. Structure modulated charge transfer in carbon atomic wires. *Scientific Reports,* 9(1):1–10, **2019**. doi:10.1038/s41598-018-38367-9.

155. A. Milani, A. Lucotti, V. Russo, M. Tommasini, F. Cataldo, A. Li Bassi, and C. S. Casari. Charge transfer and vibrational structure of sp-hybridized carbon atomic wires probed by surface enhanced Raman spectroscopy. *The Journal of Physical Chemistry C,* 115(26):12836–12843, **2011**. doi:10.1021/jp203682c.

156. F. Cataldo, O. Ursini, A. Milani, and C. S. Casari. One-pot synthesis and characterization of polyynes end-capped by biphenyl groups (α, α-biphenylpolyynes). *Carbon,* 126:232–240, **2018**. doi:10.1016/j.carbon.2017.09.098.

157. A. Lucotti, M. Tommasini, D. Fazzi, M. Del Zoppo, W. A. Chalifoux, M. J. Ferguson, G. Zerbi, and R. R. Tykwinski. Evidence for solution-state nonlinearity of sp-carbon chains based on IR and Raman spectroscopy: violation of mutual exclusion. *Journal of the American Chemical Society,* 131(12):4239–4244, **2009**. doi:10.1021/ja078198b.

158. N. R. Agarwal, A. Lucotti, D. Fazzi, M. Tommasini, C. Castiglioni, W. A. Chalifoux, and R. R. Tykwinski. Structure and chain polarization of long polyynes investigated with infrared and Raman spectroscopy. *Journal of Raman Spectroscopy,* 44(10):1398–1410, **2013**. doi:10.1002/jrs.4300.

159. P. Marabotti, A. Milani, A. Lucotti, L. Brambilla, M. Tommasini, C. Castiglioni, P. Mecik, B. Pigulski, S. Szafert, and C. S. Casari. Vibrational and nonlinear optical properties of amine-capped push-pull polyynes by infrared and Raman spectroscopy. *Carbon Trends,* 5:100115, **2021**. doi:10.1016/j.cartre.2021.100115.

160. A. Campion, and P. Kambhampati. Surface-enhanced Raman scattering. *Chemical Society Reviews,* 27(4):241–250, **1998**. doi:10.1039/A827241Z.

161. K. Kneipp, M. Moskovits, and H. Kneipp. *Surface-Enhanced Raman Scattering: Physics and Applications,* vol. 103, Springer Science & Business Media, 2006. doi:10.1007/3-540-33567-6.

162. K. Kneipp, Y. Wang, H. Kneipp, L. T. Perelman, I. Itzkan, R. R. Dasari, and M. S. Feld. Single molecule detection using surface-enhanced Raman scattering (SERS). *Physical Review Letters,* 78(9):1667, **1997**. doi:10.1103/PhysRevLett.78.1667.

163. L. D'Urso, G. Grasso, E. Messina, C. Bongiorno, V. Scuderi, S. Scalese, O. Puglisi, G. Spoto, and G. Compagnini. Role of linear carbon chains in the aggregation of copper, silver, and gold nanoparticles. *The Journal of Physical Chemistry C,* 114(2):907–915, **2010**. doi:10.1021/jp9096309.

164. A. Lucotti, M. Tommasini, M. Del Zoppo, C. Castiglioni, G. Zerbi, F. Cataldo, C. S. Casari, A. L. Bassi, V. Russo, M. Bogana, *et al.* Raman and SERS investigation of isolated sp carbon chains. *Chemical Physics Letters,* 417(1–3):78–82, **2006**. doi:10.1016/j.cplett.2005.10.016.

165. F. Cataldo. Spectroscopical characterization of carbonaceous matter prepared through the Glaser coupling reaction route. *Carbon,* 37(1):161–163, **1999**. doi:10.1016/S0008-6223(98)90106-3.

166. K. Hanamura, M. Fujii, T. Wakabayashi, and S. Hayashi. Surface-enhanced Raman scattering of size-selected polyynes (C_8H_2) adsorbed on silver colloidal nanoparticles. *Chemical Physics Letters,* 503(1–3):118–123, **2011**. doi:10.1016/j.cplett.2010.12.078.

167. D. A. Long. *Raman Spectroscopy,* McGraw-Hill *New York,* 1, 1977.

168. A. C. Albrecht. On the theory of Raman intensities. *The Journal of Chemical Physics,* 34(5):1476–1484, **1961**. doi:10.1063/1.1701032.

169. S. A. Asher. UV resonance Raman studies of molecular structure and dynamics: applications in physical and biophysical chemistry. *Annual Review of Physical Chemistry,* 39(1):537–588, **1988**. doi:10.1146/annurev.pc.39.100188.002541.

170. S. Kang, J. H. Ryu, B. Lee, K. H. Jung, K. B. Shim, H. Han, and K. M. Kim. Laser wavelength modulated pulsed laser ablation for selective and efficient production of graphene quantum dots. *RSC Advances,* 9(24):13658–13663, **2019**. doi:10.1039/C9RA02087J.

171. S. Kang, K. H. Jung, S. Mhin, Y. Son, K. Lee, W. R. Kim, H. Choi, J. H. Ryu, H. Han, and K. M. Kim. Fundamental understanding of the formation mechanism for graphene quantum dots fabricated by pulsed laser fragmentation in liquid: experimental and theoretical insight. *Small,* 16(38):2003538, **2020**. doi:10.1002/smll.202003538.

172. D. Zhang, B. Gokce, and S. Barcikowski. Laser synthesis and processing of colloids: fundamentals and applications. *Chemical Reviews,* 117(5):3990–4103, **2017**. doi:10.1021/acs.chemrev.6b00468.

173. M. Tsuji, S. Kuboyama, T. Tsuji, and T. Hamagami. Formation of $C_{2n}H_2$ polyynes by laser ablation of graphite, coal or C60 particles suspended in selected solvents. *Polyynes,* 147–174, **2005**.

174. R. Matsutani, T. Kakimoto, K. Wada, T. Sanada, H. Tanaka, and K. Kojima, *et al.* Preparation of long-chain polyynes $C_{18}H_2$ and $C_{20}H_2$ by laser ablation of pellets of graphite and perylene derivative in liquid phase. *Carbon,* 46:1091–1109, **2008**. doi:10.1016/j.carbon.2008.03.009.

175. A. Zaidi, A. Hu, M. Wesolowski, X. Fu, J. Sanderson, Y. Zhou, and W. Duley. Time of flight mass spectrometry of polyyne formation in the irradiation of liquid alkanes with femtosecond laser pulses. *Carbon,* 48(9):2517–2520, **2010**. doi:10.1016/j.carbon.2010.03.026.

176. M. Wesolowski, S. Kuzmin, B. Moores, B. Wales, R. Karimi, A. Zaidi, Z. Leonenko, J. Sanderson, and W. Duley. Polyyne synthesis and amorphous carbon nano-particle formation by femtosecond irradiation of benzene. *Carbon,* 49(2):625–630, **2011**. doi:10.1016/j.carbon.2010.10.008.

177. Y. Taguchi, H. Endo, T. Kodama, Y. Achiba, H. Shiromaru, T. Wakabayashi, B. Wales, and J. Sanderson. Polyyne formation by ns and fs laser induced breakdown in hydrocarbon gas flow. *Carbon,* 115:169–174, **2017**. doi: 10.1016/j.carbon.2017.01.003.

178. N. Takizawa, T. Kodama, H. Shiromaru, T. Wakabayashi, S. Al-Tuairqi, Q. Wang, H. Zhang, and J. Sanderson. Efficient polyyne formation by ns and fs laser-induced breakdown in ethylene and acetylene gas flow. *Carbon,* 152:372–375, **2019**. doi:10.1016/j.carbon.2019.06.013.

179. N. R. Arutyunyan, P. V. Fedotov, and V. V. Kononenko. Laser synthesis and stability of one-dimensional polyynic carbon chains in liquid media. *Journal of Nanophotonics,* 10(1):012519, **2015**. doi:10.1117/1.JNP.10.012519.

180. H. Tabata, M. Fujii, and S. Hayashi. Laser ablation of diamond nanoparticles suspended in solvent: synthesis of polyynes. *Chemical Physics Letters,* 395:138–142, **2004**. doi:10.1016/j.cplett.2004.07.063.

181. S. Shin, J. Song, and S. Park. Preparation of polyynes by laser ablation of graphite in aqueous media. *App. Surf. Sci.,* 257:5156–5158, **2011**. doi:10.1016/j.apsusc.2010.10.074.

182. Y. K. Choi, J. K. Song, and S. M. Park. Production of hydrogen-capped polyynes by laser ablation of graphite in neat water. *Bulletin of the Korean Chemical Society,* 30(12):3073–3074, **2009**. doi:10.5012/bkcs.2009.30.12.3073.

183. D. Nishide, H. Dohi, T. Wakabayashi, E. Nishibori, S. Aoyagi, M. Ishida, S. Kikuchi, R. Kitaura, T. Sugai, M. Sakata, *et al.* Single-wall carbon nanotubes encaging linear chain $C_{10}H_2$ polyyne molecules inside. *Chemical Physics Letters,* 428(4–6):356–360, **2006**. doi:10.1016/j.cplett.2006.07.016.

184. H. Tabata, M. Fujii, and S. Hayashi. Synthesis of polyynes by laser ablation of diamond nanoparticles suspended in solution. *The European Physical Journal D,* 34:223–225, **2005**. doi:10.1140/epjd/e2005-00147-0.

185. K. S. Khashan. Synthesis of polyynes by laser ablation of graphite in ethanol. *Iraqi Journal of Physics,* 11(21):37–47, **2013**. doi:10.30723/ijp.v11i21.365.

186. R. Matsutani, F. Ozaki, R. Yamamoto, T. Sanada, Y. Okada, and K. Kojima. Preparation of polyynes up to $C_{22}H_2$ by liquid-phase laser ablation and their immobilization into SiO_2 gel. *Carbon,* 47(7):1659–1663, **2009**. doi:10.1016/j.carbon.2009.02.026.

187. Y. E. Park, S. K. Shin, and S. M. Park. The solvent effects on the formation of polyynes by laser ablation. *Bulletin of the Korean Chemical Society,* 33(7):2439–2442, **2012**. doi:10.5012/bkcs.2012.33.7.2439.

188. M. Condorelli, L. D'Urso, G. Compagnini, E. Fazio, F. Neri, V. Litvinenko, E. Kurmaev, and I. Zhidkov. The catalytic role of platinum nanoparticles in laser generated nanocarbons. *Applied Surface Science,* 558:149890, **2021**. doi:10.1016/j.apsusc.2021.149890.

189. M. Condorelli, A. Speciale, F. Cimino, C. Muscara, E. Fazio, L. D'Urso, C. Corsaro, G. Neri, A. M. Mezza-salma, G. Compagnini, *et al.* Nano-hybrid Au@LCCs systems displaying anti-inflammatory activity. *Materials,* 15(10):3701, **2022**. doi:10.3390/ma15103701.

190. C. Corsaro, M. Condorelli, A. Speciale, F. Cimino, G. Forte, F. Barreca, S. Spadaro, C. Muscara, M. D'Arrigo, G. Toscano, *et al.* Nano-hybrid Ag@LCCs systems with potential wound-healing properties. *Materials,* 16(6):2435, **2023**. doi:10.3390/ma16062435.

191. C. Wu, S. Chen, and P. Shen. Polyynes and flexible Si-H doped carbon nanoribbons by pulsed laser ablation of graphite in tetraethyl orthosilicate. *Carbon,* 67:27–37, **2014**. doi:10.1016/j.carbon.2013.09.049.

192. C. S. Casari, V. Russo, A. Li Bassi, C. E. Bottani, F. Cataldo, A. Lucotti, M. Tommasini, M. Del Zoppo, C. Castiglioni, and G. Zerbi. Stabilization of linear carbon structures in a solid Ag nanoparticle assembly. *Applied Physics Letters,* 90(1):013111, **2007**. doi:10.1063/1.2430676.

193. F. Goncalves, K. Hamano, and J. Sengers. Density and viscosity of tetralin and trans-decalin. *International Journal of Thermophysics,* 10:845–856, **1989**. doi:10.1007/BF00514480.

194. A. A. Maryott, and E. R. Smith. *Table of Dielectric Constants of Pure Liquids,* vol. 514, US Government Printing Office 1951.

195. M. Benoliel. *Investigation of Naphthalene and its Hydrogenated Derivatives in the Formation Process of Polyynes by Pulsed Laser Ablation in Liquid.* Master's thesis, Materials Engineering and Nanotechnology, 2021.

196. A. P. Altshuller. The dielectric constants, polarizations and dipole moments of some alkylbenzenes. *The Journal of Physical Chemistry,* 58(5):392–395, **1954**. doi:10.1021/j150515a003.

197. A. Kucherik, S. Arakelian, T. Vartanyan, S. Kutrovskaya, A. Osipov, A. Povolotskaya, A. Povolotskii, and A. Man'Shina. Laser-induced synthesis of metal-carbon materials for implementing surface-enhanced Raman scattering. *Optics and Spectroscopy,* 121(2):263–270, **2016**. doi:10.1134/S0030400X16080105.

198. B. Pan, J. Xiao, J. Li, P. Liu, C. Wang, and G. Yang. Carbyne with finite length: the one-dimensional sp carbon. *Science Advances,* 1(9):e1500857, **2015**. doi:10.1126/sciadv.1500857.

199. H. Kim, P. Tarakeshwar, N. M. Fujikado, K. Evraets, A. K. Jones, M. Meneghetti, P. R. Buseck, and S. G. Sayres. Pseudocarbynes: linear carbon chains stabilized by metal clusters. *The Journal of Physical Chemistry C,* 124(35):19355–19361, **2020**. doi:10.1021/acs.jpcc.0c05014.

200. H. Kim, P. Tarakeshwar, M. Meneghetti, P. R. Buseck, and S. G. Sayres. Formation of Au-pseudocarbynes by self-assembly of carbon chains and gold clusters. *Carbon,* 205:546–551, **2023**. doi:10.1016/j.carbon.2023.01.060.

201. Y. Jiang, P. Liu, Y. Liang, H. Li, and G. Yang. Promoting the yield of nanoparticles from laser ablation in liquid. *Applied Physics A,* 105(4):903–907, **2011**. doi:10.1007/s00339-011-6557-z.

202. F. Yang, C. Li, J. Li, P. Liu, and G. Yang. Carbyne nanocrystal: one-dimensional van der Waals crystal. *ACS Nano,* 15(10):16769–16776, **2021**. doi:10.1021/acsnano.1c06863.

203. W. Cao, H. Xu, P. Liu, Y. He, and G. Yang. The kinked structure and interchain van der Waals interaction of carbyne nanocrystals. *Chemical Science,* **2023**. doi:10.1039/D2SC04926K.

204. P. Tarakeshwar, P. R. Buseck, and H. W. Kroto. Pseudocarbynes: charge-stabilized carbon chains. *The Journal of Physical Chemistry Letters,* 7(9):1675–1681, **2016**. doi:10.1021/acs.jpclett.6b00671.

205. S. Okada, M. Fujii, and S. Hayashi. Immobilization of polyynes adsorbed on Ag nanoparticle aggregates into poly (vinyl alcohol) films. *Carbon,* 49(14):4704–4709, **2011**. doi:10.1016/j.carbon.2011.06.074.

206. R. Sata, H. Suzuki, N. Ueno, Y. Morisawa, M. Hatanaka, and T. Wakabayashia. UV-polarizing linear polyyne molecules aligned in PVA. *Chinese Journal of Chemical Physics,* 32(2):175–181, **2019**. doi:10.1063/1674-0068/ cjcp1812273.

207. L. Malard, D. Nishide, L. Dias, R. B. Capaz, A. Gomes, A. Jorio, C. Achete, R. Saito, Y. Achiba, H. Shinohara, *et al.* Resonance Raman study of polyynes encapsulated in single-wall carbon nanotubes. *Physical Review B,* 76(23):233412, **2007**. doi:10.1103/PhysRevB.76.233412.

208. L. Moura, C. Fantini, A. Righi, C. Zhao, H. Shinohara, and M. Pimenta. Dielectric screening in polyynes encapsulated inside double-wall carbon nanotubes. *Physical Review B,* 83(24):245427, **2011**. doi:10.1103/PhysRevB.83.245427.

209. S. Mallakpoura, and A. Ezhieh. Preparation and characterization of chitosan-poly(vinyl alcohol) nanocomposite films embedded with functionalized multi-walled carbon nanotube. *Carbohydrate Polymers,* 166:377–386, **2017**. doi:10.1016/j.carbpol.2017.02.086.

210. S. Pandey, S. K. Pandey, V. Parashar, G. K. Mehrotra, and A. C. Pandey. Ag/PVA nanocomposites: optical and thermal dimensions. *Journal of Materials Chemistry,* 21:17154–17159, **2011**. doi:10.1039/ c1jm13276h.

211. S. Pecorario, A. D. Scaccabarozzi, D. Fazzi, E. Gutierrez-Fernandez, V. Vurro, L. Maserati, M. Jiang, T. Losi, B. Sun, R. R. Tykwinski, *et al.* Stable and solution-processable cumulenic sp-carbon wires: a new paradigm for organic electronics. *Advanced Materials,* 34(15):2110468, **2022**. doi:10.1002/adma.202110468.

212. A. D. Scaccabarozzi, A. Milani, S. Peggiani, S. Pecorario, B. Sun, R. R. Tykwinski, M. Caironi, and C. S. Casari. A field-effect transistor based on cumulenic sp-carbon atomic wires. *The Journal of Physical Chemistry Letters,* 11(5):1970–1974, **2020**. doi:10.1021/acs.jpclett.0c00141.

Chapter 4

Discovery of Carbon Nanowires and Mass Production of Single-Walled Carbon Nanowires by Polyyne Fusion

Xinluo Zhao and Weiwei Chang
Department of Physics, Shanghai University,
Shanghai, 200444, China
Institute of Low-dimensional Carbons and Device Physics,
Shanghai University, Shanghai, 200444, China
xlzhao@shu.edu.cn

Carbon nanowires (CNWs, LLCCs@CNTs), made of long linear carbon chains (LLCCs, i.e., carbyne) inserted inside carbon nanotubes (CNTs), are a novel 1-D carbon allotrope, and possess excellent physical properties. This chapter will firstly introduce the preparation of polyynes by submerged arc-discharge, and their purification and storage. Secondly, we will primarily describe the formation and theoretical feasibility of single-walled CNWs (SWCNWs, LLCCs@SWCNTs) by polyyne fusion inside single-walled CNTs (SWCNTs). HRTEM observation and Raman characterization of SWCNWs are

Emerging Carbyne: Truly One-Dimensional Form of Carbon
Edited by Yahachi Saito

ISBN 978-981-5129-15-1 (Hardcover), 978-1-003-51036-9 (eBook)
www.jennystanford.com

also reviewed, and the temperature effect on Stokes and anti-Stokes of SWCNWs are briefly mentioned. Finally, the remaining challenges and prospects are addressed to advance the field of carbyne and CNWs.

4.1 Introduction

4.1.1 Discovery of Carbon Nanowires (CNWs)

In order to search for short linear carbon chains (LCCs) relevant to the origin of life in the universe in the fields of astrophysics and astrochemistry, Kroto *et al.* serendipitously discovered the fullerene C_{60} in 1985 [1], and their mass-production method by direct current (DC) arc discharge evaporation of carbon electrodes in helium gas led to the finding of one-dimensional (1-D) carbon nanotubes (CNTs) by Iijima in 1991 [2]. In 2003, it was a complete surprise for Zhao *et al.* to discover by chance a novel 1-D carbon allotrope, carbon nanowires (CNWs), comprised of long LCCs (LLCCs) inserted inside CNTs, in the multi-walled CNTs (MWCNTs) samples obtained by hydrogen arc discharge method [3].

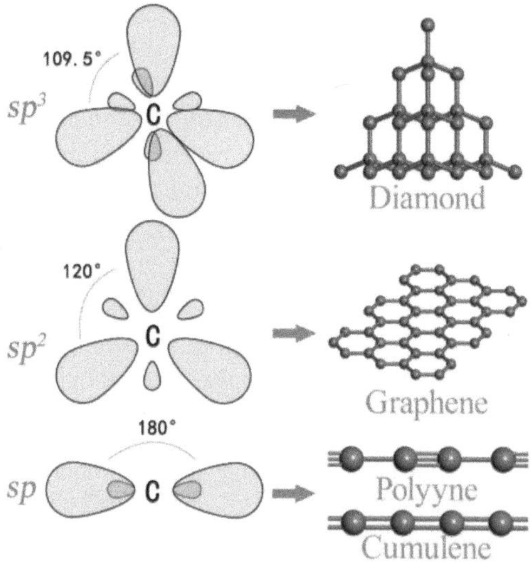

Figure 4.1 *sp*^{3}-, *sp*^{2}-, and *sp*-hybridization of carbon, and their corresponding allotropes [4]. © 2022 by Chinese Physics B.

As we all known, there are three ways that carbon could hybridize into bonds: sp^3, sp^2, and *sp* hybridization, as shown in Fig. 4.1 [4], the most familiar naturally occurring ones are diamond [5] that is composed entirely of sp^3 hybrid orbitals, and graphite [6] with sp^2 hybridization, but the simplest form of carbon bonding, *sp*, is still rarely understood. Since the bond angle of *sp* orbitals hybridized into bonds is 180°, the nanostructure and molecules formed by them are linear, so this kind of researches could be called as LCC researches. In fact, the earliest studies of *sp* carbon went back to 1885, Baeyer began trying to prepare 1-D polymers of carbon [7, 8], but the failure of that experiment led researchers to believe that this type of carbon would be extremely difficult to make for a long time. It was not until 1953 that the researchers discovered the molecular structures containing carbon chains in plants and fungi, which gave the research field renewed hope [9–12]. In 1968, the statement of discovering linear carbon crystals in Ries Crater in Germany made *sp* carbon a hot topic [13], however, the evidence from relevant experimental studies is insufficient to rigorously prove the exact structure of these *sp* carbon materials, and the finding was questioned. As mentioned above, during the quest for LCCs the fullerene C_{60} was discovered by using laser ablation of graphite [1] and their mass-preparation method by DC arc-discharge evaporation of carbon electrodes in helium gas resulted in the finding of CNTs [14], these important discoveries directly resulted in the shift from the study of *sp* carbon to that of the fullerene C_{60} [1], CNT [2], as well as graphene [15] with sp^2-hybridization.

The first observation of Raman band (~1850 cm^{-1}) of LLCCs in 1998 [16, 17] and a subsequent HRTEM study in 2003 fully demonstrated that the LLCCs with *sp* hybridization existed inside the center holes of MWCNTs with sp^2-hybridization, resulting in the formation of a new 1-D carbon nanostructure of CNWs [3]. The discovery of CNWs completely ignited people's enthusiasm for the studies on carbon nanostructures containing *sp* carbon, such as polyynes, LCCs, LLCCs or carbynes, CNWs, graphyne, and so on, as shown in Fig. 4.2 [18].

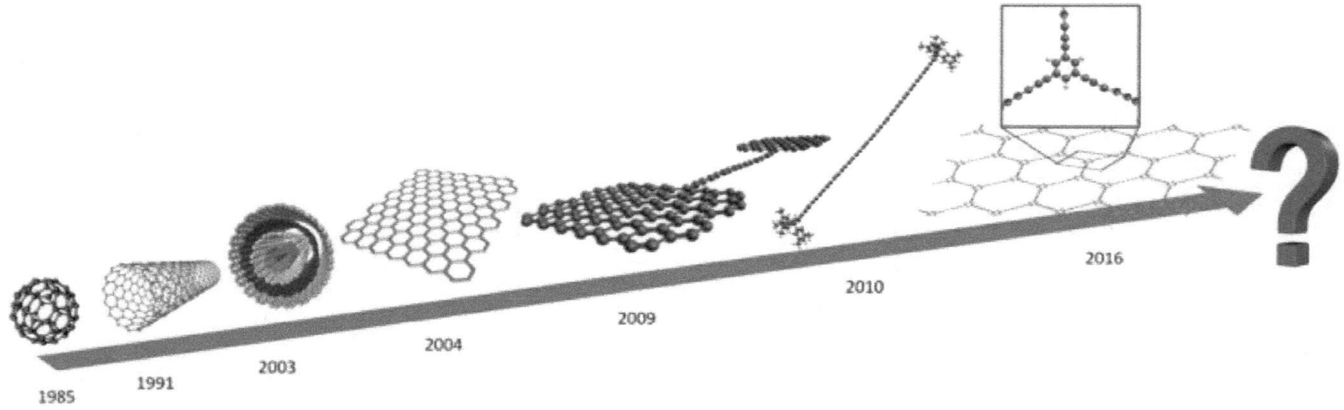

Figure 4.2 The path of the most relevant sp^2 and sp-sp^2 allotropes or carbon nanostructures discovered and or fabricated [18]. © 2018 The Materials Research Society.

4.1.2 Impact on Carbyne Research of CNWs Discovery

At present, the motivation for studying CNWs, quite interestingly, is to hunt the excellent physical properties of LLCCs or carbyne, such as the strongest mechanical properties [19], ultrahigh thermal conductivity [20], excellent electronic properties [21] and a direct band gap of 2.6–4.7 eV tunable *via* tension stress [22]. However, the LLCCs and even short LCCs are extremely unstable in air and easily transform to other carbon nanomaterials, which is the main obstacle to the researches of LCCs. Polyyne is a kind of short *sp* hybridization LCC with hydrogen atoms or other functional groups at its either end, and their properties are often determined by its length, end group, and so on [18, 23]. It was reported that the length of an isolated LCC was no more than 48 carbon atoms until now when they were end-capped by some huge functional groups [24, 25], showing that the longer of LCCs, the more difficult it is to be synthesized. The instability of LLCs and their preparation difficulties casted doubts on the longer LLCs, especially, the existence of LLCCs or carbyne, had a significant controversy for a long time since 1960s [26–30], therefore, the discovery of CNWs (LLCCs@CNTs) not only proved the real presence of LLCCs or carbyne with *sp*-hybridization, but also indicated their stable existence inside CNTs for scientific study [3].

Recently, some outstanding works on *sp* carbon has been reported, for example, the LCCs inside CNTs [3, 16, 17], between graphene layers [31] or CNTs [32], and white carbon with Au nanoparticles [33], cyclocarbon [34], as well as graphdiyne [35]. Of all these works, only the CNWs contain the LLCCs that are more than a hundred of carbon atoms in length [3], especially, Shi *et al.* prepared an extra-long LCC with more than 6000 carbon atoms inside DWCNTs in 2016 [36]. Like polyynes, the properties of short LCCs should also depend on many factors, such as length, end group, environmental, and so on, but some theoretical calculations have shown that their physical properties are almost independent of their length when the length of LLC is long enough, even though the number of carbon atoms in a LCC (about 60 to 80) is not very precise, depending on the calculation methods employed [25, 37–39]. Therefore, we have called a LCC whose properties did not vary with its length as the LLCC or carbyne, and the recently experimental results of tip-enhanced Raman scattering (TERS) measurement on 14 different LLCCs inside DWCNWs by Heeg *et al.* also supports our denominating [37]. Obviously, the LLCCs with L-band of ~1850 cm^{-1} inside CNTs are all long enough to be considered as the LLCCs or carbyne. Based on the above research results, it is clear that the CNWs samples have the most potential to be one of the best options for studying the physical properties and crystal structure of the third allotrope of carbon, carbyne, with *sp*-hybridization.

4.1.3 Classification of CNWs and Preparation Methods

The CNWs can be divided into three categories according to the type of CNT outside of LLCCs. As shown in Fig. 4.3, Fig. 4.3a is the atomic structure models of multi-walled CNWs (MWCNWs), i.e., LLCC inside multi-walled CNTs (LLCCs@MWCNTs), Fig. 4.3b is the double-walled CNWs (DWCNWs, LLCCs@DWCNTs), and Fig. 4.3c is the single-walled CNWs (SWCNWs, LLCCs@SWCNTs). This classification method was proposed by Zhang *et al.* in 2018 [40], it is consistent with the development process of CNWs, and now has been widely recognized.

Since the discovery of CNWs in 2003, three types of CNWs have been successively synthesized in different ways. In 1997, Zhao *et al.* developed hydrogen arc discharge technique in order to prepare high-quality MWCNTs [41], accidentally observed the L-band of 1850 cm^{-1} for the first time in 1998 [16], and confirmed the 1-D crystal structure of MWCNWs (LLCCs@MWCNTs) by means of high-resolution transmission electron microscopy (HRTEM) observations, measurements of Raman spectroscopy and theoretical calculations in 2003. After then, DC arc discharge was the main method for preparing MWCNWs samples. In simple terms, two graphite rods were used as the electrodes, and the distance between the electrodes was 1 mm. The arc (DC 50 A) started between cathode (d = 10 mm) and anode (d = 6 mm) for 1 minute at a pressure of 50 Torr in pure hydrogen atmosphere, and finally the MWCNWs were formed in the cathode deposits. Besides MWCNWs (LLCCs@MWCNTs) and MWCNTs, the smallest CNTs with the diameters of 0.4 and 0.3 nm were also found in the cathode deposits, respectively [42, 43]. Moreover, in 1997 the multi-layers graphene, i.e., petal-like graphene sheets or carbon nanowalls, were serendipitously observed on the cathode surface around the deposits [44]. At present, researchers are able to prepare MWCNWs samples by various methods [3], such as arc-discharge in helium [45], in liquid nitrogen [46], in argon gas [47], laser irradiating CNT [48], heat treatment CNT [49], and so on. Zhang *et al.* have controlled the plasma in the arc discharge area by cooling ambient gas and applying magnetic field, greatly expanded the conditions for preparing MWCNWs, and improved their yield and purity [40, 50].

It was in 2011 that the DWCNWs (LLCCs@DWCNTs) were successfully synthesized and confirmed by heating DWCNTs with a small inner diameter of 0.7 nm [51] or by fusion reactions of polyyne molecules, $C_{10}H_2$, in the ultrathin 1-D nano-space of DWCNTs [52], enriching the types of CNWs. After more than a decade of difficult exploration, the SWCNWs (LLCCs@SWCNTs) were finally synthesized [53, 54], especially, Chang *et al.* mass-produced SWCNWs samples with high yield by a controllable fusion reaction of polyyne molecules, $C_{2n}H_2$, inside ultrathin SWCNTs [55].

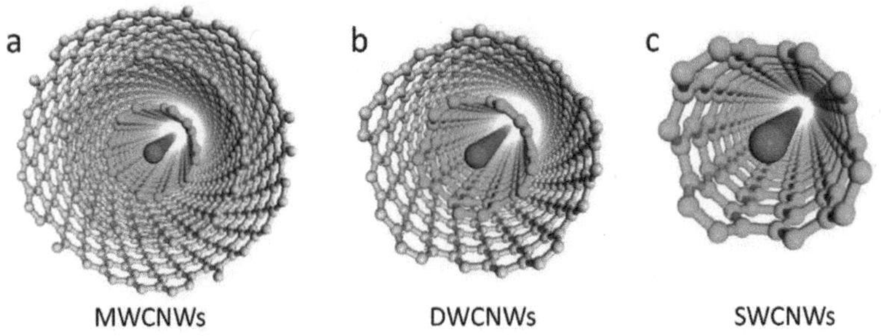

Figure 4.3 Atomic structure models of three different types of carbon nanowires, (a) MWCNWs, (b) DWCNWs, and (c) SWCNWs.

4.1.4 Research Progress of CNWs

In recent years, some great progresses have been achieved in the research field of CNWs. Firstly, in terms of CNWs preparation, as mentioned above three types of CNWs have been successfully synthesized. Secondly, some new advances have been made in the characterization of CNWs. Raman spectroscopy plays an important role in the characterization of CNWs samples [3, 16, 17, 36], the Raman shift between 1900 to 2300 cm^{-1} is considered to be due to a short LCCs (polyynes, $C_{2n}H_2$) [31, 36–39], and that located at ~1780

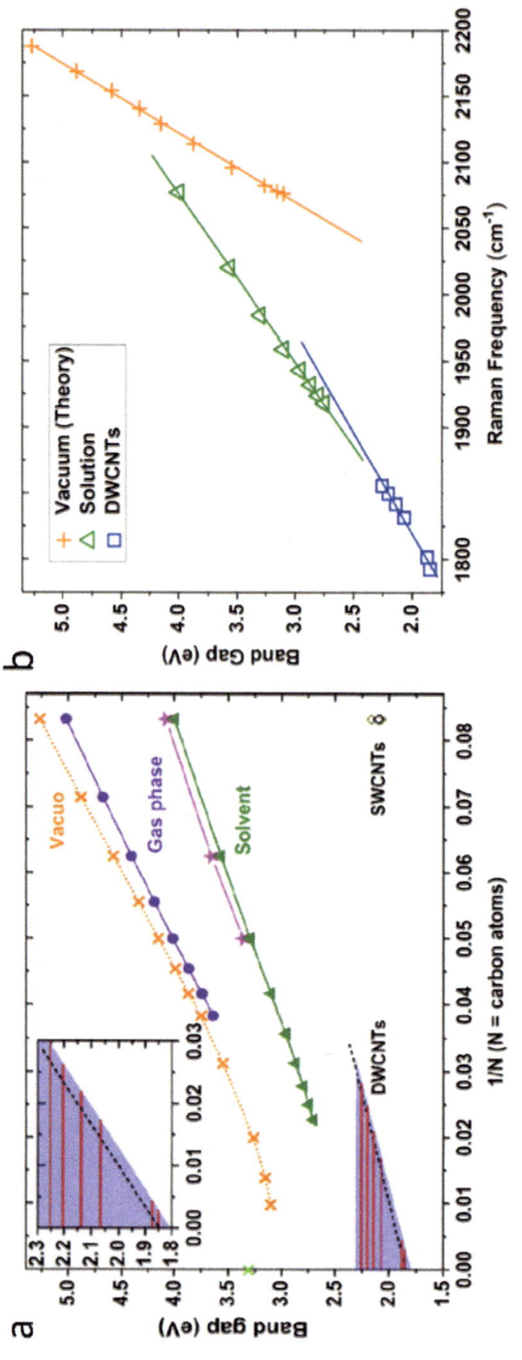

Figure 4.4 (a) Relationship between band gap and length of LCCs in different environments. (b) Relationship between band gap of LCCs and Raman characteristic peak position [58]. © 2017 by American Chemical Society.

to 1880 cm^{-1} is the Raman characteristic peaks, i.e., L-band, of LLCCs inside CNTs [3, 36, 56, 57]. The band gap width of LLCCs inside CNTs is 1.8~2.6 eV measured by resonance Raman scattering [56–60], and it has been said that the band gap of LCCs depends on the length and environment of it (Fig. 4.4a). Moreover, it has been found that there is a linear relationship between the band gap of LCC and the frequency of its Raman characteristic peak (Fig. 4.4b). Subsequently, the Stokes and anti-Stokes studies on a single DWCNWs (LLCCs@ DWCNTs) by TERS have shown that the LLCCs have great potential for the application to nanoscale temperature sensors [61, 62].

From the first observation of LLCCs inside MWCNTs by means of HRTEM [3], HRTEM are being used to directly confirm and accurately distinguish the types of CNWs [36, 55]. Especially, Andrade *et al.* used HRTEM with electron energy loss spectroscopy (EELS) to analyze the cross-section of MWCNWs in 2015, unambiguously confirmed the existence of a 1-D LLCC structure present within the innermost tubes of MWCNTs, and put an end to the controversy about whether the LLCCs existed in the center of CNTs or between of them [59].

4.1.5 The Content of This Chapter

In this chapter, we firstly introduce the preparation of polyyne molecules, $C_{2n}H_2$, by submerged arc-discharge in organic solvents, and their concentration, purification and storage. Subsequently, we describe in detail the formation of SWCNWs by the fusion reaction of polyyne molecules inside SWCNTs with diameters ranging from 0.7 to 0.95 nm, and the theoretical feasibility of this method is also expounded. Moreover, the HRTEM observation and Raman characterization of SWCNWs are also reviewed, and some research results of the temperature effect on Stokes and anti-Stokes of SWCNWs are briefly mentioned. Finally, the remaining challenges and prospects are presented. Our purpose is to let more people know about SWCNWs (LLCCs@SWCNTs) and participate in the research field of SWCNWs through the subsequent sections, so as to reveal the scientific secrets of *sp* carbon and realize the industrial application of 1-D CNWs in the near future.

4.2 Preparation of Polyyne Molecules, $C_{2n}H_2$, by Submerged Arc-Discharge in Organic Solvents, Their Purification and Storage

4.2.1 Preparation Methods of Polyyne Molecules, $C_{2n}H_2$, by Submerged Arc-Discharge

The aforementioned polyynes were originally synthesized by traditional chemical methods [63–68], which usually involved more complex organic chemical reactions processes, intermediate products and byproducts. In recent years, the linear polyyne molecules, $C_{2n}H_2$, were prepared by laser ablation of graphite [69–73] or submerged arc-discharge in organic solvents [74–78]. For the laser ablation of graphite, the obtained polyyne molecules have an advantage in length and purity [70, 72, 73], while the technique of submerged arc-discharge in organic solvents can be used to produce the samples of polyyne molecules in large-scale [74, 76]. From now on, we mainly introduce the preparation of polyyne molecules, $C_{2n}H_2$, by submerged arc-discharge in organic solvents method and the characteristics of obtained polyyne solution.

Figure 4.5a is the schematic drawing of the apparatus employed for mass-producing polyyne molecules, $C_{2n}H_2$, by submerged arc-discharge method [76]. The apparatus consists of three parts, the container that provides DC arcing region containing organic solvents, the cooling system, and the welding machine. When arcing, the temperature of cooling system is set to −20 °C, and the arcing current is generally about 5 to 20 A. In the arcing zone of container, two pure graphite rods with a diameter of 10 mm could be arranged in a "V" geometry with a spacing of less than 2 mm. Generally, in order to ensure that the arc area of graphite rod was completely under the organic solvent, the organic solvent should reach about two-thirds of the container that was used. The arcing time depended on the size of container, and for the 5 L three-necked flask, the average arcing time was about 2 h, with an appropriate stop in the middle to ensure that the temperature in the arcing area is not too high. Figure 4.5b shows the UV absorption spectra of polyyne solution obtained by Wu *et al.* using submerged arc-discharge in cyclopentane at different arcing times. The concentration of polyyne solution increased with the increase of arcing time, and the concentration of polyyne molecules with more carbon atoms was lower than that of polyyne molecules with fewer carbon atoms. At present, it is generally believed that the formation mechanism of polyyne molecules was explained by the difference in the competition of polymerization and hydrogenation [79, 80]. Figure 4.5c is the UV absorption spectra of C_6H_2, C_8H_2, $C_{10}H_2$ and $C_{12}H_2$, which are all separated by using high performance liquid chromatography (HPLC). The surface enhanced Raman spectra (SERS) of C_8H_2, $C_{10}H_2$ and $C_{12}H_2$ in acetonitrile are shown in Fig. 4.5d, being measured in the presence of Ag colloid. This suggests that the polyyne molecules with different lengths have different UV absorption spectra and SERS, which provide a basis for the detection of polyyne molecules. Recently, it was reported that the polyynes-based materials showed great application prospects in the classification of living cells, with high throughput diagnosis and screening [81].

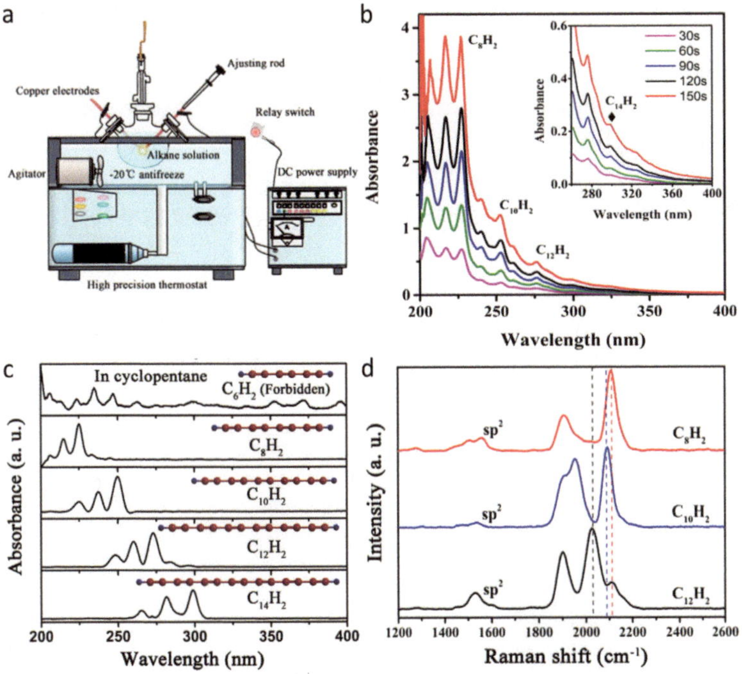

Figure 4.5 (a) Schematic drawing of apparatus for preparing polyyne molecules by submerged arc-discharge method. (b) UV absorption spectra of the obtained polyyne solution by submerged arc-discharge in cyclopentane for different time. (c) UV absorption spectra of polyyne samples separated by HPLC. (d) SERS spectra of C_8H_2, $C_{10}H_2$, and $C_{12}H_2$ in acetonitrile, which were measured in the presence of Ag colloid [76]. © 2019 by Elsevier.

4.2.2 Purification and Concentration of Polyyne Solution via Carbides

The linear polyyne molecules, $C_{2n}H_2$, are considered to be an important molecular model, and the precursors for the growth of LLCCs or carbyne inside both DWCNTs and SWCNTs [52, 55]. The raw solution of polyynes prepared by submerged arc-discharge method usually has more impurity and low concentration, and a high-purity and high-concentration polyyne solution is highly required for scientific researches. In 2021, Fang *et al.* developed a purification method of polyyne solution *via* carbides, and the schematic diagram is shown in Fig. 4.6a [82]. Firstly, Tollen's reagent is added to the original polyyne solution to form carbides, this process is continued to stir until the reaction is completed. Secondly, the sediment obtained in the previous step is centrifuged and cleaned with pure water and ethanol to obtain solid Ag_2C_{2n}. Due to the possibility of explosion, argon gas should be used to protect the drying process, and then a solution of n-hexane was added to a test tube containing Ag_2C_{2n}. Finally, the purified polyyne solution was obtained by a chemical reaction between carbides and hydrochloric acid. Figure 4.6b is the UV absorption spectra of polyyne solution before

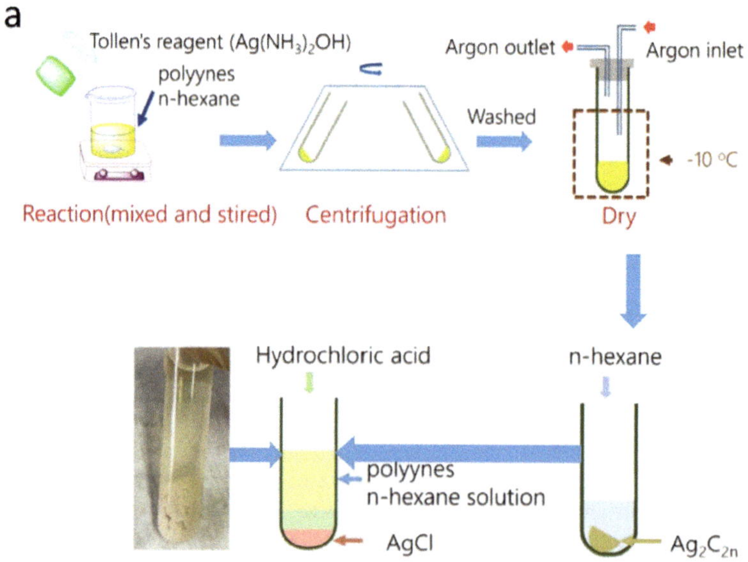

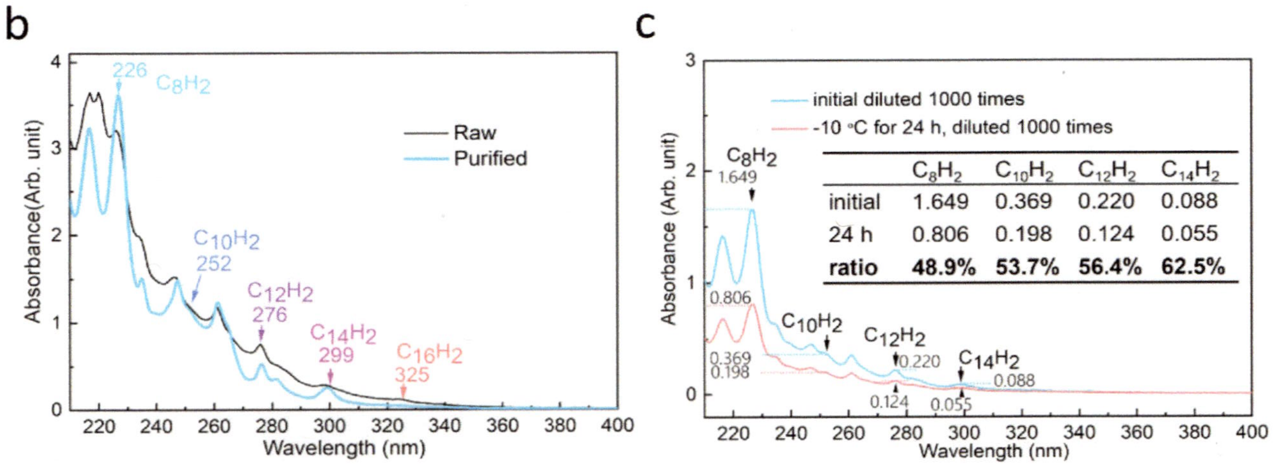

Figure 4.6 (a) Schematic diagram of high-purity and high-concentration polyyne solution obtained by purification technique *via* carbides. (b) UV absorption spectra of raw and purified polyyne solutions. (c) UV absorption spectra of initial purified polyyne solution and the purified polyyne solution after stored for 24 h at −10 °C [82]. © 2021 by Elsevier.

and after purification. It can be seen that the peak of UV absorption spectra of purified polyynes solution becomes significantly sharper than the original one. This indicates that the purity and concentration of polyyne solution are improved greatly after purification, especially, the concentration of C_8H_2 has increased significantly. Figure 4.6c shows the UV absorption spectra of initial purified polyyne solution and that after 24 h of storage at −10 °C. It can be seen that the concentration of purified polyyne solution decreases significantly after a period of time, but the purified polyyne solution is still at a higher concentration. This demonstrates that our purification process for obtaining a high-purity and high-concentration polyyne solution by chemical reaction of solid metal carbides precursors with hydrogen chloride is an easy and feasible technique.

4.2.3 Long-Term Storage of Polyyne Molecules, $C_{2n}H_2$, in Paraffin

After the mass-production, purification and concentration of polyyne solution have been completed, the remaining problem is the stockpile of polyyne molecules. In order to store polyyne solution, it is necessary to prevent the cross-linking reactions between polyyne molecules. Introducing large functional groups at the ends of polyynes can create spatial barriers between polyynes [29], and the large functional groups will cause the polyynes to bend. Encapsulation of polyyne molecules into CNTs can also create spatial barriers to completely prevent their cross-linking reaction [56, 83–88], but it is difficult or almost impossible to extract polyyne molecules again from CNTs for a reuse. Even though the cold storage at low temperature could slow down the cross-linking reaction between polyynes to some extent, the effect is still far from satisfactory. Therefore, in order to meet the rising demand for their fundamental researches and practical applications, developing a way to store polyynes stably and extract them for reuse becomes urgent.

In 2022, Luo *et al.* took advantage of the fact that paraffin wax is solid at room temperature, which maybe can provide a spatial barrier between polyyne molecules and prevent further cross-linking, and authenticated that the polyyne molecules with high-purity and high-concentration could be stockpiled in solid paraffin and taken out whenever [89]. The specific steps of this method, which have three principal sections: purification, stockpile, and extraction, are shown in Fig. 4.7. In the purification section, we had used the aforementioned purification technique of polyynes *via* carbides to obtain high-purity and high-concentration polyyne *n*-hexane solution [82, 89]. It should be pointed out that here the pristine polyynes solution were prepared by submerged arc-discharge method in organic solvent. In the stockpile section, the polyyne *n*-hexane solution was first mixed with paraffin granules and dried in air at room temperature to obtain a solid paraffin containing polyyne molecules. In the extraction section, by dissolving the paraffin containing polyyne molecules in acetone and using the same purification technique *via* carbides, the polyyne molecules could be replaced in *n*-hexane solution.

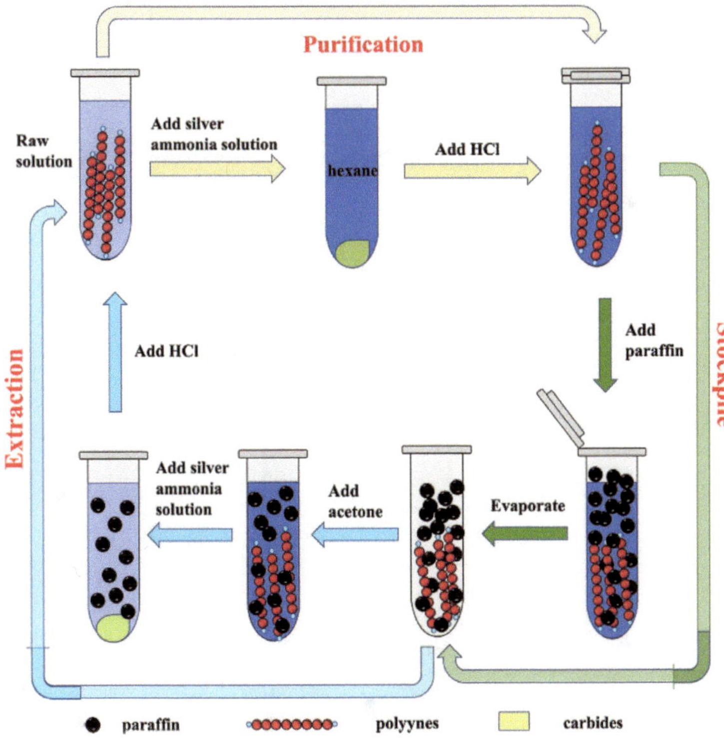

Figure 4.7 (a) Process diagram of storing and extracting polyyne molecules in solid paraffin at room temperature [89]. © 2022 by Elsevier.

Figure 4.8a shows the SERS of polyyne molecules stored in solid paraffin at different time, from top to bottom, initial, 7 days,

14 days, 21 days, and 28 days. Although the Raman intensity of polyynes decreases gradually with the increase of storage time, it can be seen that a visible Raman intensity remains after storage for 28 days, indicating that the polyynes samples still possess a certain concentration. Figure 4.8b is the Raman spectrum of SWCNWs (LLCCs@SWCNTs) obtained by the fusion reaction of polyyne molecules inside SWCNTs [55, 89]. Here, the extracted polyyne molecules from paraffin wax have been used, indicating that the polyyne molecules can be extracted intact after being stored in solid paraffin for some time.

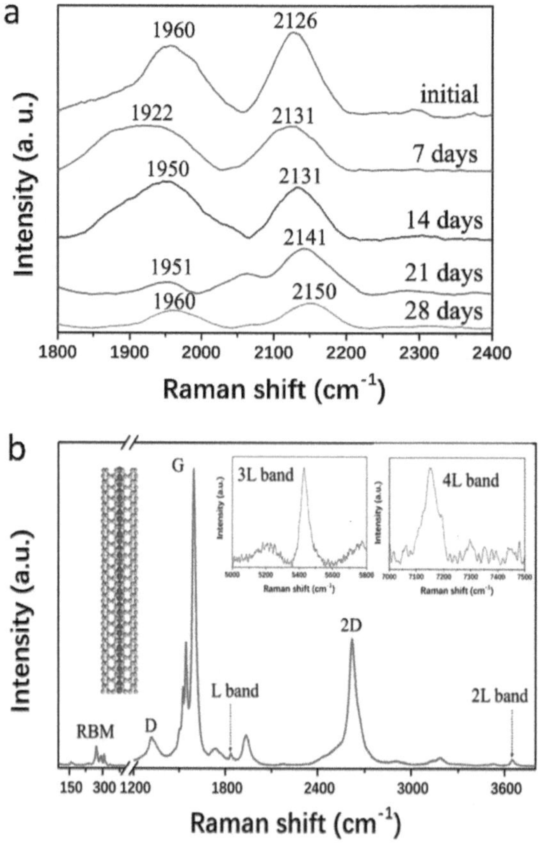

Figure 4.8 (a) SERS of polyynes stored in paraffin, from top to bottom, initial, 7 days, 14 days, 21 days, and 28 days. (b) Raman spectrum of LLCCs@SWCNTs synthesized by using extracted polyynes from paraffin. Both of them were measured with an excitation wavelength of 532 nm [89]. © 2022 by Elsevier.

4.3 Preparation of SWCNWs by Fusion Reaction of Polyyne Molecules inside SWCNTs

4.3.1 Synthesis of SWCNWs (LLCCs@SWCNTs)

Comparing with MWCNWs and CWCNWs, the SWCNWs have only one layer of graphene outside LLCCs, and the graphene possesses a light transmittance of ~97.7%, therefore by using SWCNWs samples, it is relatively easy to detect the physical and chemical properties of both LLCCs and CNWs. Until now, as the simplest and smallest CNWs, the structure of SWCNWs (LLCCs@SWCNTs) has been only considered as the model of CNWs in the theoretical calculations [90–93], and it has been predicted that they may have superconductivity and ferromagnetism [94], and have much larger elastic modulus than the isolated LLCCs [95]. It is expected that these theoretical predictions can be directly verified by the experimental measurements of SWCNWs, but the SWCNWs samples had been difficult to be obtained for a long time mainly due to their own structural characteristics. Firstly, the traditional methods for preparing SWCNTs do not contain LLCCs inside. Secondly, there are some problems in preparing SWCNWs by converting amorphous carbon inside SWCNTs into LLCCs. Because the amorphous carbon inside DWCNTs is converted into LLCCs with *sp* hybridization at 1200–1500 °C [36, 49], while SWCNTs tend to be converted to other forms of carbon at 1200 °C [96]. Therefore, it is believed that the synthesis of SWCNWs by traditional heat treatment method, which has been used to prepare DWCNWs [36, 51] or MWCNWs [49], is not possible. With the deepening of researches in recent years, there are three unconventional approaches developed for obtaining SWCNWs structure until now, which are the tip-ultrasonic and density gradient ultracentrifugation, vacuum discharge and polyynes fusion method. Next, we will introduce in detail these three preparation techniques of SWCNWs, respectively.

In 2018, Shi *et al.* successfully extracted extremely small amount of SWCNWs from DWCNWs samples using the method of tip-ultrasonic and density gradient ultracentrifugation [53]. In simple terms, it was divided into two steps: Firstly, the DWCNTs synthesized by chemical vapor deposition were heat-treated at 1500 °C under high vacuum for 30 mins to obtain DWCNWs samples [36, 51]. Secondly, the outer CNTs of DWCNWs were removed by tip-ultrasonication method so as to obtain SWCNWs structure, and then the SWCNWs were separated from the DWCNWs, DWCNTs, empty and water-filled SWCNTs by using density gradient centrifugation method (DUG) [97, 98]. Even though the SWCNWs obtained by this method are of small amount, poor crystallinity, and have a lot of surfactants on their surface, it is the direct experimental evidence of the existence of SWCNWs structure, greatly encouraging researchers to study enthusiastically SWCNWs.

The vacuum discharge method proposed by Toma *et al.* in 2019 [54] is used to prepare SWCNWs under high vacuum condition by changing the crystal structure of CNTs film with the energy released at the moment of field emission (FE). In short, a typical SWCNTs film with an area of 5 mm × 5 mm and a thickness of dozens of microns as a cathode, which faces to an anode grid, the gap between cathode and anode was about 2 mm. Then the FE current and voltage were controlled until the SWCNTs film was broken down, and finally a small amount of SWCNWs, DWCNWs, and MWCNWs mixture were found at the damaged edges of the SWCNTs film. This method has the advantage of being quick and easy, and its disadvantage is that the SWCNWs cannot be prepared in large scale and the proportion of SWCNWs in the mixture is uncontrollable. From now on, we will mainly describe the synthesis method of SWCNWs by fusion reaction of polyyne molecules inside SWCNTs.

4.3.1.1 A controlled approach for mass-producing SWCNWs: polyynes fusion method

The polyynes fusion method for mass-producing SWCNWs samples was developed by Chang *et al.* in 2021 [55], which utilized the characteristics of polyyne molecules moving along the axial direction inside the SWCNTs to promote polyynes fusion into LLCCs. It was inspired by the previous work of DWCNWs (LLCCs@DWCNTs) preparation in 2011, in which the linear polyyne molecules, $C_{10}H_2$, were encapsulated inside DWCNTs and the DWCNWs (LLCCs@ DWCNTs) were prepared by heat treatment [52]. The schematic diagram of preparation process for SWCNWs is shown in Fig. 4.9, and it can be described as follows: Firstly, the pristine SWCNTs were heated in dry air to remove their caps; Secondly, both the open-ended SWCNTs and the high-concentrations polyyne solution were placed in a high-pressure reactor, so that the linear polyyne molecules, $C_{2n}H_2$, could be encapsulated inside SWCNTs to form polyynes@SWCNTs structure by keeping their temperature at 80 °C for 24 hours. Finally, the polyynes@SWCNTs were heat-treated at the temperature of 700 °C in Ar gas for 3 hours by using a tubular furnace, the polyyne molecules then grew into the LLCCs within the narrow space of SWCNTs to form the SWCNWs (LLCCs@SWCNTs) samples [55].

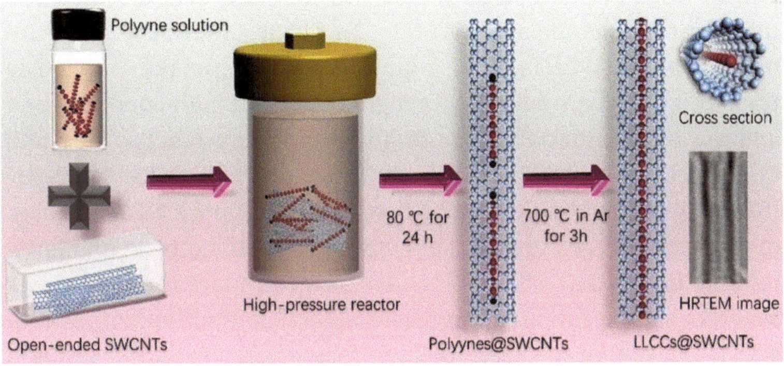

Figure 4.9 Schematic diagram of the preparation process for SWCNWs by polyynes fusion method [55]. © 2021 by Elsevier.

4.3.1.2 Removal of end-caps of ultrathin SWCNTs

One difficulty point, which we must pay special attention to, in the preparation of SWCNWs by polyynes fusion method is to remove the end-caps of ultrathin SWCNTs with the diameters about 0.7~0.95 nm. The previous studies have shown that the CNTs have a protective effect on LLCCs, that is, there is an optimal distance between the LLCCs and the wall of CNTs, which is the key to the stable existence of LLCCs inside CNTs [3, 36, 40, 50]. This optimal distance is thought to be close to the interlayer spacing between graphite (~0.34 nm) [3], which corresponds exactly to a SWCNT with a diameter about 0.7 nm. The curvature of the end cap of SWCNTs with such a small diameter is very close to that of the side wall of SWCNTs, therefore, it is possible to damage the SWCNTs while removing the end-cap of SWCNTs, meaning that the removal conditions of the end-cap of ultrathin SWCNTs will be very strict. Luckily, Chang *et al.* found that the heat treatment of SWCNTs with the diameters of about 0.7 nm at 300~600 °C for 0.2~2 hours in dry air could remove the end-caps of SWCNTs. The successful solution of opening problem of ultrathin SWCNTs makes it possible to synthesize SWCNWs by polyynes fusion method.

4.3.1.3 Fusion reaction of polyyne molecules inside SWCNTs

As mentioned above, the second step of preparing SWCNWs by polyynes fusion method is to insert linear polyyne molecules into the ultrathin channels of SWCNTs. To put it simply, the structure of polyynes@SWCNTs were obtained by placing the previously prepared polyyne solution with high-purity and high-concentration and the open-ended SWCNTs in a high-pressure reactor [55], and this process is a little bit similar to insert polyyne molecules $C_{10}H_2$ [56], adamantane molecules [99], or C_{60} [100] into CNT. Obviously, in order to induce the polyyne molecules to interconnect and grow into LLCCs within SWCNTs, it is necessary to place the polyynes@SWCNTs samples in an appropriate induction environment. Figure 4.10 is the Raman spectra of polyyne fusion reaction within SWCNTs by using heat induction [55], the effects of different heat treatment times (Fig. 4.10a) and temperatures (Fig. 4.10b) on the polyynes fusion inside SWCNTs have been investigated. Although the exact relationship was unknown between the Raman intensity of L-band and the proportion of LLCCs inserted inside SWCNTs, the intensity ratio between the L-band of LLCCs and the G-band of SWCNTs could be related to the relative content of LLCCs in the same SWCNWs [40, 50]. The different batches may have different filling ratios of polyynes inside SWCNTs, so we have studied the effects of temperature and time using the same batch of polyynes@SWCNTs samples. By observing the relative intensity of L-band with fusion temperature and fusion time, a rough fusion condition of 700 °C for 3 hours could be obtained from Fig. 4.10, the whole fusion process indicates that the polyyne fusion reaction needs a certain fusion time as well as the desired fusion temperature.

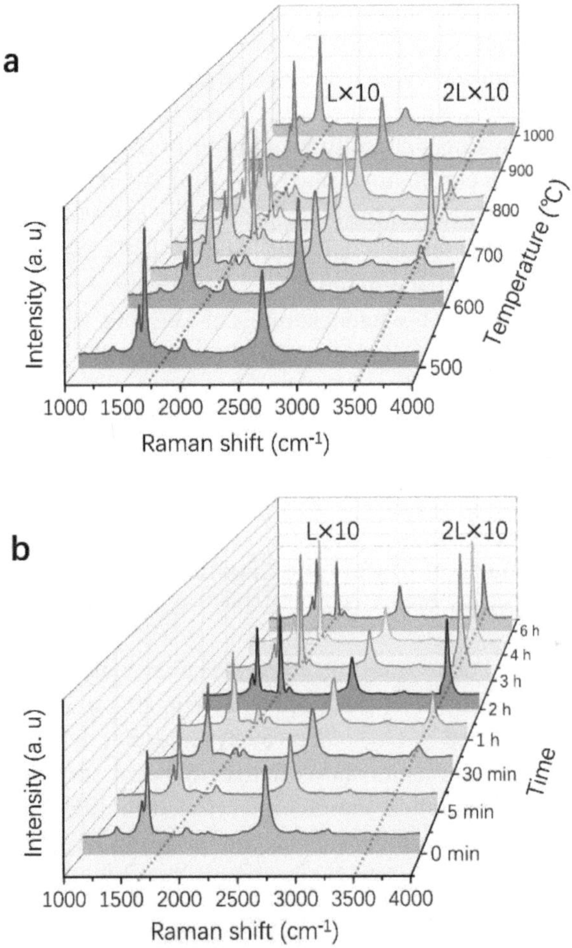

Figure 4.10 (a) Raman spectra of polyynes@SWCNTs after heat treatment from 500 to 1000 °C for 6 h. (b) Raman spectra of polyynes@SWCNTs annealed at 700 °C for various times [55]. © 2021 by Elsevier.

The advantages of polyynes fusion method for preparing SWCNWs are the controllability and mass-productivity, this is because both the polyyne solution with high-purity and high-concentration and the ultrathin SWCNTs could be synthesized in large scale. Such as, the linear polyyne molecules, $C_{2n}H_2$, were synthesized in large scale by submerged arc-discharge method in organic solvents, or commercial laser marking technology [101–104], and the technique of their purification and concentration *via* carbides were also developed [82]. On the other hand, the SWCNTs with the diameters between 0.7 and 1.0 nm could be obtained by single-chirality separation using gel chromatography method, especially, the Kataura group reported the milligram-scale single-chirality separation of SWCNTs (6, 5) using stepwise elution chromatography [105, 106]. All of this has already laid a solid foundation for the mass-production of SWCNWs samples. In conclusion, this feasible, efficient, and scalable method has the great advantage that the quantity of carbon source is controllable for growing LLCCs in the nanochannels of SWCNTs. From this point of view, the polyynes fusion method is even possible to synthesize SWCNWs close to 100% purity, therefore we hope that more and more researchers will participate in the study of SWCNWs to develop new characterization methods and look for special applications in nanoelectronics, nanomechanics, nanomaterials in the near future.

4.3.2 Mechanism of Polyynes Fusion Method for Preparing SWCNWs

Inserting short LCCs inside ultrathin SWCNTs is also theoretically feasible, the processes of polyynes C_8H_2 entering a SWCNT (6, 5) have been simulated by Liu *et al.* through molecular dynamics (MD), and the calculation results show that the process of carbon atoms entering SWCNTs is almost unhindered [107]. It could be understood that the essence of the process of polyynes entering SWCNT is caused by the attractive van der Waals interactions between more inserted carbon atoms of polyynes and the SWCNT, until the whole polyynes enter the SWCNT completely, this interaction becomes stable [107].

The polyynes will undergo a dehydrogenation fusion process inside the SWCNTs and grow into LLCCs, which can exist stably inside the SWCNTs [55]. Using ReaxFF reactive MD, we have compared the total energies of two C_8H_2 chains@SWCNT (6, 5) and a $C_{16}H_2$ chain@ SWCNT (6, 5) plus an isolated H_2 (Fig. 4.11a). The calculations show that the energy of a $C_{16}H_2$ chain inside SWCNT (6, 5) plus an isolated H_2 is 0.48 eV lower than that of two C_8H_2 chains inside SWCNT (6, 5), indicating that fused long carbon chains are thermodynamically more stable than separated short carbon chains inside SWCNT (6, 5) at 700 °C. Furthermore, the binding energy of $C_{16}H_2$@SWCNT (6, 5) (Fig. 4.11b) was −4.15 eV at an optimal distance of 3.79 Å between $C_{16}H_2$ and SWCNT (6, 5), consistent with the HRTEM observations.

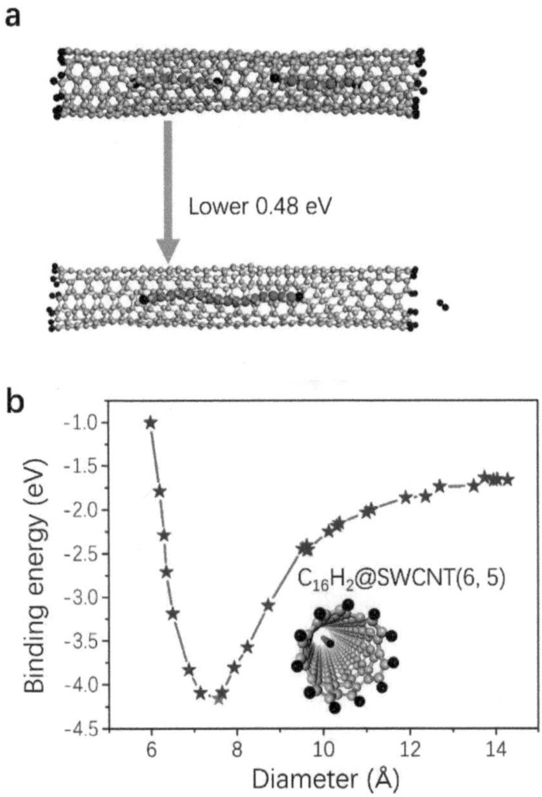

Figure 4.11 (a) The two C_8H_2 chains inside an SWCNT (6, 5) lowered energy by 0.48 eV after fusion into one $C_{16}H_2$ chain@SWCNT (6, 5). (b) The binding energy of $C_{16}H_2$ at various diameters of SWCNTs. The lowest binding energy is reached when the diameter of SWCNTs is 0.75 nm [55]. © 2021 by Elsevier.

4.4 Characterization of SWCNWs

4.4.1 Raman Spectra Measurements of SWCNWs

Raman spectroscopy is recognized as the best tool to fast identify carbon materials with different hybridization without damage. As we known, when the incident photons are inelastic scattered with the lattice of matter (phonons), Raman effects will be generated, and the wavelength of the scattered light can be larger (Stokes scattering) or smaller (anti-Stokes scattering) than that of the incident light. The anti-Stokes scattering process is related to the annihilation of phonons, so their high-energy (<-1000 cm^{-1}) Raman modes are several orders of magnitude less intense than that in Stokes scattering at room temperatures. Therefore, it is not easy to be identified because its intensity is similar to that of the noise produced in the measurement process [108, 109]. However, the LLCCs or carbyne possess a resonant Raman scattering cross section per atom exceeding that of any other known material or molecule [110], and the anti-Stokes Raman spectra of L-band from single LLCCs have been successfully obtained in the DWCNWs samples by a TERS setup [61, 62].

Figure 4.12a displays the anti-Stokes Raman spectrum (pale pink area) and Stokes spectrum (pale green area) of SWCNWs obtained by a common confocal Raman spectroscopy at 80 K, which have normalized by their intensity of G-bands, respectively [111]. The atomic structure diagram of a SWCNW is shown in the central inset in Fig. 4.12a, and a red chain of carbon atoms exists in the center of SWCNT, indicating that the SWCNWs (LLCCs@SWCNTs) are composed of both LLCCs and SWCNTs. Therefore, the Raman spectra of SWCNWs should show the Raman characteristics of both *sp*-hybridized LLCCs and *sp*^{2}-hybridized SWCNTs. For anti-Stokes, several Raman characteristic peaks are shown, the defect-induced D_A mode, the tangential mode of G_A^+ from SWCNTs [112, 113], and the out-of-phase stretching mode of L_A-band originated from the LLCCs inside SWCNTs [61, 62]. Among them, it is worth noting that the G_A^- modes were also observed, and usually they were difficult to be recognized due to their low intensity [114]. For Stokes (dark green line), it shows a richer series of characteristic peaks compared with anti-Stokes (red wine line). One noticeable difference is the iTOLA band, which is only seen in Stokes because the Stokes have stronger intensity than anti-Stokes at low temperature. In addition, the L-band from LLCCs is found to locate at around 1835 cm^{-1}, expressing that these LLCCs composed of hundreds to thousands of carbon atoms are encapsulated inside SWCNTs whose diameter is about 0.75 nm [3, 36, 55]. In the final, it should be emphasized that in the present research those faint anti-Stokes Raman signals of high-energy modes, i.e., L-band of LLCCs, as strikingly strong

Raman peaks without using SERS and amplification have been easily obtained by only used the normal Raman scattering, signifying that all of the Raman techniques can be applied to research on these SWCNWs samples.

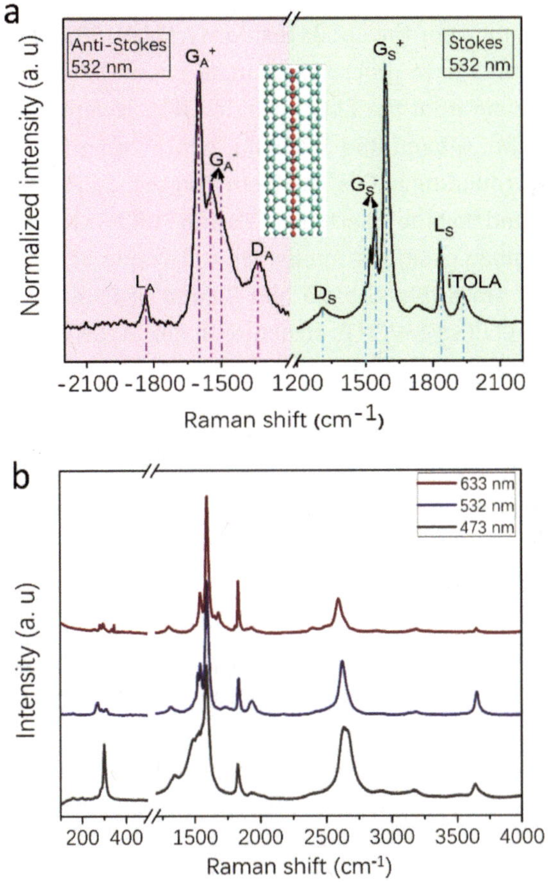

Figure 4.12 (a) The anti-Stokes and Stokes Raman spectra at 80 K, as well as atomic structure diagram of SWCNWs (LLCCs@SWCNTs). The excitation wavelength is 532 nm [111]. © 2023 by American Chemical Society. (b) Raman spectra of LLCCs@SWCNTs characterized by three different excitation wavelengths: 473 nm (grey line), 532 nm (blue line), and 633 nm (red line), respectively [55]. © 2021 by Elsevier.

To examine the resonance Raman conditions of LLCCs inside SWCNTs, the Raman spectra of SWCNWs samples were measured using lasers with excitation wavelengths of 633, 532, and 473 nm (Fig. 4.12b). Both the L and 2L bands were found in all three Raman spectra where the strongest L-bands appeared at ~1834 cm^{-1} corresponding to an excitation wavelength of 633 nm, whereas the weakest L-bands were observed at 473 nm far away from the resonant window for the LLCCs inside MWCNTs [59]. Obviously, the L and 2L bands have different resonance conditions, even though they both come from the LLCCs inside CNTs. Except L and 2L, the RBMs, D-band, G-band and 2D-band, which come from the outer SWCNTs surrounding LLCCs, can been also seen in Fig. 4.12b, and it has been found that the inserted LLCCs into SWCNTs do not markedly affect the Raman positions, intensities and shapes of outer SWCNTs. Therefore, from RBMs, G-band and L-band of Raman spectra, the presence of LLCCs@SWCNTs structure in our SWCNWs samples are easy to confirm.

4.4.2 HRTEM Observations of SWCNWs

Although Raman spectroscopy could quickly detect whether a carbon material contains CNWs, it cannot determine which kind of CNWs it is, and that's where the HRTEM comes in. Figure 4.13 shows the HRTEM images of empty SWCNTs and SWCNWs (LLCCs@SWCNTs). Comparing the empty SWCNT in Fig. 4.13a with SWCNWs in Fig. 4.13c and e, it can be clearly found that the LLCCs exist in the center of SWCNTs. In addition, the presence of LLCCs inside SWCNTs is also confirmed by comparing their line profiles of empty SWCNT (Fig. 4.13b) and SWCNWs (Fig. 4.13d, f) [55]. This means that the LLCCs could be stable within SWCNTs with the diameters ranging from 0.70 to 0.95 nm, and the diameters range of SWCNWs is obviously larger than the diameters range of the innermost CNTs of MWCNWs or DWCNWs. This phenomenon may be related to the mechanism of LLCCs growth inside deferent types of CNTs, which still needs further research.

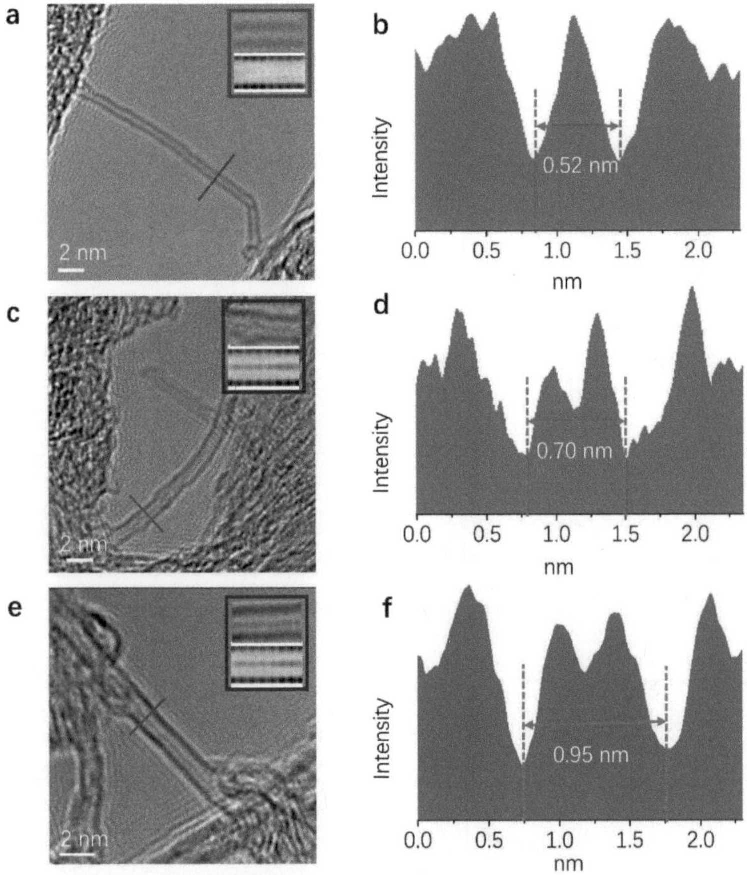

Figure 4.13 HRTEM images and the corresponding line profiles of SWCNTs and SWCNWs (LLCCs@SWCNTs). (a) An empty SWCNT with a diameter of 0.52 nm and its line profile in (b). (c) An SWCNT with a diameter of 0.7 nm containing an LLCC and its line profile in (d). (e) An SWCNT with a diameter of 0.95 nm containing an LLCC and its line profile in (f). The insets in Fig. 4.14 (a), (c) and (e) show a part of HRTEM image (top) and simulated image (bottom) of SWCNT or SWCNWs (LLCCs@SWCNTs) [55]. © 2021 by Elsevier.

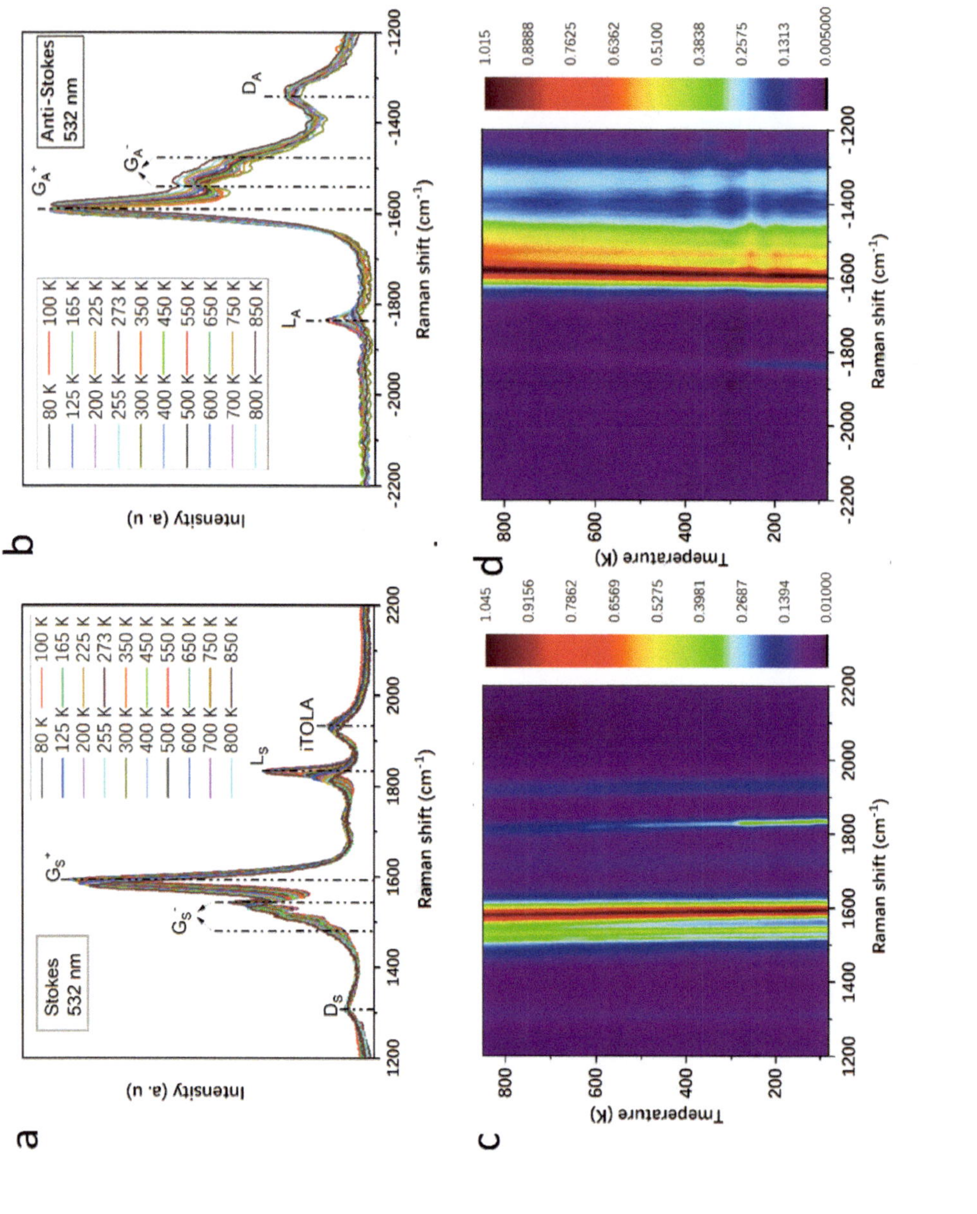
a
Stokes
532 nm
Intensity (a. u)
Raman shift (cm-1)
iTOLA
b
Anti-Stokes
532 nm
c
d
Temperature (K)
80 K
100 K
125 K
165 K
200 K
225 K
255 K
273 K
300 K
350 K
400 K
450 K
500 K
550 K
600 K
650 K
700 K
750 K
800 K
850 K

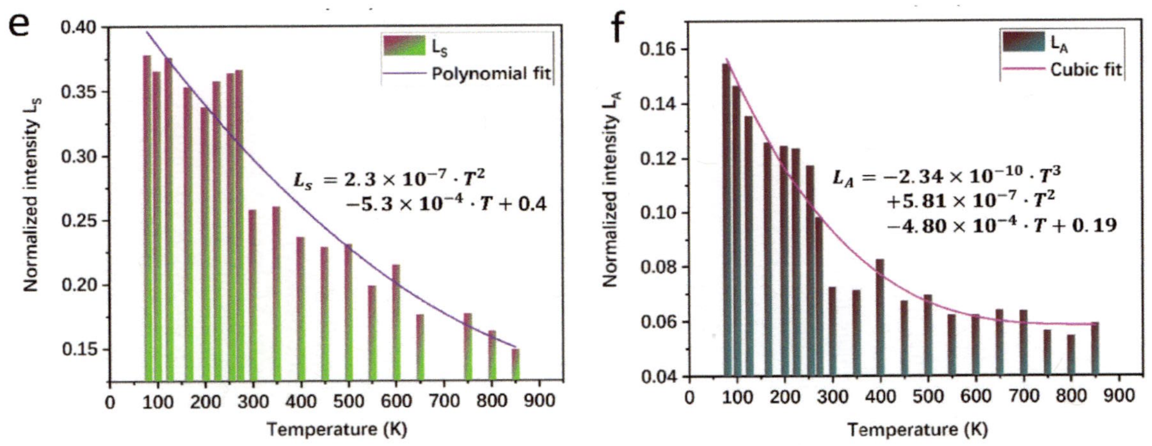

Figure 4.14 (a) Temperature-dependent Stokes Raman spectra of SWCNWs. (b) Temperature-dependent anti-Stokes spectra of SWCNWs. (c) Stokes Raman spectra contour mapping of Fig. 4.15a. (d) anti-Stokes Raman spectra contour mapping of Fig. 4.15b. (e) The intensity of normalized L_S-band varies with temperature. (f) The intensity of normalized L_A-band varies with temperature [111]. © 2023 by American Chemical Society.

4.5 Effect of Temperature on Raman Spectra of SWCNWs

Temperature could change the number of phonons and phase of matters, thus affecting the intensity and the position of its Raman peak. Figure 4.14 is the Stokes and anti-Stokes Raman scattering of SWCNWs at varying temperatures from 80 to 850 K measured by normal Raman spectroscopy. In order to more clearly observe the changes of L-band with temperature, the Raman spectra were all normalized by their G-bands. It could be seen from the Fig. 4.14a and b that the Stokes and anti-Stokes Raman peak positions and intensity of SWCNWs show certain changes except for the intensity of G-bands due to the reasons mentioned above [111]. In fact, further observation indicates that the intensity of Raman peaks decreases with the temperature increasing for Stokes, but increases for anti-Stokes, and they also include the G-bands. And beyond that, the Raman peak positions are all redshifted with increasing temperature. In order to further observe the changes of Raman intensity and position of L-band at 1834 and G-bands at 1593 cm^{-1} from SWCNWs more clearly, we have drawn the Raman contour mapping of Fig. 4.14a, b and shown them in Fig. 4.14c, d. For Stokes, the gradually changing color and slanting peaks position of L-band mean a reduction in the intensity of Raman characteristic peaks and a redshift in peaks position, while for anti-Stokes, the gradually changing color and slanting peaks position mean an increase in the intensity of Raman characteristic peaks and a redshift in peak position. But only the Raman intensity of G-bands, which were used to normalized all of Raman spectra, remains unchanged, although their redshift also occurs. Obviously, the LLCCs are more sensitive to temperature in terms of Raman intensity and peak position. When we focus on the relative intensity of L-band for LLCCs inside SWCNTs, both Stokes and anti-Stokes decrease with temperature, which obeys a quadratic polynomial fitting for Stokes (Fig. 4.14e) and cubic fitting for anti-Stokes peaks (Fig. 4.14f), while the absolute intensity of Stokes peaks decreased and that of anti-Stokes increased with the increasing temperature.

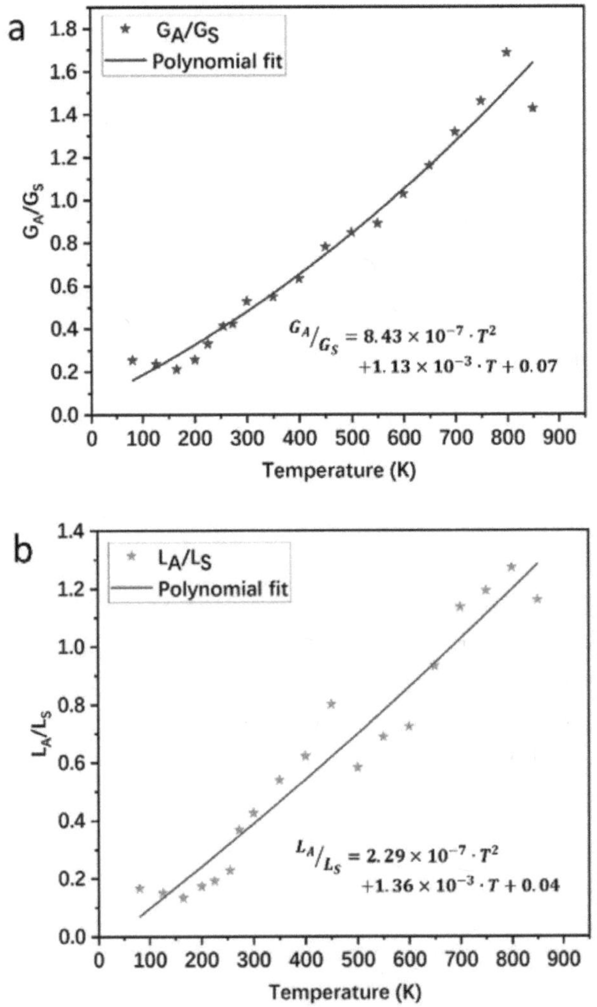

Figure 4.15 The ratio of anti-Stokes to Stokes, (a) G_A/G_S, (b) L_A/L_S [111]. © 2023 by American Chemical Society.

Figure 4.15 shows the anti-Stokes and Stokes ratios of G and L-band, which originated from the encapsulated hosting SWCNTs and the LLCCs inside SWCNTs, respectively. Both the ratios of G_A/G_S and L_S/L_A satisfy the polynomial fitting and exhibit a non-linear growth trend. From the growing trend, the ratio of L_A/L_S is obviously greater than that G_A/G_S, it also reflects that the LLCCs is more sensitive to temperature than SWCNTs. As we all know, the intensity of a Raman mode is proportional to its phonon population, which is relate to the local temperature and satisfy the Bose Einstein distribution. Therefore, the ratio of anti-Stokes to Stokes is also relate to temperature and approximately satisfy the relation $\frac{I_A}{I_S} = \exp^{-\frac{\hbar\omega}{K_B T}}$, indicating that the ratio exhibited nonlinear growth trend as temperature increased [62]. Using the ratio of anti-Stokes to Stokes to measure the temperature of objects has been reported, so the LLCCs or SWCNWs samples have great potential in the field of nanoscale temperature measurement.

Table 4.1 Temperature coefficients of G and L modes of carbonaceous materials. © 2023 by American Chemical Society

Material	Coefficients ($cm^{-1}\cdot K^{-1}$)	Source
SWCNTs	−0. 0192 (G_S^+)/0.0200 (G_A^+)	This work
LLCCs inside SWCNTs	−0. 0296 (L_S)/0.0334 (L_A)	This work
LLCCs inside DWCNTs	−0. 023 (L_S)	Ref. [61]
HOPG	−0. 011 (G_S)	Ref. [117]
Graphite	−0. 011 (G_S)	Ref. [118]
Graphene	−0. 015 (G_S)	Ref. [119]
MWCNTs	−0. 028 (G_S)	Ref. [120, 121]
SWCNTs	−0. 012 (G_S)	Ref. [115]

G_S^+: Stokes scattering of high-energy longitudinal phonon of the tangential mode of G from SWCNTs encapsulating LLCCs.
G_A^+: anti-Stokes scattering of high-energy longitudinal phonon of the tangential mode of G from SWCNTs encapsulating LLCCs.
L_S: Stokes scattering of out-of-phase stretching mode of L-band from LLCCs.
L_A: anti-Stokes scattering of out-of-phase stretching mode of L-band from LLCCs.
G_S: Stokes Raman scattering of the tangential mode of G from SWCNTs.

The temperature-dependence of Raman peak positions of L and G-band are depicted in Fig. 4.16 for Stokes and anti-Stokes, respectively. It has been found that the Raman peak positions of G and L-band are quasi-linear with temperature, the temperature coefficient of G-band is −0.0192 $cm^{-1}\cdot K^{-1}$ for Stokes and 0.0200 $cm^{-1}\cdot K^{-1}$ for anti-Stokes, while the L-band is −0.0296 $cm^{-1}\cdot K^{-1}$ for

Stokes and 0.0334 $cm^{-1}\cdot K^{-1}$ for anti-Stokes [111]. In addition, it should be noted that the small numerical differences between the Stokes and anti-Stokes temperature coefficients for the G-band of SWCNTs and the L-band of LLCCs may be caused by experimental error and average error, since they come from the same materials, respectively. It is notable that the temperature coefficient (−0.0296 $cm^{-1}\cdot K^{-1}$) of LLCCs inside SWCNTs in this work was a little higher than that (−0.023 $cm^{-1}\cdot K^{-1}$) of LLCCs inside DWCNTs [61]. This may originate from the difference of hosting CNTs, bulk and individual measurements or from the proportion of SWCNWs in the total sample, and further researches are necessary.

Table 4.1 shows the temperature coefficients of some typical carbon materials and SWCNWs, i.e., LLCCs inside SWCNTs. The LLCCs have the largest temperature coefficient, and other carbon materials have similar temperature coefficients except for MWCNTs. Terekhov *et al.* reported that the SWCNTs with different purity had different temperature coefficients, and found the temperature coefficient as low as −0.012 $cm^{-1}\cdot K^{-1}$ when the purity of SWCNTs samples was 95% [115]. The small amounts of impurities were the main reason resulting in a higher temperature coefficient. Therefore, the much higher temperature coefficients of MWCNTs than SWCNTs maybe result from their defects, amorphous carbon, and the graphite particles coexisting with the MWCNTs samples. Besides, the temperature coefficient was also related to the size and curvature [116, 117]. It has been found that linear *sp* carbon is the most sensitive to temperature, so the SWCNWs (LLCCs@SWCNTs) may be the smallest nanothermometers that can be used by peoples.

In addition, the peak position of L-band changed from 1835 to 1807 cm^{-1}, which corresponded to the band gap of LLCCs inside SWCNTs having a variation of 179 meV when the temperature changed from 80 to 850 K [58]. It indicates the possibility of controlling the crystal structure, physical properties of LLCCs by means of changing temperature. Obviously, the high symmetry, high sensibility, and high temperature coefficients of L-band are all beneficial to the applications in local temperature sensing. Moreover, the linear change of peak positions of L-band with the temperature indubitably illustrates that the electronic structure of LLCCs inside SWCNTs can be linearly controlled by changing temperature in the future.

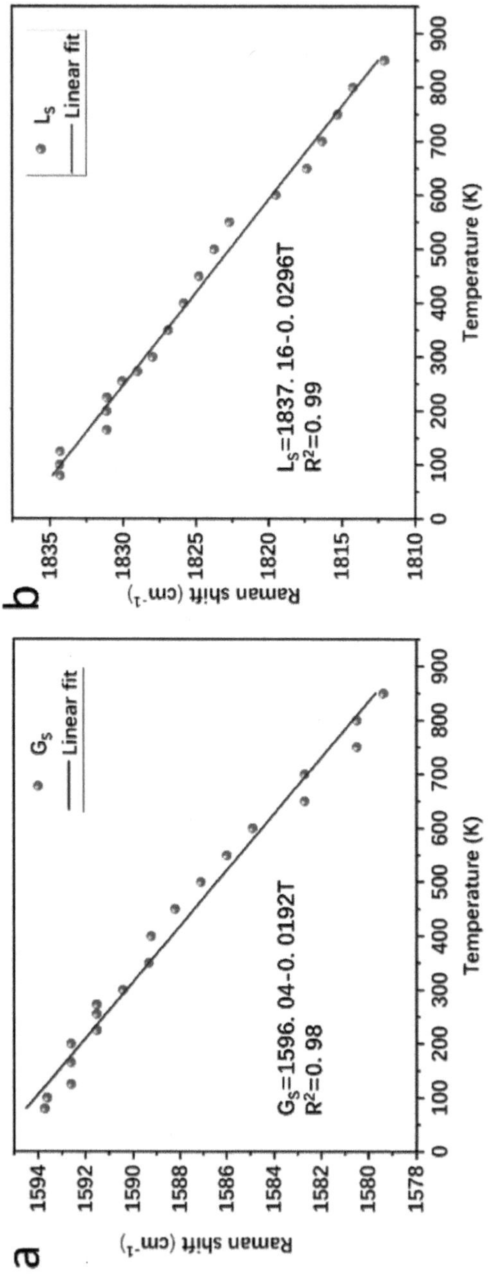
a
Raman shift (cm-1)
Temperature (K)
Gs
Linear fit
Gs=1596. 04-0. 0192T
R²=0.98
b
Raman shift (cm-1)
Temperature (K)
Ls
Linear fit
Ls=1837. 16-0. 0296T
R²=0. 99

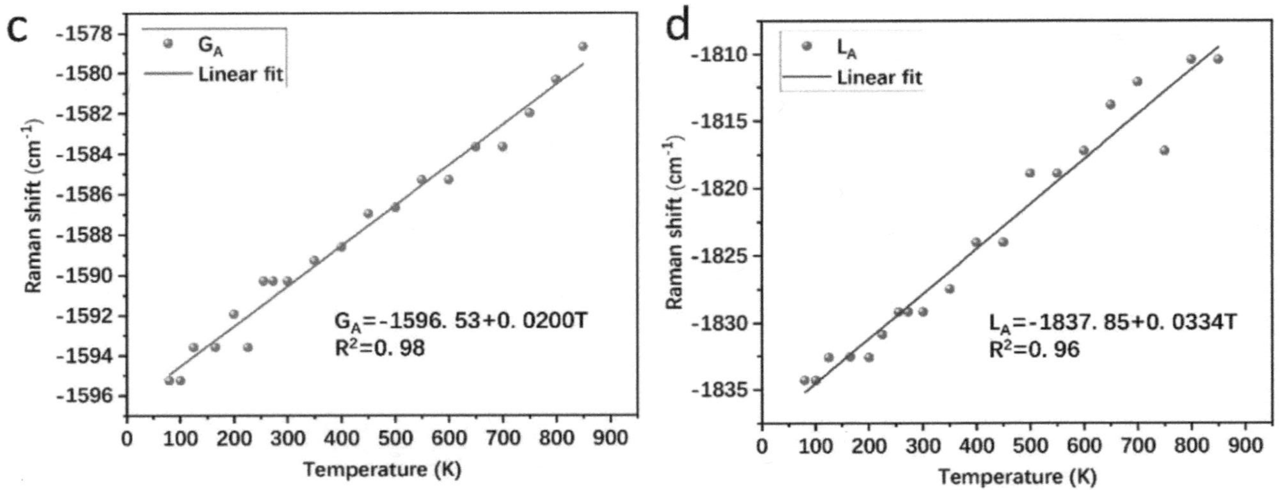

Figure 4.16 The peak positions of Stokes and anti-Stokes changes with the change of temperature from 80 to 850 K, (a) G_S^+, (b) L_S, (c) G_A^+, (d) L_A [111]. © 2023 by American Chemical Society.

4.6 Challenges and Prospects of SWCNWs

SWCNWs have been successfully prepared by tip-ultrasonic and density gradient ultracentrifugation, vacuum discharge and polyyne fusion method, especially, the polyyne fusion method can be used to feasibly and controllably mass-produce SWCNWs samples, but the challenges still exist in the synthesis of SWCNWs. First, it is the question of purity of SWCNWs samples. Since all of the current SWCNWs samples are the mixture of empty SWCNTs without LLCCs inside and SWCNWs containing LLCCs, the proportion of SWCNWs is not more than 20%, a near 100% pure sample of SWCNWs is necessary when we utilize the ultra-high mechanical properties of SWCNWs. The second challenge is the single-chirality separation of SWCNWs. As we all kwon, the LLCCs inside the SWCNTs with different diameters or chiralities will possess different band gaps [37, 122], therefore, a SWCNWs with single-chirality implies that both the LLCC and the SWCNT have the certain band gaps. Obviously, the single-chirality SWCNW is most needed for the development of scientific study on LLCCs, or carbyne, and SWCNWs, and has great potential in the industrial application of advanced semiconductor field. To date, the promising applications of SWCNWs mainly come from the theoretical predictions, in order to advance their excellent properties from theoretical prediction to practical application, it is also necessary to make breakthrough in the large-scale synthesis technology of the SWCNWs with high-purity or even single-chirality. As a novel 1-D carbon allotrope, the various new properties and applications for SWCNWs are still expected, and needed to be explored and developed by researchers.

4.7 Conclusions

The discovery of CNWs (LLCCs@CNTs) and the recent progress for SWCNWs preparation are reviewed, the process of preparing SWCNWs by polyyne fusion technique is described in detail. This is a feasible, efficient, and scalable method for synthesizing SWCNWs, and it is even possible to mass-produce the SWCNWs samples that are close to 100% purity or single-chirality. Moreover, we also introduce the growth mechanism of SWCNWs prepared by polyyne fusion technique, and their common characterization methods, i.e., Raman spectra measurements and HRTEM observations. Finally, some of the current developments regarding the temperature effect on Stokes and anti-Stokes Raman spectra of SWCNWs are presented. Although the theoretical predictions of the excellent properties of SWCNWs have led to continuous experimental breakthroughs, some problems still need to be solved, for example, the goal of producing SWCNWs with high-purity or even single-chirality in large-scale remains unaccomplished, and the practical applications that involve the outstanding properties of SWCNWs are also lacking. It can be hoped that the new nanoscience and nanotechnology relating to 1-D novel LLCCs or carbyne and SWCNWs will come after the success of great challenge for the mass-production of SWCNWs with high-purity or even single-chirality in the near future.

References

1. Kroto, H. W., Heath, J. R., O"Brien, S. C. (1985). C_{60}: buckminsterfullerene, *Nature*, **318**, 162–163.
2. Iijima, S. (1991). Helical microtubules of graphitic carbon, *Nature*, **354**, 56–58.
3. Zhao, X., Ando, Y., Liu, Y., Jinno, M., Suzuki, T. (2003). Carbon nanowire made of a long linear carbon chain inserted inside a multiwalled carbon nanotube, *Phys. Rev. Lett.*, **90**, 187401.
4. Zhang, Y. F. (2022). A review of arc-discharge method towards large-scale preparation of long linear carbon chains, *Chin. Phys. B*, **31**, 125201.
5. Prelas, M. A., Popovici, G., Bigelow, L. K. (1997). Handbook of industrial diamonds and diamond films, Marcel Dekker, New York.
6. Kelly, B. T. (1981). *Physics of Graphite*, Applied Science Publishers Ltd, London.
7. Baeyer, A. (1885). Ueber Polyacetylenverbindungen, *Ber. Dtsch. Chem. Ges.*, **18**, 674–681.
8. Baeyer, A. (1885). Ueber Polyacetylenverbindungen, *Ber. Dtsch. Chem. Ges.*, **18**, 2269–2281.
9. Anchel, M. (1953). Identification of an antibiotic polyacetylene from clitocybediatreta as a suberamic acid enediol, *J. Am. Chem. Soc.*, **75**, 4621–4622.

10. Nakaminami, G. (1963). The structures of natural acetylenic compounds and their biogenesis, *J. Syn. Org. Chem. Jpn.*, **21**, 751–765.

11. Bu'Lock, J. D. (1956). Acetylenic compounds as natural products, *Q. Rev. Chem. Soc.*, **10**, 371–394.

12. Bohlmann, F., Burkhardt, T., Zdero, C. (1973). *Naturally Occurring Acetylenes*, Academic Press, London.

13. Goresy, A. E., Donnay, G. (1968). A New allotropic form of carbon from the ries crater, *Science*, **161**, 363–364.

14. Krätschmer, W., Lamb, L. D., Fostiropoulos, K. (1990). C_{60}: a new form of carbon, *Nature*, **347**, 354–358.

15. Novoselov, K. S. (2004). Electric field effect in atomically thin carbon films, *Science*, **306**, 666–669.

16. Zhao, X., Ando Y. (1998). Raman spectra and X-ray diffraction patterns of carbon nanotubes prepared by hydrogen arc discharge, *Jpn. J. Appl. Phys.*, **37**, 4846–4849.

17. Ando, Y., Zhao, X., Kataura, H., Achiba, Y., Kaneto, K., Tsuruta, M., Uemura, S., Iijima, S. (2000). Multi-walled carbon nanotubes prepared by hydrogen arc, *Diam. Relat. Mater.*, **9**, 847–851.

18. Casari, C. S., Milani, A. (2018). Carbyne: from the elusive allotrope to stable carbon atom wires, *MRS Commun.*, **8**, 207–219.

19. Liu, M. J., Artyuhov, V. I., Lee, H., Xu, F. B., Yakobson, B. I. (2013). Carbyne from first principles: chain of C atoms, a nanorod or a nanorope, *ACS Nano.*, **7**, 10075–10082.

20. Wang, M., Lin, S. (2015). Ballistic thermal transport in carbyne and cumulene with micron-scale spectral acoustic phonon mean free path, *Sci. Rep.*, **5**, 18122.

21. Lang, N. D., Avouris, P. (1998). Oscillatory conductance of carbon-atom wires, *Phys. Rev. Lett.*, **81**, 3515.

22. Kasatochkin, V. I., Sladkov, A. M., Kudryavtsev, Y. P., Popov, N. M., Korshak, V. V. (1967). Crystalline forms of a linear modification of carbon, *Dokl. Akad. Nauk SSSR*, **177**, 358–360.

23. Zhang, K., Zhang, Y. F., Shi L. (2020). A review of linear carbon chains, *Chin. Chem. Lett.* **31**, 1746–1756.

24. Hoye, R. H., Baire, B., Niu, D., Willoughby, P. H., Woods, B. P. (2012). The hexadehydro-dielsalder reaction, *Nature*, **490**, 208–212.

25. Gao, Y. Z., Hou, Y. X., Gámez, F. G., Ferguson, M. J., Casado, J., Tykwinski, R. R. (2020). The loss of endgroup effects in long pyridyl-end capped oligoynes on the way to carbyne, *Nat. Chem.*, **12**, 1143–1149.

26. Smith, P. P. K., Buseck, P. R. (1985). Carbyne forms of carbon: evidence for their existence, *Science*, **229**, 486–487.

27. Whittaker, A. G., Kintner, P. L. (1969). Carbon: observations of new allotropic form, *Science,* **165**, 589–591.

28. Sladkov, A., Kudryavtsev, Y. (1969). Diamond, graphite, carbyne: allotropic forms of carbon, *Priroda*, **5**, 37–44.

29. Smith, P. P. K., Buseck, P. R. (1981). Graphitic carbon in the Allende meteorite: a microstructural study, *Science,* **212**, 322–324.

30. Smith, P. P. K., Buseck, P. R. (1982). Carbyne forms of carbon: do they exist? *Science,* **216**, 984–986.

31. Jin, C., Lan, H., Peng, L., Suenaga, K., Iijima, S. (2009). Deriving carbon atomic chains from graphene, *Phys. Rev. Lett.,* **102**, 205501.

32. Ben Romdhane, F., Adjizian, J. J., Charlier, J.C., Banhart, F. (2017). Electrical transport through atomic carbon chains: the role of contacts, *Carbon*, **122**, 92–97.

33. Pan, B., Xiao, J., Li, J., Liu, P., Wang, C., Yang, G. (2015). Carbyne with finite length: the one-dimensional *sp* carbon, *Science Advances*, **1**, e1500857.

34. Kaiser, K., Scriven, L. M., Schulz, F., Gawel, P., Gross, L., Anderson, H. L. (2019). An *sp*-hybridized molecular carbon allotrope, cyclo [18] carbon, *Science*, **365**, 1299–1301.

35. Li, G. X., Li, Y. L., Liu, H. B., Guo, Y. B., Li, Y. J., Zhu, D. B. (2010). Architecture of graphdiyne nanoscale films, *Chem. Commun.*, **46**, 3256–3258.

36. Shi, L., Rohringer, P., Suenaga, K., Niimi, Y., Kotakoski, J., Meyer, J. C., Peterlik, H., Wanko, M., Cahangirov, S., Rubio, A. (2016). Confined linear carbon chains as a route to bulk carbine, *Nat. Mater.,* **15**, 634–639.

37. Heeg, S., Shi, L., Poulikakos, L. V., Thomas, P., Lukas N. (2018). Carbon nanotube chirality determines properties of encapsulated linear carbon chain, *Nano Lett.,* **18**, 5426–5431.

38. Heeg, S., Shi, L., Pichler, T., Novotny, L. (2018). Raman resonance profile of an individual confined long linear carbon chain, *Carbon*, **139**, 581–585.

39. Chalifoux, W. A., Tykwinski, R. R. (2010). Synthesis of polyynes to model the *sp*-carbon allotrope carbyne, *Nat. Chem.* **2**, 967–971.

40. Zhang, Y., Zhao, J., Fang, Y., Liu, Y., Zhao, X. (2018). Preparation of long linear carbon chain inside multi-walled carbon nanotubes by cooling enhanced hydrogen arc discharge method, *Nanoscale,* **10**, 17824–17833.

41. Zhao, X., Ohkohchi, M., Wang, M., Iijima, S., Ichihashi, T., Ando, Y. (1997). Preparation of high-grade carbon nanotubes by hydrogen arc discharge, *Carbon*, **35**, 775–781.

42. Qin, L. C., Zhao, X., Hirahara, K., Miyamoto, Y., Ando, Y., Iijima, S. (2000). The smallest carbon nanotube, *Nature*, **408**, 50.

43. Zhao, X., Liu, Y., Inoue, S., Suzuki, T., Jones, R. O., Ando, Y. (2004). Smallest carbon nanotubes is 3 Å in diameter, *Phys. Rev. Lett.*, **92**, 125502.

44. Ando, Y., Zhao, X., Ohkohchi, M. (1997). Production of petal-like graphite sheets by hydrogen arc discharge, *Carbon*, ***35***, 153–158.

45. Cazzanelli, E., Castriota, M., Caputi, L. S., Cupolillo, A., Giallombardo, C., Papagno, L. (2007). High-temperature evolution of linear carbon chains inside multiwalled nanotubes, *Phys. Rev. B.*, **75**, 12145(R).

46. Scuderi, V., Scalese, S., Bagiante, S., Compagnini, G., D'Urso, L., Privitera, V. (2009). Direct observation of the formation of linear C chain/carbon nanotube hybrid systems, *Carbon*, **47**, 2134–2137.

47. Kim, Y. A., Muramatsu, H., Hayashi, T., Endo, M. (2012). Catalytic metal-free formation of multi-walled carbon nanotubes in atmospheric arc discharge, *Carbon*, **50**, 4588–4595.

48. Ha, J., Jung, H. Y., Hao, J., Li, B., Raeliarijaona, A., Alarcón, J., Terrones, H., Ajayan, P. M., Jung, Y. J., Kim, J. (2017). Ultrafast structural evolution and formation of linear carbon chains in single-walled carbon nanotube networks by femtosecond laser irradiation, *Nanoscale*, **9**, 16627–16631.

49. Sheng, L., Jin, A., Yu, L., An, K., Ando, Y., Zhao, X. (2012). A simple and universal method for fabricating linear carbon chains in multiwalled carbon nanotubes, *Mater. Lett.*, **81**, 222–224.

50. Zhang, Y., Chang, W., Liu, Y., Maruyama, T., Zhao, X. (2020). High-yield growth of multi-walled carbon nanowires by magnetic field controlled arc technique, *Carbon*, **158**, 672–680.

51. Shi, L., Sheng, L., Yu, L., An, K., Ando, Y., Zhao, X. (2011). Ultra-thin double-walled carbon nanotubes: a novel nanocontainer for preparing atomic wires, *Nano Res.*, **4**, 759–766.

52. Zhao, C., Kitaura, R., Hara, H., Irle, S., Shinohara, H. (2011). Growth of linear carbon chains inside thin double-wall carbon nanotubes, *J. Phys. Chem. C*, **115**, 13166–13170.

53. Shi, L., Yanagi, K., Cao, K., Kaiser, U., Ayala, P., Pichler, T. (2018). Extraction of linear carbon chains unravels the role of the carbon nanotube host, *ACS Nano*, **12**, 8477–8484.

54. Toma, S., Asaka, K., Irita, M., Saito, Y. (2019). Bulk synthesis of linear carbon chains confined inside single-wall carbon nanotubes by vacuum discharge, *Surf. Interface Anal.*, **51**, 131–135.

55. Chang, W. W., Liu, F., Liu, F. Y., Zhu, T. X., Fang, L., Li, Q. Q., Liu, Y., Zhao, X. L. (2021). Smallest carbon nanowires made easy: long linear carbon chains confined inside single-walled carbon nanotubes, *Carbon,* 183, 571–577.

56. Daisuke, N., Hirofumi, D., Tomonari, W., Nishibori, E., Aoyagi, S., Ishida, M., Kikuchi, S., Kitaura, R., Sugai, T., Kitaura, R., Sugai, T., Sakata, M. (2006). Single-wall carbon nanotubes encaging linear chain $C_{10}H_2$ polyyne molecules inside, *Chem. Phys. Lett.*, **428**, 356–360.

57. Malard, L. M., Nishide, D., Dias, L. G., Capaz, R. B., Gomes, A. P., Jorio, A., Achete, C. A., Saito, R., Achiba, Y., Shinohara, H. (2007). Resonance Raman study of polyynes encapsulated in single-wall carbon nanotubes, *Phys. Rev. B*, **76**, 233412.

58. Shi, L., Rohringer, P., Wanko, M., Rubio, A., Waßerroth, S., Reich, S., Cambré, S., Wenseleers, W., Ayala, P., Pichler, T. (2017). Electronic band gaps of confined linear carbon chains ranging from polyyne to carbyne, *Phys. Rev. Mater.*, **1**, 075601.

59. Andrade, N. F., Vasconcelos, T. L., Gouvea, C. P., Archanjo, B. S., Achete, C. A., Kim, Y. A., Endo, M., Fantini, C., Dresselhaus, M. S., Filho, A. G. S. (2015). Linear carbon chains encapsulated in multiwall carbon nanotubes: resonance Raman spectroscopy and transmission electron microscopy studies, *Carbon,* **90**, 172–180.

60. Fantini, C., Cruz, E., Jorio, A., Terrones, M., Terrones, H., van Lier, G., Charlier, J. C., Dresselhaus, M. S., Saito, R., Kim, Y. A., Hayashi, T., Muramatsu, H., Endo, M., Pimenta, M. A. (2006). Resonance Raman study of linear carbon chains formed by the heat treatment of double-wall carbon nanotubes, *Phys. Rev. B*, **73**, 193408.

61. Tschannen, C. D., Frimmer, M., Grodeev, G., Vasconcelos, T. L., Shi, L., Pichler, T., Reich, S., Heeg, S., Novotny, L. (2021). Anti-Stokes Raman scattering of single carbyne chains, *ACS Nano,* **15**, 12249–12255.

62. Tschannen, C. D., Frimmer, M., Vasconcelos, T. L., Shi, L., Pichler, T., Novotny, L. (2022). Tip-enhanced Stokes–anti-Stokes scattering from carbyne, *Nano Lett.,* **22**, 3260–3265.

63. Eastmond, R., Johnson, T. R., Walton, D. R. M. (1972). Silylation as a protective method for terminal alkynes in oxidative couplings: a general synthesis of the parent polyynes $H(\equiv C)_nH$(n = 4–10, 12), *Tetrahedron*, **28**, 4601–4616.

64. Lei, A., Srivastava, M., Zhang, X. (2002). Transmetalation of palladium enolate and its application in palladium-catalyzed homocoupling of alkynes: a room-temperature, highly efficient route to make diynes, *J. Org. Chem.*, **67**, 1969–1971.

65. Chalifoux, W. A., Ferguson, M. J., Tykwinski, R. R. (2007). Tris(biphenyl-4-yl)silyl-endcapped polyynes, *Eur. J. Org. Chem.*, **6**, 1001–1006.

66. Marsden, J. A., Miller, J. J., Haley, M. M. (2004). Let the best ring win: selective macrocycle formation through Pd-catalyzed or Cu-mediated alkyne homocoupling, *Angew. Chem. Int. Ed.*, **43**, 1694–1697.

67. Li, J. H., Liang, Y., Xie, Y. X. (2005). Efficient palladium-catalyzed homocoupling reaction and Sonogashira cross-coupling reaction of terminal alkynes under aerobic conditions, *J. Org. Chem.*, **70**, 4393–4396.

68. Chalifoux, W. A., Tykwinski, R. R. (2009). Synthesis of extended polyynes: toward carbine, *C. R. Chimie*, **12**, 341–358.

69. Tsuji, M., Tsuji, T., Kuboyama, S., Yoon, S. H., Korai, Y., Tsujimoto, T., Kubo, K., Mori, A., Mochida, I. (2002). Formation of hydrogen-capped polyynes by laser ablation of graphite particles suspended in solution, *Chem. Phys. Lett.*, **355**, 101–108.

70. Matsutani, R., Inoue, K., Sanada, T., Wada, N., Kojima, K. (2012). Preparation of long-chain polyynes of $C_{28}H_2$ and $C_{30}H_2$ by liquid-phase laser ablation, *J. Photochem. Photobiol. A*, **240**, 1–4.

71. Matsutani, R., Inoue, K., Wada, N., Kojima, K. (2011). Wavelength dependence of polyyne preparation by liquid-phase laser ablation using pellet targets, *Commun. Chem.*, **47**, 5840–5842.

72. Inoue, K., Matsutani, R., Sanada, T., Kojima, K. (2010). Preparation of long-chain polyynes of $C_{24}H_2$ and $C_{26}H_2$ by liquid-phase laser ablation in decalin, *Carbon*, **48**, 4209–4211.

73. Matsutani, R., Kakimoto, T., Wada, K., Sanada, T., Tanaka, H., Kojima, K., Preparation of long-chain polyynes $C_{18}H_2$ and $C_{20}H_2$ by laser ablation of pellets of graphite and perylene derivative in liquid phase, *Carbon*, **46**, 1103–1106.

74. Cataldo, F. (2003). Simple generation and detection of polyynes in an arc discharge between graphite electrodes submerged in various solvents, *Carbon*, **41**, 2671–2674.

75. Cataldo, F. (2005). A methods for synthesizing polyyne in solution, *Carbon*, **43**, 2792–2800.

76. Wu, Y., Zhang, Y., Zhu, T. X., Li, H., Zhao, X. (2019). Effects of precursor molecules on polyyne formation by arc discharge between two copper electrodes, *Chem. Phys. Lett.*, **730**, 64–69.

77. An, K., Wei, G. T., Qi, G. M., Sheng, L. M., Yu, L. M., Ren, W., Zhao, X. L. (2015). Stability improvement of C_8H_2 and $C_{10}H_2$ embedded in poly(vinyl alcohol) films with adsorption on gold nanoparticles, *Chem. Phys. Lett.*, **637**, 71–76.

78. Li, Y., An, K., Sheng, L., Yu, L., Ren, W., Zhao, X. (2015). Surface-enhanced Raman scattering investigations of $C_{2n}H_2$ (n = 4–6) in as-prepared and dried Ag colloid, *Chem. Phys. Lett.*, **631**, 12–15.

79. Heath, J. R., Zhang, Q., O'Brien, S. C., Curl, R. F., Kroto, H. W., Smalley, R. E. (1987). The formation of long linear carbon chain molecules during laser vaperation of graphite, *J. Am. Chem. Soc.*, **109**, 359–363.

80. Gaumet, J. J., Wakisaka, A., Shimizu, Y., Tamori, Y. (1993). Carbon clusters vaporized directly from graphite through laser vaporization, *J. Chem. Soc., Chem. Commun.*, **89**, 347–348.

81. Hu, F., Zeng, C., Long, R., Miao, Y., Wei, L., Xu, Q., Wei, M. (2018). Super multiplexed optical imaging and barcoding with engineered polyynes, *Nat. Methods*, **15**, 194–200.

82. Fang, L., Zhu, T. X., Chang, W. W., Liu, Y., Zhao, X. (2021). Purification of polyynes via carbides, *Carbon*, **179**, 28–32.

83. Kuwahara, R., Kudo, Y., Morisato, T., Ohno, K. (2011). Encapsulation of carbon chain molecules in single-walled carbon nanotubes, *J. Phys. Chem. A*, **115**, 5147–5156.

84. Moura, L. G., Malard, L. M., Carneiro, M. A., Venezuela, P., Capaz, R. B., Nishide, D., Achiba, Y., Shinohara, H., Pimenta, M. A. (2009). Charge transfer and screening effects in polyynes encapsulated inside single-wall carbon nanotubes, *Phys. Rev. B*, **80**, 161401.

85. Wakabayashi, T., Murakami, T., Nagayama, H., Nishide, D., Kataura, H., Achiba, Y., Tabata, H., Hayashi, S., Shinohara, H. (2009). Raman spectral features of longer polyynes $HC_{(2n)}H$ (n = 4–8) in SWNTs, *Eur. Phys. J. D*, **52**, 79–82.

86. Malard, L. M., Nishide, D., Dias, L. G., Capaz, R. B., Gomes, A. P., Jorio, A., Achete, C. A., Saito, R., Achiba, Y., Shinohara, H. (2007). Resonance Raman study of polyynes encapsulated in single-wall carbon nanotubes, *Phys. Rev. B*, **76**, 233412.

87. Nishide, D., Wakabayashi, T., Sugai, T., Kitaura, R., Kataura, H., Achiba, Y., Shinohara, H. (2007). Raman spectroscopy of size-selected linear polyyne molecules $C_{2n}H_2$ (n = 4–6) encapsulated in single-wall carbon nanotubes, *J. Phys. Chem. C*, **111**, 5178–5183.

88. Moura, L. G., Fantini, C., Righi, A., Zhao, C., Shinohara, H., Pimenta, M. A. (2011). Dielectric screening in polyynes encapsulated inside double-wall carbon nanotubes, *Phys. Rev. B*, **83**, 245427.

89. Luo, T., Chang, W., Zhu, T., Liu, Y., Zhao, X. (2022). Synthesis and stockpile of polyynes in paraffin as well as extraction for preparing single-walled carbon nanowires (LLCCs@SWCNTs), *Chem. Phys.*, **563**, 111688.

90. Yang, W., Huang, Y., Yang, B., Liu, R. (2006). Structural and electronic properties of carbon nanowires made of linear carbon chains enclosed inside zigzag carbon nanotubes, *Carbon*, **44**, 456–462.

91. Tapia, A., Aguilera, L., Cab, C., Medina-Esquivel, R. A., de Coss, R., Ganto, G. (2010). Density functional study of the metallization of a linear carbon chain inside single wall carbon nanotubes, *Carbon*, **48**, 4057–4062.

92. Bonabi, F., Brun, S. J., Pedersen, T. G. (2017). Excitonic optical response of carbon chains confined in single-walled carbon nanotubes, *Phys. Rev. B*, **96**, 155419.

93. Deng, Y., Cranford, S. W. (2019). Mapping temperature and confinement dependence of carbyne formation within carbon nanotubes, *Carbon*, **141**, 209–217.

94. Chen, J., Yang, L., Yang, H., Dong, J. (2003). Electronic and transport properties of a carbon-atom chain in the core of semiconducting carbon nanotubes, *Phys. Lett. A*, **316**, 101–106.

95. Yang, W., Huang, Y., Yang, B., Liu, R. (2008). Crystal orbital study on carbon chains encapsulated in armchair carbon nanotubes with various diameter, *Carbon*, **46**, 276–284.

96. Endo, M., Kim, Y. A., Hayashi, T., Muramatsu, H., Terrones, M., Satio, R., Villalpando-Paez, F., Chou, S. G., Dressehaus, M. S., Nanotube coalescence-inducing mode: a novel vibrational mode in carbon systems, *Small*, **2**, 1031–1036.

97. Yanagi, K., Udoguchi, H., Sagitani, S., Oshima, Y., Takenobu, T., Kataura, H., Ishida, T., Matsuda, K., Maniwa, Y. (2010). Transport mechanisms in metallic and semiconducting single–wall carbon nanotube networks, *ACS Nano*, **4**, 4027–4032.

98. Rohringer, P., Shi, L., Liu, X., Yanagi, K., Pichler, T. (2014). Purification, separation and extraction of inner tubes from double–walled carbon nanotubes by tailoring density gradient ultracentrifugation using optical probes, *Carbon*, **74**, 282–290.

99. Zhang, J., Feng, Y., Ishiwata, H., Miyata, Y., Kitaura, R., Dahl, J. E. P., Carlson, R. M. K., Shinohara, H., Tomanek, D. (2012). Synthesis and transformation of linear adamantane assemblies inside carbon nanotubes, *ACS Nano*, **6**, 8674–8683.

100. Hirahara, K., Suenaga, K., Bandow, S., Kato, H., Okazaki, T., Shinohara, H., Iijima, S. (2000). One-dimensional metallofullerene crystal generated inside single-walled carbon nanotubes, *Phys. Rev. Lett.*, **85**, 5384–5387.

101. Cataldo, F. (2004). Synthesis of polyynes in a submerged electric arc in organic solvents, *Carbon,* **42**, 129–141.

102. Cataldo, F., Ornella, U., Giancarlo, A. (2007). Kinetics of polyynes formation with the submerged carbon arc, *J. Electroanal. Chem.*, **60**, 282–290.

103. Li, H., Wu, Y. L., Zhang, Y. F., Zhu, T. X., Maruyama, T., Liu, Y., Zhao, X. L. (2020). Submerged carbon arc in deionized water: a green route for scalable preparation of gas containing polyynes, *Chem. Phys.*, **535**, 110804.

104. Fang, L., Xie, Y. P., Sun, S. J., Zi, W. (2022). Large-scale synthesis of polyynes with commercial laser marking technology, *Chin. Phys. B*, **31**, 126803.

105. Wei, X., Tanaka, T., Hirakawa, T., Wang, G., Kataura, H. (2017). High-efficiency separation of (6,5) carbon nanotubes by stepwise elution gel chromatography, *Phys. Status Solidi B*, **254**, 1700279.

106. Wei, X., Tanaka, T., Hirakawa, T., Hirakawa, T., Tsuzuki, M., Wang, G., Yomogida, Y., Hirano, A., Kataura, H. (2018). High-yield and high-throughput single-chirality enantiomer separation of single-wall carbon nanotubes, *Carbon*, **132**, 1–7.

107. Liu, F., Wang, Q. Q., Tang, Y. C., Du, W., Chang, W. W., Fu, Z. W., Zhao, X. L., Liu. Y. (2023). Carbon nanowires made by the insertion and fusion method toward carbon–hydrogen nanoelectronics, *Nanoscale,* **15**, 6143–6155.

108. Brown, S. D. M., Corio, P., Marucci, A., Dresselhaus, M. S. (2000). Anti-Stokes Raman spectra of single-walled carbon nanotubes, *Phys. Rev. B,* **61**, 5317–5340.

109. Souza Filho, A. G., Jorio, A., Hafner, J. H., Lieber, C. M., Saito, R., Pimenta, M. A., Dresselhaus, G., Dresselhaus, M. S. (2001). Electronic transition energy E_{ii} for an isolated (n,m) single-wall carbon nanotube obtained by anti-Stokes/Stokes resonant Raman intensity ratio, *Phys. Rev. B,* **63**, 241404.

110. Tschannen, C. D., Grodeev, G., Reich, S., Shi, L., Pichler, T., Frimmer, M., Novotny, L., Heeg, S. (2020). Raman scattering cross section of confined carbyne, *Nano Lett.*, **20**, 6750–6755.

111. Chang, W. W., Luo, T., Zhu, T. X., Li, Q. Q., Lu, B., Deng, Z. Y., Liu, Y., Maruyama, T., Zhao, X. L. (2023). Anti-Stokes Raman scattering of the smallest carbon nanowires made of long linear carbon chains inserted inside single-walled carbon nanotubes, *J. Phys. Chem. C*, **127**, 12068–12076.

112. Fantini, C., Jorio, A., Souza, M., Ladeira, L. O., Pimenta, M. A., Souza Filho, G., Saito, R., Samsonidze, Ge. G., Dresselhaus, G., Dresselhaus, M. S. (2004). One-dimensional character of combination modes in the resonance Raman scattering of carbon nanotubes, *Phys. Rev. Lett.*, **93**, 087401.

113. Dresselhaus, M. S., Dresselhaus, G., Saito, R., Jorio, A. (2005). Raman spectroscopy of carbon nanotubes, *Phys. Rep.*, **409**, 47–99.

114. Gordeev, G., Jorio, A., Kusch, P., Vieira, B. G. M., Flavel, B., Krupke, R., Barros, E. B., Reich, S. (2017). Resonant anti-Stokes Raman scattering in single-walled carbon nanotubes, *Phys. Rev. B*, **96**, 245415.

115. Terekhov, S. V., Obraztsova, E. D., Lobach, A. S., Konov, V. I. (2002). Laser heating method for estimation of carbon nanotube purity, *Appl. Phys. A*, **74**, 393–396.

116. Liu, H. N., Cong, X., Lin, M. L., Tan, P. H. (2019). The intrinsic temperature-dependent Raman spectra of graphite in the temperature range from 4 K to 1000 K, *Carbon*, **152**, 451–458.

117. Tan, P. H., Deng, Y. M., Zhao, Q. (1998). Temperature-dependent Raman spectra and anomalous Raman phenomenon of highly oriented pyrolytic graphite, *Phys. Rev. B*, **58**, 5435–5439.

118. Tan, P. H., Deng, Y. M., Zhao, Q., Cheng, W. (1999). The intrinsic temperature effect of the Raman spectra of graphite, *Appl. Phys. Lett.*, **74**, 1818–1820.

119. Berciaud, S., Han, M. Y., Mak, K. F., Brus, L. E., Kim, P., Heinz, T. F. (2010). Electron and optical phonon temperatures in electrically biased graphene, *Phys. Rev. Lett.*, **104**, 227401.

120. Huang, F. M., Yue, K. T., Tan, P. H., Zhang, S. L., Shi, Z. J., Zhou, X. H., Gu, Z. N. (1998). Temperature dependence of the Raman spectra of carbon nanotubes, *J. Appl. Phys.*, **84**, 4022–4024.

121. Tsujimoto, M., Tanimura, M., Tachibana, M. (2020). Temperature dependence of the Raman spectra of multilayer graphene nanoribbons fabricated by unzipping method, *Diam. Relat. Mater.*, **109**, 108047.

122. Shi, L., Senga, R., Suenaga, H., Kataura, H., Saito, T., Paz, A. P., Rubio, A., Ayala, P., Pichler, T. (2021). Toward confined carbyne with tailored properties, *Nano Lett.*, **21**, 1096–1101.

Chapter 5

Synthesis of Long Linear Carbon Chains Confined in Multi-Walled Carbon Nanotubes

Yifan Zhang[a] and Xinluo Zhao[b]
[a]*School of Engineering, Huzhou University,*
759 Erhuandong Road Huzhou, 313000, China
[b]*Department of Physics, Shanghai University,*
99 Shangda Road Shanghai, 200444, China
yifan_zhang@zjhu.edu.cn, xlzhao@shu.edu.cn

Carbyne as the third allotrope of carbon, attracts much attention from researchers. However, it remains controversial for a long time. In 2003, Zhao *et al.* discovered that long linear carbon chains (LLCCs) as the true 1-D form of carbyne stably exist inside the hollow space of the innermost tube of multi-walled carbon nanotubes (MWCNTs) *via* hydrogen arc-discharge method. The whole structure was named as "carbon nanowires" (CNWs) by Zhao *et al.* This chapter is intended for readers using the arc-discharge method to prepare CNWs. The apparatus, parameters and characterization are reviewed. The

Emerging Carbyne: Truly One-Dimensional Form of Carbon
Edited by Yahachi Saito

ISBN 978-981-5129-15-1 (Hardcover), 978-1-003-51036-9 (eBook)
www.jennystanford.com

growth mechanism of LLCCs is discussed. Lastly, several challenges and prospects are suggested towards futural development of CNWs and arc-discharge method.

5.1 Introduction to the Discovery of MWCNWs

The story of the discovery of CNWs is like the discovery of fullerene, where Kroto *et al.* was looking for carbyne but coincidentally found fullerene [1]. In 2003, Zhao *et al.* discovered that LLCCs with length of more than one hundred atoms can be encapsulated inside MWCNTs in the product of hydrogen arc-discharge method [2], when looking for the thinnest carbon nanotubes [3, 4]. Afterwards, Heeg *et al.* proved that the LLCCs with L-band of ~1850 cm^{-1} have no evident length-dependent property, which suggests that LLCCs inside CNTs meet the definition of carbyne in 1-D form [5].

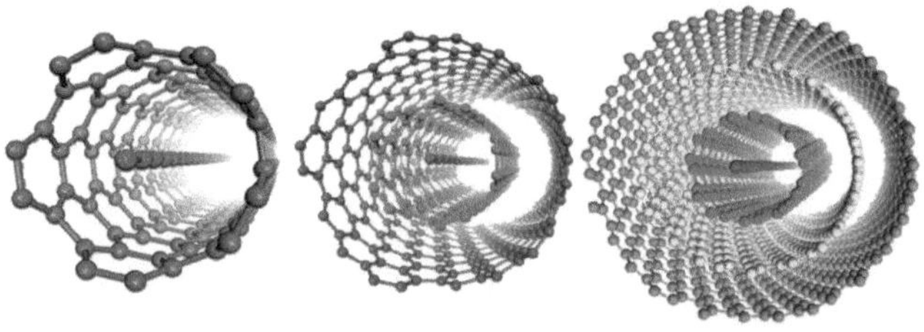

Figure 5.1 The three kinds of CNWs depending on the type of CNTs.

Figure 5.1 shows three kinds of CNWs depending on the type of CNTs. There are two customary terms used for the new structure. At the beginning, the novel 1-D carbon nanostructure made of the LLCCs inserted inside carbon nanotubes (CNTs) was firstly named "carbon nanowires" (CNWs) as a new carbon allotrope by Zhao *et al.* The extended terms SWCNWs, DWCNWs and MWCNWs are also used to distinguish the LLCCs inside single-walled, double-walled, and multi-walled CNTs, respectively. The other term "LLCC inside carbon nanotubes" (LLCC@CNTs) is also used in relevant works, which may appear as the similar variants such as LCC@CNTs, CC@CNTs, carbyne@CNTs or C_n@CNTs. In order to represent different types of CNTs, LLCC@MWCNTs, LLCC@DWCNTs and LLCC@SWCNTs can be used. The CNWs represent the whole 1-D carbon allotrope composed of LLCCs with *sp* and CNTs with sp^2 bonding. While LLCC@CNTs give an intuitive expression of the component, compared with other CNT-filling structures in the view of 1-D van der Waals heterostructure.

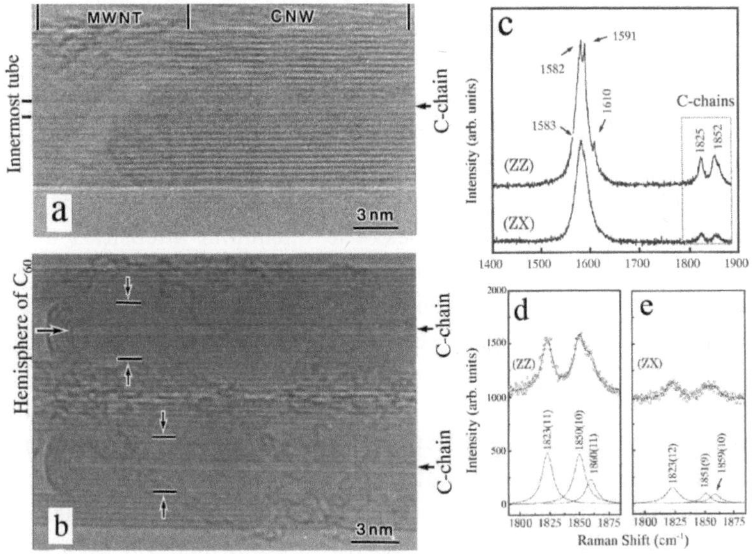

Figure 5.2 (a) and (b) HRTEM micrographs showing MWCNTs containing LLCCs, (c)–(e) Polarized Raman spectra of a straight MWCNT bundle containing LLCCs. Figure reprinted with permission from Ref. [2], ©2003 by American Physical Society.

The impact of CNW discovery is that the micrographic and spectral evidence are well-combined, giving the first intuitive recognition of the existence of LLCCs and providing a reliable new thought to obtain carbyne. In that work, three HRTEM micrographs were presented by Zhao *et al.* As shown in Fig. 5.2a, it is a typical HRTEM micrograph displaying an MWCNT partially filled by an LLCC (right side). The MWCNT has a seamless tubular structure of 13 rolled-up graphitic sheets. Furthermore, a gradual decrease of the diameter of innermost tube was found 1.0 nm to 0.7 nm from the left to the right of the MWCNT. The LLCC appears as the new straight line with the length of 13.5 nm (marked by an arrow) between the two lines of the innermost tube, increasing the number of parallel lines from even (26) to odd (27). It was noticed that, the atomic line shows large contrast possibly results from the high density of π-electrons (two per carbon atom) in the LLCC. In Fig. 5.2b, the HRTEM micrograph of two horizontally aligned MWCNTs are presented. Both innermost tubes of two MWCNTs are filled by LLCCs in the center. The inserted LLCCs have a length of 20 nm, shall be more than 100 atoms even if assuming the maximum C–C bond length to be 0.1389 nm.

Meanwhile, a new significant Raman band consisting of several peaks was found near 1850 cm^{-1} as shown in Fig. 5.2c, which is distinguished from known graphite and diamond. The excitation laser of 1.75, 2.41, 2.54 and 2.71 eV were used to prove that these peaks cannot be attributed to the second-order harmonics or combination modes in sp^2 carbon allotropes. The maximum intensities were observed near 2.5 eV, in agreement with the transition energy for LLCCs. The new peaks were finally attributed to the stretching mode of LLCC, though it is slightly redshifted, compared to the theoretical calculations and experimental measurements at that time. The redshift of Raman frequency is an important phenomenon in this system, directly related to the physical property of inserted carbyne, which is the most basic indicator now.

The correspondence between the new structure in HRTEM micrographs and its Raman spectra is confirmed *via* polarized Raman measurement. The spectra in Fig. 5.2c–e were taken from a straight bundle of MWCNTs containing LLCCs. The bundle oriented along the Z direction and the light propagated along the Y direction. Raman spectra of (ZZ) and (ZX) polarization geometry are presented. The G-band modes of MWCNTs can be observed both in the (ZZ) and (ZX) geometry with slight difference, indicating that the G band of MWCNTs contains several modes with different symmetries. The spectra of LLCCs (marked by dotted rectangle) can also be seen in both the (ZZ) and (ZX) geometries. However, the decrease of peak intensities in the (ZX) geometry is significant, which can be explained by the depolarization effects due to highly oriental 1-D structure of LLCCs along the axial direction of MWCNTs. Generally, the Raman band can be splitted to several peaks *via* Lorentzian function. The full width at half maximum (FWHM) of each peak is near 10 cm^{-1}, which agree with the typical value for the width of phonons in crystal. The origin of multiple peaks will be discussed in the section of characterization later.

The discovery also provides a reliable idea to obtain stable 1-D carbyne in solid state under normal condition. The innermost tubes of CNTs can not only provide a well-confined 1-D sub-nanometer space for the growth of LLCCs avoided from bending or twisting, but also perfectly protect LLCCs from crosslink and oxidation. The reactivity and aging effect of carbyne in these samples hasn't been observed. Cui *et al.* further proved LLCCs can be stable inside CNTs even over hundreds degree centigrade until the CNTs outside were oxidized under air condition [6].

To date, various methods have been developed to synthesize LLCCs inside CNTs, as summarized previously [7, 8]. Besides the direct preparation of MWCNWs from graphite, several indirect methods provide the possibility to synthesize LLCCs inside different types of CNTs. For instance, heat-treatment method can produce carbyne inside DWCNTs *via* annealing proper SWCNTs and DWCNTs. The world's longest carbon chain observed by Shi *et al.*, owning more than 6000 carbon atomic length in DWCNTs, was obtained through high-temperature heat-treatment of DWCNTs in high vacuum [9]. The nanoreactor method is based on the confined chemical reaction in the limited sub-nanometer space of CNTs, which can produce SWCNWs now by Chang *et al.* [10]. Another interesting and efficient method recently reported to synthesize CNWs is field electron discharge [11], presented by Toma *et al.*

By contrast, the arc-discharge method is a classic but still a unique method being able to produce LLCCs and surrounding MWCNTs simultaneously from graphite, which is the main difference compared to the other methods. This method has the advantages of low cost and simple set. Moreover, the LLCCs can be generated just in seconds. Thus, arc-discharge method can be one of the applicable ways for large-scale researches of carbon chains at present. In the following sections, the arc-discharge method to produce MWCNWs is introduced. The present state of researches on the instruments and experimental parameters is reviewed in detail. The most common characterization techniques and corresponding ideal criterion are summarized, which may help to conveniently evaluate the products. The growth mechanism is discussed based on arc-discharge process.

5.2 The Preparation of MWCNWs via Arc-Discharge Method

The arc-discharge method is a physical way to produce nanostructures, including various carbon allotropes such as fullerene, multi-layers graphene, and CNTs [12–15]. The preparation of MWCNWs depends on the preparation of high-quality MWCNTs with similar apparatus but stricter parameters. HRTEM and Raman spectroscopy are recommended to characterize and evaluate the obtained samples.

5.2.1 The Apparatus and Conditions of Arc-Discharge Method to Prepare MWCNWs

The common and core parts of apparatus of arc-discharge method are anode and cathode electrodes connected with a current supply, and a proper chamber. Figure 5.3a is the illustration of arc-discharge equipment, which only shows the main part of the equipment. Under real circumstance, the whole chamber is attached to a sturdy bracket, and the bottom is connected to the vacuum pump. Gas path and cooling pipe are also fixed on the chamber. Both sides of the chamber have a quartz window. The anode and cathode are vertically placed on the top and bottom of the right cylinder chamber, connected with an arc generator. In the case of the first synthesis of MWCNWs, the electrodes were high-purity graphite rods connected with direct current (DC) arc generator. The chamber was vacuumized to lower than 10^{-4} Pa and was then injected with 60 Torr of hydrogen gas. The DC was applied between two vertically placed electrodes to generate a high-temperature arc plasma in the gap of 1–2 mm for tens of seconds. After the arc-discharge process, the anode can be observed ablated and a deposit with a black core surrounded by a grey shell was obtained on the top surface of cathode, shown in Fig. 5.3d. The MWCNWs can be found in the black core, while the grey shell only contains disorder graphite.

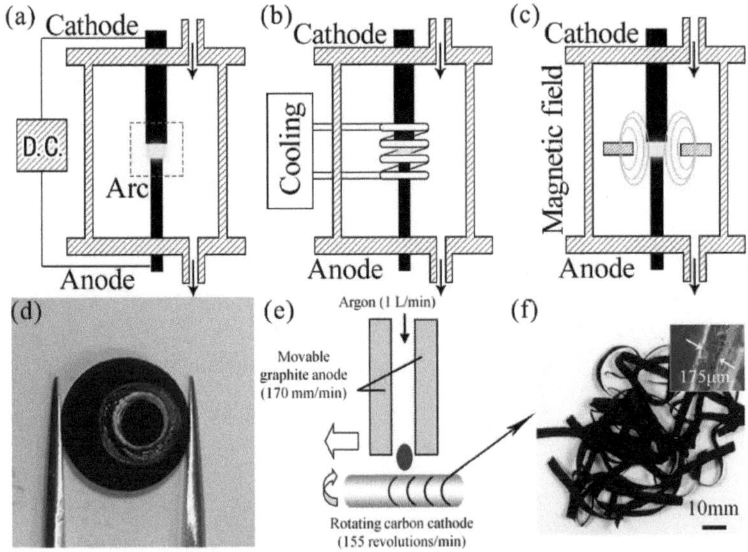

Figure 5.3 (a) The simplified illustration of arc-discharge apparatus, modified by gas cooling (b) and magnetic field (c). (d) The typical circular cathode deposit obtained by (a)–(c). (e) Rotating arc-discharge apparatus and (f) the obtained tape-like deposit. Panel (a) to (c) are reprinted with permission from Ref. [8], ©2023 by Chinese Physical Society. Panel (d) is reprinted with permission from Ref. [20], ©2020 by Elsevier. Panel (e) and (f) are reprinted with permission from Ref. [26], ©2012 by Elsevier.

The preparing condition of MWCNWs applied for the first time was the same as the one used to prepare the high-purity and high-crystalline MWCNTs. Hydrogen provided a reductive atmosphere and was thought to increase the temperature of arc plasma. The parameter window to grow MWCNWs under hydrogen arc was thought very narrow at the beginning. But further researches have largely extended the applicable parameter window, which benefit the study on growth mechanism and property of LLCCs. According to the basic set of equipment and method, the main adjustable factors are usually the material of electrode (including rod material and catalyst), the background gas and pressure for arc, and the current. As listed in Table 5.1, those previous works provided different parameters and modifications of arc-discharge method to prepare MWCNWs.

Table 5.1 Possible arc-discharge parameters to prepare MWCNWs

Electrodes	Catalyst	Current (A)	Gas	Pressure (Torr)	Modification	Ref.
Graphite	/	50–60	H_2/D_2	50–70		[2, 16]
Graphite	/	50	H_2–N_2	70		[16]
Graphite	Ni–Cr	60	He	375		[17, 18]
Graphite	/	50–80	H_2	50	Gas cooled by coolant	[19]
Graphite	/	90	H_2	90–100	Gas cooled by liquid nitrogen	[19]
Graphite	/	30–60	H_2	50	Magnetic field	[20]
Graphite	/	80	N_2	/	Immersed in liquid nitrogen	[21, 22]
Graphite	B_2O_3	100	Ar	760	Rotating electrode in air condition	[23]

In Zhao's experiment, pure graphite without catalyst was used as the proper electrode material for the preparation of MWCNWs. Cazzanelli *et al.* filled Ni–Cr alloy with weight proportion 8:2 into the anode by drilling a 1 mm diameter hole. The deposit shows irregular shape when the Ni–Cr catalytic mixture was irregularly inserted in the anode. Meanwhile, the diameter of deposits prepared by anodes with catalyst is slightly smaller than that of without catalyst. Further measurements confirmed the existence of MWCNWs in all deposits obtained with or without catalyst. So, the authors thought that the presence of Ni–Cr catalyst does not appear correlated with the formation of MWCNWs. In another report, Kang *et al.* used boron trioxide as a chain promoter by feeding argon gas containing boron trioxide into the hollow tube of graphite anode. But, the boiling point of boron oxide is over 1800 °C, so it is better to classify that it acts as a catalyst rather than background gas in this review, though everything will be vaporized in arc plasma. Kang *et al.* reported that boron trioxide helps to largely increase the productivity of nanotube tape (the shape of their cathode deposit) up to 10 times, and it became easier to find LLCCs under HRTEM. The authors thought that the existence of boron atoms into carbon cluster may lead to the reduction in the surface tension and then increase the sustained time of the viscous state, which promote the growth of both MWCNTs and LLCCs. To sum up, the catalyst for MWCNWs growth hasn't been totally experimented, especially compared to the big number of catalysts for the growth of CNTs. It is expected to find the way to promote the growth of MWCNWs *via* catalyst like boron.

The background gas plays an important role in the preparation of MWCNTs as well as MWCNWs. The composition of gas will largely affect the property of arc plasma by different ionization energy and thermal conductivity. At the beginning, Jinno *et al.* reported to produce MWCNWs in the mixture of H_2–N_2. The electrodes and pressure were controlled to be the same as Zhao's work. The mixing ratio of H_2 and N_2 gases was changed in the range from 0% to 100%. According to the signal of LLCCs in Raman spectra, it was found that the LLCC signals disappeared when nitrogen was over 20%, where the growth of MWCNTs was not affected much. But no obvious change was found when 100% deuterium was used to replace hydrogen in that experiment. Afterwards, helium and argon have been tried to successfully produce MWCNWs in the experiments using Ni–Cr and boron trioxide as catalysts, respectively. Moreover, Scuderi *et al.* reported an interesting experiment using liquid nitrogen as the arc medium. The electrodes were immersed into liquid nitrogen with a gap of 1 mm. By applying current up to 70 A and 80 A, MWCNWs were obtained. The authors found the size of anode and cathode also became a decisive factor in their experiments.

Fortunately, the composition of gas is not so critical for the growth of MWCNWs. However, it is noted that the corresponding optimal combinations of current and pressure parameter under different gas environment are different. The applicable pressure and current in the case of hydrogen can be lower than 50 Torr and 50 A, respectively, while 1 atm and 100 A of that is needed in the case of argon. As mentioned above, those parameters largely affect the property of arc plasma. So, it can be assumed that to obtain a suitable arc plasma is the essential condition to prepare LLCCs and the most important parameter of plasma is temperature, because it directly controls the ablation of anode and the temperature of deposit. Thus, in order to promote the growth of MWCNWs, an alternative approach is to increase the arc temperature of plasma. Besides adjusting the background gas and exploring the correct combination of pressure and current, the temperature of plasma can also be changed by applying gas cooling and magnetic field. The two approaches were early experimented [24, 25] to increase the purity of MWCNTs, which is also available to promote the growth of MWCNWs in our previous works. In the gas cooling experiment, hydrogen gas was used as the arc medium just like the first work. As shown in Fig. 5.3b, a cooling system was inserted into the hydrogen arc-discharge apparatus, consisting of a coil of hollow copper pipe and a refrigerator. The refrigerator supplied circulating cooling liquid from –20 °C to 20 °C to cool down hydrogen gas in the chamber. We also applied liquid nitrogen by a self-pressurized liquid nitrogen container. In this work, the combinations of different cooling temperature and arc current were experimented. The optimal current parameter was found largely moving to higher value with the decrease of cooling temperature. In the case of liquid nitrogen, the optimal pressure was changed from 50 to 100 Torr. Besides, the suitable current range to produce LLCCs was largely extended. Figure 5.3c shows the other modification of the apparatus by introducing a static magnetic field around arc plasma. In this experiment, four rectangular Sm_2Co_{17} magnets were used to generate an external magnetic field. It must be careful to control the distance between the magnets and arc plasma. Though the working temperature of Sm_2Co_{17} can be over 300 °C and the copper films were used to block the strong thermal radiation from plasma to the magnets, the temperature of plasma generated by high current can obviously degauss the magnets after arc-discharge process, in which the highest current in this work was 70 A. Contrary to the cooling experiment, magnetic field promote the optimal current parameter to lower value. Surprisingly, MWCNWs can be found in a very tiny deposit obtained under 30 A. With the decrease of current, the plasma became much weaker and the ablation of anode was slowed down, which cause the significant shrink of deposit. In addition, Kim *et al.* modified the arc-discharge apparatus by combining a movable and hollow anode and a rotating cathode, as shown in Fig. 5.3e, to realize the continuous production of MWCNTs in the shape of tape (Fig. 5.3f) under atmospheric condition. This indicates the potential towards the industrial preparation of high-crystalline MWCNTs [26] and Kang *et al.* realizes the preparation of MWCNWs *via* this method [23].

5.2.2 The Characterization of MWCNWs Obtained via Arc-Discharge Method

The common ways to understand the nanostructure and property of MWCNWs are limited to Raman spectroscopy and HRTEM in the present researches.

5.2.2.1 HRTEM observations on MWCNWs

As mentioned above, in the first observation of MWCNWs, HRTEM gave the most intuitive sight of this 1-D carbon nanostructure. However, limited by the development of HRTEM technique, the resolution of those images cannot reach to a perfect level to avoid any doubt, because the other contents such as amorphous carbon and even virtual image inside the innermost tube of CNTs were very common at that time. So, the new model of inserted structure was not accepted by all researchers at the beginning. Fortunately, more and more researchers have obtained clearer micrographs and well proved the existence of LLCCs, as collected in Fig. 5.4. The main features that can be observed *via* HRTEM are the length of LLCCs, the diameter of surrounding CNTs, and the filling ratio. The length of LLCCs is an important feature to distinguish the 1-D structure between carbyne and short polyynes. For the definition of carbyne, the chain length must be long enough to reduce the length-dependent effect such as the change of bond-length alternation (BLA). The observed LLCCs are usually over 20 nm, being longer than one hundred carbon atoms according to relevant reports, which is mainly limited by the field of view under high magnification but not the chains themself. It is not applicable to observe LLCCs across the influence of multiple graphite walls under low magnification, so a possible way to measure whole long chains *via* HRTEM might be image stitching. Nevertheless, the observed chain length should already satisfy the definition of carbyne, according to the study on long polyynes protected by functional group, where the carbon chains longer than 44–48 atoms show little length-dependent effect [27, 28].

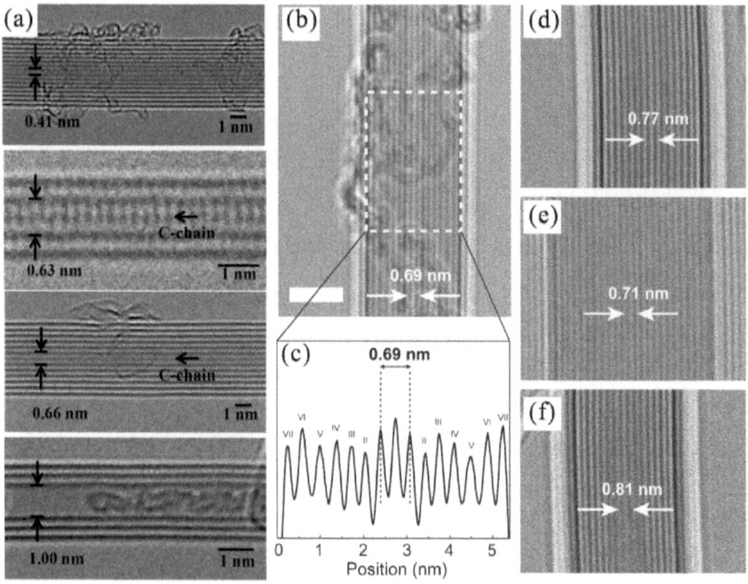

Figure 5.4 The HRTEM images presented by (a) Kang *et al.* and (b) to (f) Andrade *et al.* Panel (a) is reprinted with permission from Ref. [23], ©2016 by Elsevier. Panel (b) to (f) are reprinted with permission from Ref. [29], ©2015 by Elsevier.

Generally, the innermost CNTs surrounding LLCCs show similar diameter distribution. Zhao *et al.* observed the diameter of surrounding CNTs is near 0.7 nm. Interestingly, the innermost tube partly containing LLCC shows a shrinkage of diameter from the empty to the filled side. Kang *et al.* presented two pictures that the LLCCs are filled inside the innermost tube with diameter of 0.63 nm and 0.66 nm, while the innermost tube with diameter of 0.41 contains nothing. The innermost tube with diameter of 1.0 nm containing amorphous carbon species was observed [23]. Scuderi *et al.* show the LLCC inside the innermost tube with diameter of 0.71 nm [21]. Additionally, the statistical distribution of the innermost diameter was obtained *via* HRTEM analysis. The diameters of innermost tubes in MWCNTs produced *via* arc discharge in liquid nitrogen were relatively narrow, where the ratio of CNTs with the innermost diameter lower than 1 nm was 82.2%. The MWCNTs containing LLCCs accounted for 64.3% of all MWCNTs and they were all found inside the innermost tube with diameter lower than 1 nm. Andrade *et al.* presented similar result with a series of HRTEM images [29], as shown in Fig. 5.4b to f. The authors noted that those HRTEM images were taken under a defocus condition, so the bright contrast is assigned to carbon walls. The innermost tubes with diameter of 0.69, 0.77, 0.71, and 0.81 nm were observed being able to contain LLCCs.

However, Andrade *et al.* suggested that the acquired image was not a direct representation of the structure in real space, because the HRTEM image with the magnification and specimen thickness was basically a phase-contrast image. Thus, they provided a more direct confirmation of presence of an LLCC inside the CNT through observing the cross-section of an individual MWCNW with scanning TEM (STEM). As shown in Fig. 5.5a, the high-angle annular dark field (HAADF) image clearly shows the walls of CNT without interference artifact. A digital treatment was performed using a fast Fourier transform (FFT) band pass mask on the image and shown in Fig. 5.5c to make the wall contrast clearer. As a result, Fig. 5.5d was finally obtained. The walls of CNT show the innermost tube that owns a diameter of 0.82 nm and LLCC can also be clearly distinguished in the center of innermost tube. The image has unambiguously proved the existence of LLCC inside the innermost tube of MWCNT with an unprecedented clarity. In addition, the authors acquired an electron energy loss spectrum (EELS) along the diameter of this tube in order to analyze the plasmon absorption across the cross-section, however, no influence of LLCC can be observed on the EELS.

In conclusion, HRTEM helps to directly understand the crystal structure of MWCNWs. The diameter distribution of the innermost tubes surrounding LLCCs agrees with a van der Waals interaction distance between carbon chains and CNT walls, which is one of the basic mechanisms for the existence of LLCC and will be discussed in later section. It is noted that the statistical study *via* HRTEM is still the first choice to quantitively get the filling ratio of MWCNWs, though it is low-efficiency and subjective. Another remaining question is that the shrinkage of the diameter of CNT from 1.0 to 0.7 nm when inserted with LLCC has been observed, but it hasn't been fully understood. Similarly, we suppose that LLCCs may be able to expand the innermost CNTs smaller than 0.7 nm in diameter.

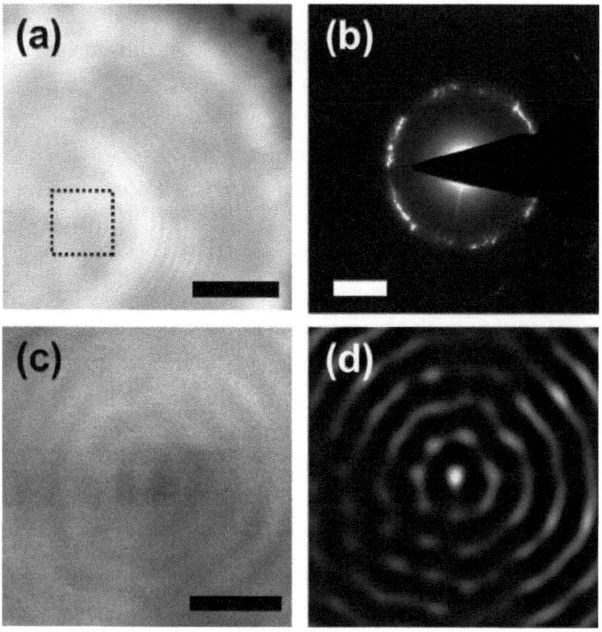

Figure 5.5 (a) HAADF image of the cross section of an MWCNT containing LLCC in the center. (b) Selected area electron diffraction of a MWCNT cross section. The dashed square in (a) is zoomed in (c). (d) The treated image of (c). Figure is reprinted with permission from Ref. [29], ©2015 by Elsevier.

5.2.2.2 Raman spectroscopic studies on MWCNWs

Besides the direct observation *via* HRTEM, Raman spectroscopy is considered as the best way to characterize MWCNWs because of the high efficiency and accuracy. As shown in Fig. 5.2d and Fig. 5.6a, LLCCs inside MWCNWs own unique signals locating from 1780 to 1880 cm^{-1}, differing from the signal of MWCNTs. Those signals were named "L", "C_n", "C-chain", etc. in different papers. Here we use L-band. The deposit obtained after arc-discharge process can be easily measured without any post-treatment to check the growth of LLCCs *via* those characteristic peaks. As a fast qualitative tool, it has been widely used to confirm the formation of MWCNWs rather than HRTEM image. The relative intensity of the peaks normalized by G-band of MWCNTs can be a macroscopical and statistical indicator to qualitatively evaluate the filling ratio of MWCNWs between homogeneous samples under the same measuring parameters such as the measuring temperature, the wavelength and the power of laser. The typical examples can be seen in our experiments modifying the instrument of arc-discharge *via* gas cooling and magnetic field, where the L-bands of MWCNW samples obtained under a series of similar conditions were normalized by G-band and then were used to compare the growth yield of MWCNWs.

Moreover, Raman spectroscopy can provide a deeper understanding of the property of LLCCs and the interaction between LLCCs and CNTs. The L-band usually consists of multiple peaks, which needs to be fitted *via* Lorentzian function and normalized before analysis. The distribution of L-bands in previous works has similar range limitation from ~1780 to 1880 cm^{-1}, though the shapes of L-bands are different. The frequency of L-band is higher than the G-band of CNTs, indicating the higher phonon energy of stretching vibrational mode generated from LLCCs. Meanwhile, the frequency is red shifted by hundreds of wavenumbers compare to polyyne molecules [30, 31]. Further resonant Raman spectroscopy shows that the resonance window of a certain LLCC peak is in direct proportion to its Raman shift, which means the optical band gap of LLCC inside CNTs can be directly calculated from its Raman frequency [32].

Among previous papers, two viewpoints were supposed to explain the multiple peaks in L-band as well as different band gap of LLCCs inside CNTs: the length-dependent effect and the influence of CNTs. The former one was easy to be accepted early, according to the known experimental and theoretical researches on polyynes at that time. The study on polyynes shows obvious length-dependent behavior and the Raman signals are red shifted with the increase of chain length. Theoretical calculations depending on different parameters indicate the inversed relation between BLA and length of carbon chains, which cause the change of band gap and phonon energy, shifting the resonance window and Raman frequency, respectively [33]. Based on this viewpoint, Andrade *et al.* measured the Raman spectra of MWCNWs under high pressure and supposed the coalescence between two or more chains caused the pressure-induced red shifts in the vibrational bands of carbon chain after the pressure cycle [34].

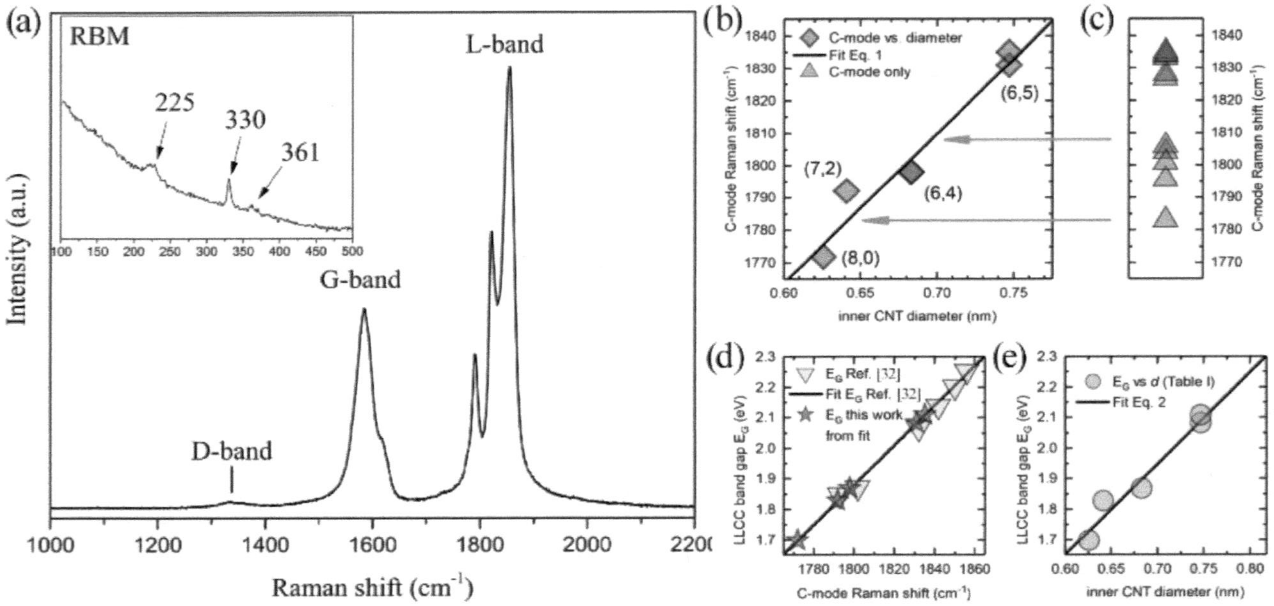

Figure 5.6 (a) Typical Raman spectrum of MWCNWs obtained by hydrogen arc discharge. (b) Raman frequency of LLCCs inside DWCNTs depends on inner tube's diameter. (c) Raman frequencies of LLCCs for which no RBM from the inner tube could be detected. (d) Band gap E_G linearly depends on Raman frequency. (e) E_G linearly depends on diameter of innertube. Panel (a) is reprinted with permission from Ref. [19], ©2018 by Royal Society of Chemistry. (b) to (e) are adapted with permission from Ref. [5], ©2018 by American Chemical Society.

However, Heeg *et al.* revealed that the diameter of inner tubes surrounding LLCCs determined the vibronic and electronic properties of the encapsulated LLCCs *via* tip-enhanced Raman scattering (TERS) measurement on 14 different LLCCs inside DWCNWs [5]. As a result, they concluded three distinct characteristics. Firstly, it is important to know that an individual LLCC only causes a single Raman frequency with FWHM lower than 10 cm^{-1}, so that they can combine the single Raman frequency of each individual LLCC and CNT together. Secondly, they revealed that the chains inside inner tubes with same chirality show similar L-band frequencies. The length-dependent effect was not found, indicating the LLCCs inside CNTs satisfy the definition of carbyne. The chirality of inner tube uniquely determines the BLA and hence its vibronic and electronic structure. Thirdly, the L-band frequency is inversely proportional to RBM frequency.

Thus, the linear correlation between diameter of surrounding CNT, the frequency of L-band, and the band gap of LLCC can be unified, as shown in Fig. 5.6b to e, given by linear functions:

$$\omega_c(d) = a + b \cdot d \tag{5.1}$$

$$E_G(d) = c + f \cdot d \tag{5.2}$$

where d is the diameter of inner tube surrounding LLCC. $\omega_c(d)$ and $E_G(d)$ are the diameter-dependent Raman frequency and band gap of LLCC, respectively. The empirical constants a, b, c and d are from experiment, equaling 1487 ± 36 cm^{-1}, 462 ± 41 cm^{-1} nm^{-1}, −0.14 ± 0.25 eV and 2.98 ± 0.37 eV nm^{-1}, respectively. We suppose that the empirical constants obtained *via* DWCNT system might need adjustment when applying to MWCNT system.

Though the analysis on band gap of LLCC is not affected, the meaning of the two viewpoints is totally different, which decide whether we synthesize LLCCs with different length distribution or LLCCs inside different innermost tubes. It will also affect how we figure out the growth mechanism. Moreover, it directly influences the investigation on the property of carbyne. For instance, Sharma *et al.* explored the universal pressure-dependent responses for LLCCs' Young's modulus, Grüneisen parameter, and strain *via* analyzing L-band of MWCNWs under high pressure [35]. The authors considered the Raman frequency is length-dependent and stated that LLCCs possessed distinct Young's modulus depending on the number of carbon atoms. The LLCCs with a variety of lengths (number of carbon atoms) display mechanical properties that result in a pressure-independent γ/E. Costa *et al.* used the L-band in MWCNW samples to evaluate the thermodynamical properties of LLCCs in terms of the number of carbons atoms and the temperature [36]. The experiments are explained based on the length-dependent effect of L-band in the two works. Maybe, it is possible to come up with more conclusions if considering the diameter-dependent effect.

Besides the most significant L-band, there are other new peaks arising in MWCNWs. Jinno *et al.* noticed broad peaks near 612 cm^{-1} and 660 cm^{-1} arising with the appearance of L-band [16]. Moura *et al.* have reported that the peak around 1637 cm^{-1} can be assigned to the longitudinal acoustic (LA) phonon of LLCCs [37]. Cupolillo *et al.* observed that the Raman frequency of the overtone of L-band are slightly lower than the exact doubling of the one-phonon Raman frequency [19, 38]. The new Raman signals may bring a deeper understanding on the LLCCs inside CNTs in the future.

5.3 The Mechanism of MWCNW Growth in Arc-Discharge Process

The comprehending of the growth mechanism of MWCNWs helps to understand the property of carbyne and to develop better synthetic method. Though it hasn't been fully explained, this section reviews several thoughts towards the mechanism. Generally, the main topics about the mechanism of the formation of MWCNWs are the detailed growth process and the required conditions.

In the arc-discharge process, a nearly cylindroid plasma zone is generated between the cathode and anode. The thermodynamics of plasma is complicated. Briefly, the temperature in the center of arc plasma is usually over 5000 K according to optical emission spectra measurements and theoretical models [39, 40]. The thermal conduction through the carbon electrode is much higher than the convection near plasma, leading to a non-uniform temperature distribution along the axis of the electrode gap. Furthermore, the ablation and solidification of carbon, radiation, and emission of thermal electrons also have a large influence on the temperature of the electrode and the arc plasma [41]. As a result, the temperature of anode is higher than the melting point of graphite and the graphitic anode vaporize to form carbon species such as tiny C_n clusters or even carbon ion into the arc plasma, where the densities of carbon species can be described by the Saha ionization equation. A part of carbon species will diffuse into background gas and condense to form other carbon materials. In the case of hydrogen arc discharge, the hot carbon species can also react with hydrogen to form hydrocarbons [42]. The rest of the carbon species migrate to the surface of the cathode and form the cathode deposit, which is usually near half of the anode mass loss. In this depositing process, the LLCCs grow together with MWCNTs. Kang *et al.* discussed the detailed process of the crystallization of LLCCs and suggested that the close-end growth model of metal-catalyst free MWCNTs can describe the growth process of MWCNWs, rather than the open-end growth model, where the remaining carbon atoms inside the proper innermost tube generated LLCCs through a super-cooling process from the outer surface towards the internal part [23].

Previous works including the adjusting growth condition, characterization and theoretical calculations show several requirements for the growth of MWCNWs.

The first aspect is the diameter of hosting innermost tube of MWCNTs. As mentioned above, all HRTEM observations indicate that LLCCs tend to exist in the innermost tube with narrow diameter near 0.7 nm, agreeing with the van der Waals radius of carbon atom. Liu *et al.* provided density functional calculations of the energies, structures, and vibration frequencies of carbon chains, rings, graphitic sheets, bowls, cages, and tubes inside SWCNT with different diameter, among which LCCs can be inserted coaxially into 0.7 nm CNTs [43]. It reveals confinement of proper CNT can stabilize carbyne and no covalent bond is needed between carbon chain and wall. Thus, the growth of narrow innermost tubes with diameter near 0.7 nm is one of the necessary conditions for the existence of LLCCs.

Secondly, our previous researches on modifying instrument *via* gas cooling and magnetic field show that the quality of MWCNTs is also important to the existence of MWCNWs. As shown in Fig. 5.7, the relative intensity of D-band to G-band (I_D/I_G) is an indicator to evaluate the amount of disorder on CNTs, meanwhile the relative intensity of L-band to G-band (I_L/I_G) shows the filling ratio of

MWCNWs. Thus, we established the correlation between the relative intensity of L-band and D-band to figure out the relation between the growth of MWCNWs and quality of MWCNTs. The scatter symbols are original data picked from normal, cooling and magnetic conditions from our previous works. Those data can roughly fit with an inverse relationship denoted as the red line. Strong L-band typically corresponds to weak D-band at the upper-left corner, indicating the well growth of MWCNWs depends on the well growth of MWCNTs.

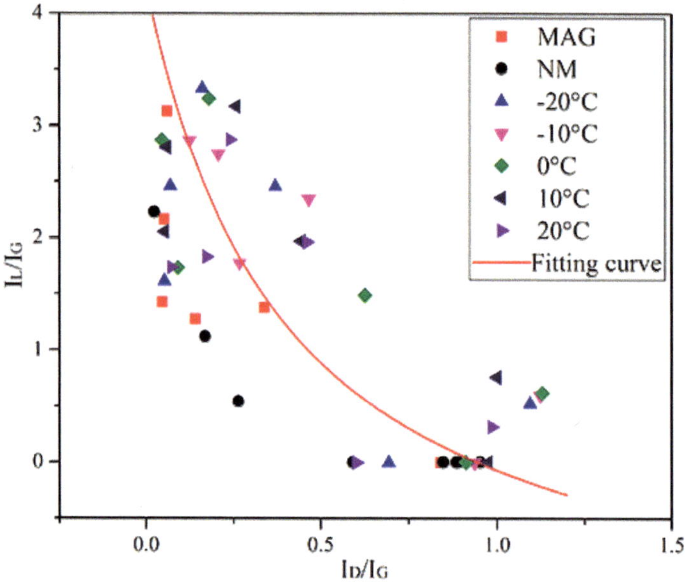

Figure 5.7 The relation between the relative intensity of the L-band and the disorder of MWCNTs. The data points come from our previous works, where NM is normal condition, MAG is magnetic field-enhanced condition and the others are cooling-enhanced conditions.

The corresponding SEM images of those samples show the same appearance, seen in Figs. 5.8 and 5.9. Though SEM morphology cannot characterize LLCCs inside MWCNTs as HRTEM and Raman spectroscopy, it helps to observe the growth yield and purity of MWCNTs in the case of hydrogen arc-discharge method. The original appearance of deposit can be directly observed without any post-treatment.

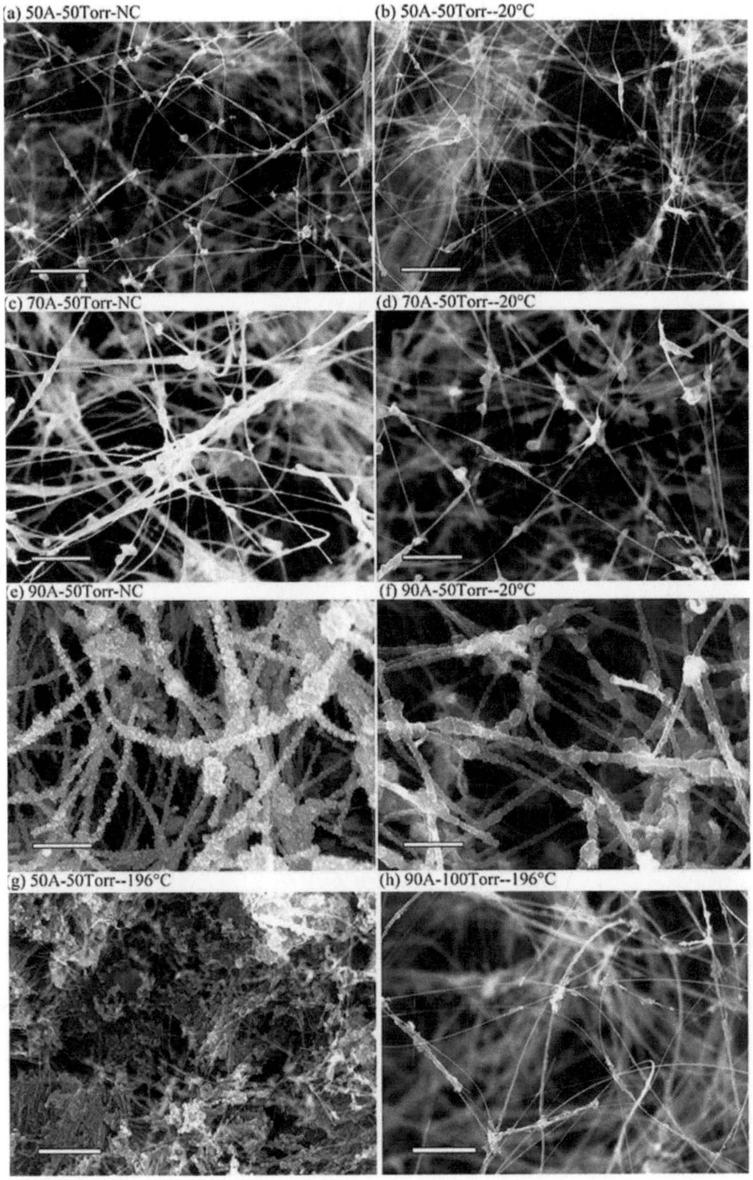

Figure 5.8 SEM images of MWCNTs prepared under normal and cooling-enhanced conditions. The term xA-yTorr-z°C expresses current, pressure and cooling temperature used for the growth of LLCCs. NC means no cooling. The scale bar is 1 μm. Figure is reprinted with the permission from Ref. [19], ©2018 by Royal Society of Chemistry.

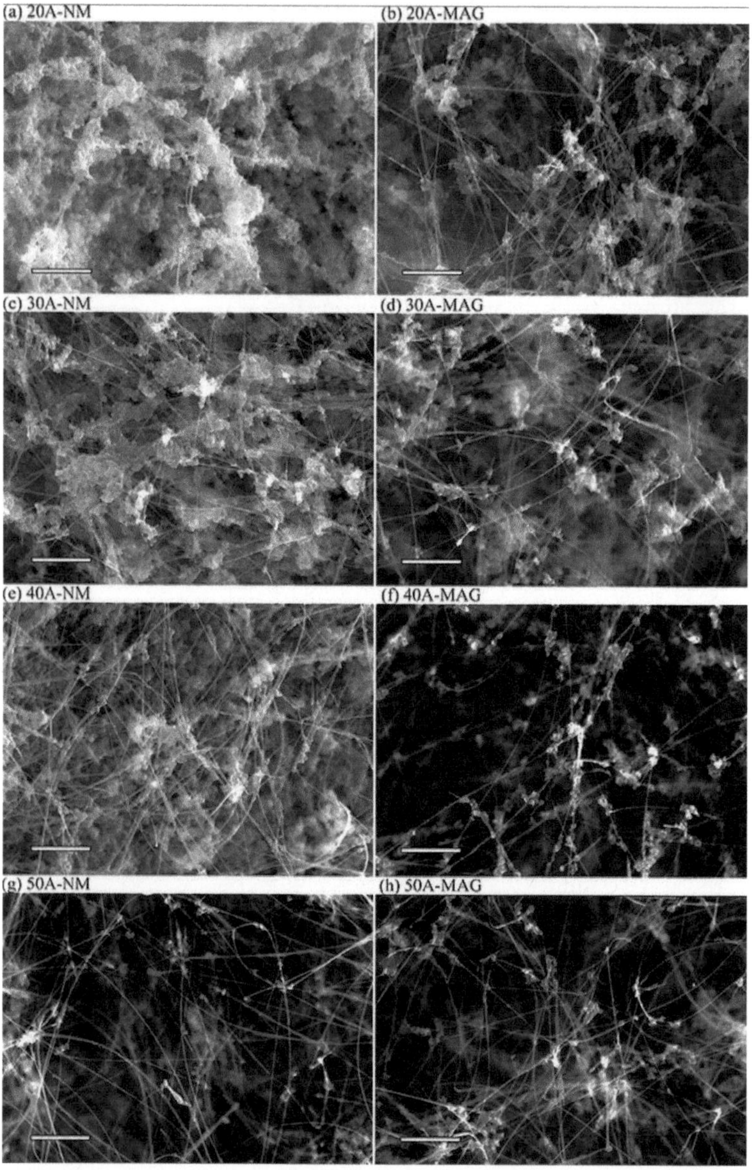

Figure 5.9 SEM images of MWCNTs prepared under normal and magnetic field-enhanced conditions. The term xA-NM/MAG expresses current condition without/with magnetic field used for the growth of LLCCs. The scale bar is 1 µm. Figure is reprinted with the permission from Ref. [20], ©2020 by Elsevier.

Figure 5.8 shows the SEM images from the samples obtained by hydrogen arc discharge with/without cooling modification. Very clear and high-density MWCNT bundles with length longer than tens of micrometers can be observed in Figs. 5.8a and b, corresponding to the optimal current under conventional condition. However, the impurities obviously appear with the increase of current, leaving from the optimal condition. As shown in Fig. 5.8e, the impurities totally cover the surface of MWCNTs when the current reaches 90 A, in which case no L-band can be detected in the corresponding Raman spectrum. The minus 20 °C gas cooling is effective to improve the purity of MWCNTs obtained at higher current. Figure 5.8d is corresponding to the optimal current under cooling condition, indicating a clearer surface with fewer impurities than Fig. 5.8c. The impurities are also reduced at 90 A in Fig. 5.8f, meanwhile, the growth of MWCNWs is still possible at such high current owing to cooling enhancement. In the case of cooling by liquid nitrogen, MWCNTs cannot grow well and stack with graphite sheets under low current and pressure. The optimal pressure and current are moved to 100 Torr and 90 A, respectively, where the growth of MWCNTs becomes much better, as shown in Fig. 5.8h. The correlation between purity of MWCNTs observed *via* SEM and the growth of MWCNWs is well applied to the magnetic experiment. Figure 5.9 shows the SEM images from the samples obtained with/without magnetic assistance. As mentioned above, the cooling enhancement helps to promote the growth of LLCCs inside MWCNTs in high arc current range, while magnetic field helps to improve the yield of LLCCs in low arc current range. The density of MWCNTs decreases obviously at over low current as shown in Fig. 5.9a, in which the current became unstable and the deposit became very thin. When the external magnetic field was applied, the arc plasma became stable and formed better deposit. Both the density and the purity of MWCNTs were improved, also allowing the growth of MWCNWs. The enhancement can be easily observed by comparing the right column to the left column.

The third aspect is the temperature of arc plasma. Both the purity of MWCNTs and the yield of LLCCs rely on the temperature of arc plasma. The influence of cooling and magnetic enhancement on the arc plasma was indirectly studied and discussed *via* finite elemental simulation. The precise calculation of plasma is very complicated, including a series of dynamic problems such as the ablation of anode, the depositing process, the density of state in plasma, ionization and deionization, and so on. However, our simulations do not aim to solve the whole dynamic process, instead of which the simulations only need to show the possible way to influence the arc plasma. The temperature increase of arc plasma under modified condition is supposed through phenomenological explanation.

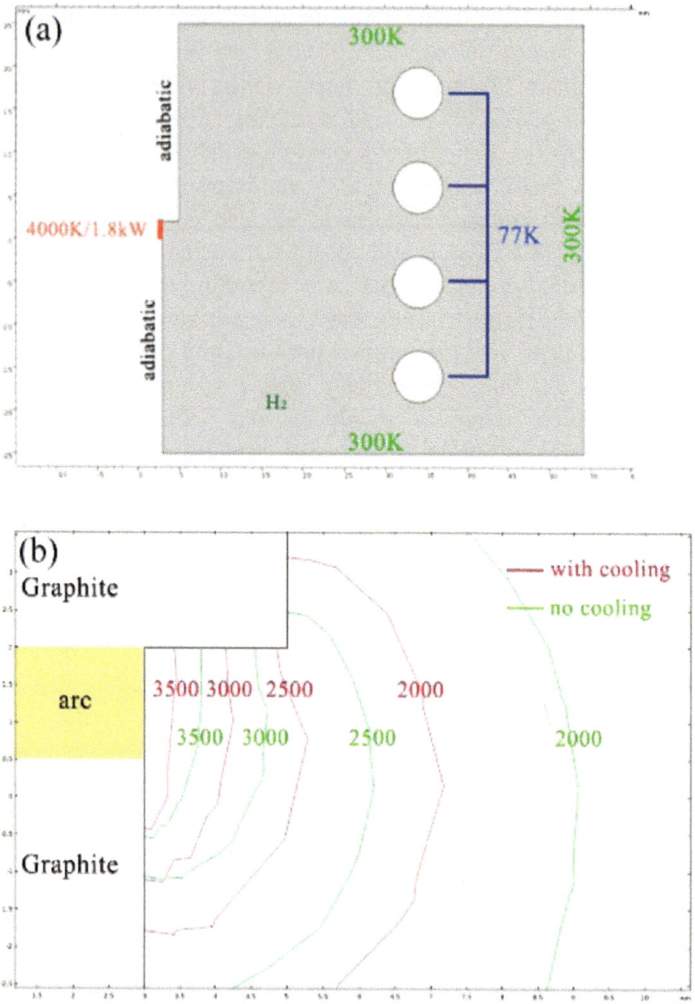

Figure 5.10 (a) The model for finite elemental simulation of temperature gradient in the arc-discharge chamber. (b) The isotherms of simulated temperature (K) distribution near arc plasma with (red line) or without (green line) gas cooling.

In the case of cooling enhancement, Fig. 5.10 shows the model and result to calculate the temperature gradient ranging from the arc to the cold pipe (77 K) or to the chamber wall (room temperature, 300 K), only considering laminar flow, gas heat transfer and radiation as main parameters. The ablation and deposition on electrodes were not considered in this model. So, they were set to be statical and adiabatic to simplify the simulation. The model of the chamber including the boundary of electrodes, plasma and cooling pipe was simplified to two-dimension with real size parameters. The temperature of the upper, bottom and side walls of chamber were fixed to 300 K, being in correspondence with the constant water-cooling protection on the walls. The four hollow circles were the cross section of cooling pipe and were fixed to 77 K. 50 Torr of hydrogen was filled into the chamber. The boundary of arc plasma was treated as a fixed heat source, where the mass transfer with gas or electrodes, the ionization and deionization on the boundary were ignored. This boundary was set to be 4000 K with 1800 W of constant output power. As a result, the temperature gradient becomes lager when starting from the arc to the cold pipe instead of to the chamber wall. The isotherms show the temperature distribution near the arc plasma, which become closer to the arc plasma boundary when applying cooling pipe as shown in the red line of Fig. 5.10b. The plasma boundary is built by the balance of ionization and deionization, which is closely related to the temperature. Thus, the larger temperature gradient leads to a fast energy dissipation of arc plasma, causing the shrink of the arc boundary and leading to the increase of current density. Then the electron temperature increases. Meanwhile, the potential drop near the surface of electrodes becomes larger to maintain the arc plasma, increasing electron emission and electron collision, as well as the temperature near the electrodes [44]. Besides, Akita *et al.* supposed that over high current can cause over supplying carbon vapor ablated from anode, which is much cooler than the arc plasma, finally decreasing the temperature of plasma [39]. Keidar *et al.* calculated the existence of the highest temperature of arc plasma and cathode under certain current and pressure [45]. The densities of C^+ and C become higher with the increase of plasma temperature. The higher ratios of the densities of C^+ and C compared to that of C_2 help to increase the purity of MWCNTs [25, 40]. It can be supposed that the higher density of smaller carbon monomers with high energy also benefit the growth of LLCCs as well as narrow innermost tubes at the same time. Those mechanisms suggest that the growth of LLCCs needs a relatively high temperature of arc plasma, and explain the existence of different optimal current-pressure combination for the growth of LLCCs using different background gas.

In the case of magnetic enhancement, the mechanism is similar. The magnetic field also helps to improve the temperature of the arc plasma. Different from cooling enhancement, magnetic field performed better effect in low current range. The model for finite elemental simulation is presented in Fig. 5.11a. The magnetic field generated by 40 A of direct current was calculated to be 26.6 Oe, encircling the arc area. The magnetic conductivity and magnetization of magnets is near 1.03 H/m and 450 kA/m, respectively. The simulated static magnetic field generated by the magnets near the gap of electrodes is vertically downward and is about 95.5 Oe. Thus, the total magnetic field is coupled by direct current and external magnetic field, which is downward with spiral near the arc area and is about 98.7 Oe, as shown in Fig. 5.11b. The boundary of arc plasma is affected by the self-generated magnetic field of the arc current because of the Lorentz force. The additional magnetic field enhances the confinement effect, resulting in the shrink of plasma. The optical image of arc in the case of 40 A without and with magnetic field-enhancement are shown in Fig. 5.11c and d. The diameter of the observed arc boundary was shrunk by about 9.8% owing to the additional magnetic field, which will improve the current density in the arc plasma as well as the temperature. As mentioned above, it increases the density of high-energy carbon monomers and benefits the growth of MWCNWs.

In conclusion, the high-quality growth of MWCNTs containing innermost tubes with diameter of 0.7 nm is required for the growth of LLCCs. To ensure a high temperature of arc plasma in arc-discharge method benefits the high-yield and high-purity preparation of both MWCNTs and MWCNWs.

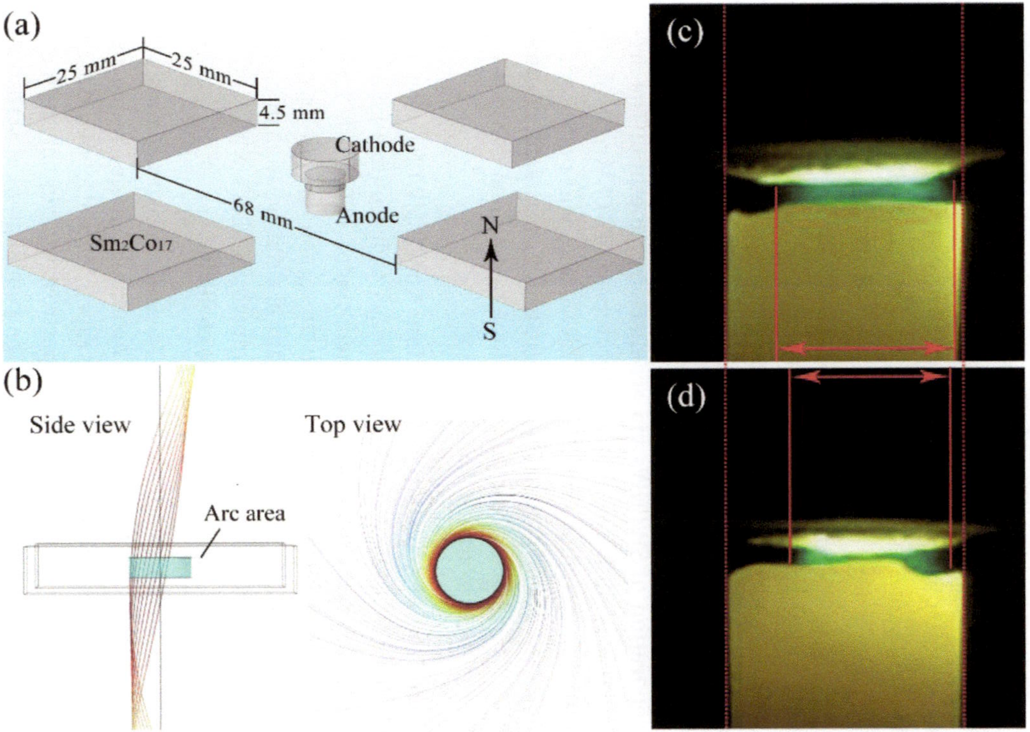

Figure 5.11 (a) The model of magnetic field-enhanced arc-discharge method. (b) The magnetic field calculated by finite elemental simulation. The photos of arc plasma (c) without and (d) with the confinement of additional magnetic field. All panels are reprinted with the permission from Ref. [20], ©2020 by Elsevier.

5.4 The Challenges and Prospects of MWCNWs and Arc-Discharge Method

The arc-discharge method to prepare MWCNWs is reviewed in this chapter. The basic apparatus is introduced in detail. Multiple adjustments on gas, current, pressure or even additional physical field widely extend the optimal conditions and realize the large-scale preparation of MWCNWs. The HRTEM and Raman spectroscopy are recommended as two best tools to evaluate the samples and investigate the nanostructure and property of LLCCs. The studies on growth mechanism aim to offer a better guide to the development of arc-discharge method. Finally, we present several challenges as well as prospects:

1. Preparation of SWCNWs via arc-discharge method: The arc-discharge method is the only way to prepare MWCNWs simply in one step from graphite. However, MWCNWs is also the only product that this method can prepare, compare to heating treatment or other methods. It will be a great progress if the preparation of SWCNWs can be realized *via* arc-discharge method.

2. Separation or selective preparation: Not all MWCNTs contain LLCCs and the as-grown MWCNWs always own multiple kinds of band gaps. Besides distinguishing them under TERS, it is highly expected to develop a way to separate empty and filled CNTs, or to separate LLCCs with different band gaps like the isolation of chiral CNTs. The method towards selective growth of LLCCs with certain band gaps is also expected to solve this challenge.

3. Determination of filling ratio: To date, statistical study *via* HRTEM or TERS is the only way to quantitatively determine the filling ratio of MWCNWs. However, it is expensive and low-efficiency. On the other hand, normal Raman spectroscopy can only provide qualitative comparison. We suppose the combination of statistical study and normal Raman spectroscopy may provide empirical formula to determine the filling ratio of CNWs.

4. Modeling of growth process: The growth process of MWCNWs from arc plasma is very complicated, including multiple physical parameters. Only discussion based on circumstantial evidence or phenomenological model cannot show the precise growth mechanism. More experiments and theoretical simulation are required to build the precise growth model.

5. Development of application: Theoretical studies show the outstanding physical properties and the bright prospects of LLCCs, i.e., carbyne [46]. However, the potential usage of LLCCs is not presented experimentally. The highly expected applications such as nanodevices need to be further investigated based on CNWs.

References

1. Kroto, H. W., Heath, J. R., O"Brien, S. C., Curl, R. F., and Smalley, R. E. (1985). C60: buckminsterfullerene, *Nature*, **318**, 162–163.
2. Zhao, X., Ando, Y., Liu, Y., Jinno, M., and Suzuki, T. (2003). Carbon nanowire made of a long linear carbon chain inserted inside a multiwalled carbon nanotube, *Phys. Rev. Lett.*, **90**, 187401.
3. Qin, L., Zhao, X., Hirahara, K., Miyamoto, Y., Ando, Y., and Iijima, S. (2000). The smallest carbon nanotube, *Nature*, **408**, 50.
4. Zhao, X., Liu, Y., Inoue, S., Suzuki, T., Jones, R. O., and Ando, Y. (2004). Smallest carbon nanotube is 3 Å in diameter, *Phys. Rev. Lett.*, **92**, 125502.
5. Heeg, S., Shi, L., Poulikakos, L. V., Pichler, T., and Novotny, L. (2018). Carbon nanotube chirality determines properties of encapsulated linear carbon chain, *Nano Lett.*, **18**, 5426–5431.
6. Cui, W., Saito, T., Ayala, P., Pichler, T., and Shi, L. (2019). Oxidation stability of confined linear carbon chains, carbon nanotubes, and graphene nanoribbons as 1-D nanocarbons, *Nanoscale*, **11**, 15253–15258.
7. Zhang, K., Zhang, Y., and Shi. L. (2020). A review of linear carbon chains, *Chinese Chem. Lett.*, **31**, 1746–1756.
8. Zhang, Y. (2022). A review of arc-discharge method towards large-scale preparation of long linear carbon chains, *Chin. Phys. B*, **31**, 125201.
9. Shi, L., Rohringer, P., Suenaga, K., Niimi, Y., Kotakoski, J., Meyer, J. C., Peterlik, H., Wanko, M., Cahangirov, S., Rubio, A., and Lapin, Z. J. (2016). Confined linear carbon chains: a route to bulk carbyne, *Nat. Mater.*, **15**, 634–639.
10. Chang, W., Liu, F., Liu, Y., Zhu, T., Fang, L., Li, Q., Liu, Y., and Zhao, X. (2021). Smallest carbon nanowires made easy: long linear carbon

chains confined inside single-walled carbon nanotubes, *Carbon*, 183, 571–577.

11. Toma, S., Asaka, K., Irita, M., and Saito, Y. (2018). Bulk synthesis of linear carbon chains confined inside singlewall carbon nanotubes by vacuum discharge, *Surf. Interface Anal.*, **51**, 131–135.
12. Ando, Y., Zhao, X., and Ohkohchi, M. (1997). Production of petal-like graphite sheets by hydrogen arc discharge, *Carbon*, **35**, 153–158.
13. Ando, Y., and Zhao, X. (2006). Synthesis of carbon nanotubes by arc-discharge method, *New Diam. Front. C. Tec.*, **16,** 123–137.
14. Chen, Y., Zhao, H., Sheng, L., Yu, L., An, K., Xu, J., Ando, Y., and Zhao, X. (2012). Mass-production of highly-crystalline few-layer graphene sheets by arc discharge in various H_2–inert gas mixtures, *Chem. Phys. Lett.*, **538,** 72–76.
15. Farhat, S., and Scott, C. D. (2006). Review of the arc process modeling for fullerene and nanotube production, *J. Nanosci. Nanotechno.*, **6,** 1189–1210.
16. Jinno, M., Bandow, S., and Ando, Y. (2004). Multiwalled carbon nanotubes produced by direct current arc discharge in hydrogen gas, *Chem. Phys. Lett.*, **398,** 256–259.
17. Cazzanelli, E., Castriota, M., Caputi, L. S., Cupolillo, A., Giallombardo, C., and Papagno, L. (2007). High-temperature evolution of linear carbon chains inside multiwalled nanotubes, *Phys. Rev. B*, **75**, 121405.
18. Cupolillo, A., Castriota, M., Cazzanelli, E., Caputi, L., Giallombardo, C., Mariotto, G., and Papagno, L. (2008). Second-order Raman scattering from linear carbon chains inside multiwalled carbon nanotubes, *J. Raman Spectrosc.*, **39**, 147–152.
19. Zhang, Y., Zhao, J., Fang, Y., Liu, Y., and Zhao, X. (2018). Preparation of long linear carbon chain inside multi-walled carbon nanotubes by cooling enhanced hydrogen arc discharge method, *Nanoscale*, **10,** 17824–17833.
20. Zhang, Y., Chang, W., Liu, Y., Maruyama, T., and Zhao, X. (2020). High-yield growth of multi-walled carbon nanowires by magnetic field controlled arc technique, *Carbon*, **158,** 672–680.
21. Scuderi, V., Scalese, S., Bagiante, S., Compagnini, G., D'Urso, L., and Privitera, V. (2009). Direct observation of the formation of linear C chain/carbon nanotube hybrid systems, *Carbon*, **47**, 2112–2142.
22. Scalese S., Scuderi V., Bagiante S., Simone, F., Russo, P., D'Urso, L., Compagnini, G., and Privitera, V. (2010). Controlled synthesis of carbon nanotubes and linear C chains by arc discharge in liquid nitrogen, *J. Appl. Phys.*, **107**, 014304.

23. Kang, C. S., Fujisawa, K., Ko, Y. I., Muramatsu, H., Hayashi, T., Endo, M., Kim, H. J., Lim, D., Kim, J. H., Jung, Y. C., Terrones, M., Dresselhaus, M. S., and Kim, Y. A. (2016). Linear carbon chains inside multi-walled carbon nanotubes: growth mechanism, thermal stability and electrical properties, *Carbon*, **107**, 217–224.

24. Akita, S., Kamo, S., and Nakayama, Y. (2002). Diameter control of arc produced multiwall carbon nanotubes by ambient gas cooling, *Jpn. J. Appl. Phys.*, **41**, 487–489.

25. Anazawa, K., Shimotani, K., Manabe, C., Watanabe, H., and Shimizu, M. (2002). High-purity carbon nanotube synthesis method by an arc discharging in magnetic field, *Appl. Phys. Lett.*, **81**, 739–741.

26. Kim, Y. A., Muramatsu, H., Hayashi, T., and Endo, M. (2012). Catalytic metal-free formation of multi-walled carbon nanotubes in atmospheric arc discharge, *Carbon*, **50**, 4588–4595.

27. Chalifoux, W. A., and Tykwinski, R. R. (2010). Synthesis of polyynes to model the *sp*-carbon allotrope carbyne, *Nat. Chem.*, **2**, 967–971.

28. Gao, Y. Z., Hou, Y. X., Gámez, F. G., Ferguson, M. J., Casado, J., and Tykwinski, R. R. (2020). The loss of endgroup effects in long pyridyl-endcapped oligoynes on the way to carbyne, *Nat. Chem.*, **12**, 1143–1149.

29. Andrade, N. F., Vasconcelos, T. L., Gouvea, C. P., Archanjo, B. S., Achete, C. A., Kim Y. A., Endo, M., Fantini, C., Dresselhaus, M. S., and Souza Filho, A. G. (2015). Linear carbon chains encapsulated in multiwall carbon nanotubes: resonance Raman spectroscopy and transmission electron microscopy studies, *Carbon*, **90**, 172–180.

30. Tabata, H., Fujii, M., Hayashi, S., Doi, T., and Wakabayashi, T. (2006). Raman and surface-enhanced Raman scattering of a series of size-separated polyynes, *Carbon*, **44**, 3168–3176.

31. Haque, Md. M., Yin, L., Nugraha, A. R. T., and Saito, R. (2011). Vibrational and NMR properties of polyynes, *Carbon*, **49**, 3340–3345.

32. Shi, L., Rohringer, P., Wanko, M., Rubio, A., and Pichler, T. (2017). Electronic band gaps of confined linear carbon chains ranging from polyyne to carbyne, *Phys. Rev. Mater.*, **1**, 075601.

33. Mostaani, E., Monserrat, B., Drummond, N. D., and Lambert, C. J. (2016). Quasiparticle and excitonic gaps of one-dimensional carbon chains, *Phys. Chem. Chem. Phys.*, **18**, 14810–14821.

34. Andrade, N. F., Aguiar, A. L., Kim, Y. A., Endo, M., Freire, P. T. C., Brunetto, G., Galvão, D. S., Dresselhaus, M. S., and Souza Filho, A. G. (2015). Linear carbon chains under high-pressure conditions, *J. Phys. Chem. C*, **119**, 10669–10676.

35. Sharma, K., Costa, N., Kim, Y. A., Muramatsu, H., Neto, N. M. B., Martins, L. G. P., Kong, J., Paschoal, A. R., and Araujo, P. T. (2020). Anharmonicity and universal response of linear carbon chain mechanical properties under hydrostatic pressure, *Phys. Rev. Lett.,* **125**, 105501.

36. Costa, N., Sharma, K., Kim, Y. A., Choi, G. B., Endo, M., Neto, N. M. B., Paschoal, A. R., and Araujo, P. T. (2021). Thermodynamics of linear carbon chains, *Phys. Rev. Lett.*, **126**, 125901.

37. Moura, T. A., Neves, W. Q., Alencar, R. S., Kim, Y. A., Endo, M., Vasconcelos, T. L., Costa, D. G., Candiotto, G., Capaz, R. B., Araujo, P. T., Souza Filho, A. G., and Paschoal, A. R. (2023). Resonance Raman spectroscopy characterization of linear carbon chains encapsulated by multi-walled carbon nanotubes, *Carbon*, **212**, 118123.

38. Castriota, M., Cazzanelli, E., Caputi, L., Cupolillo, A., Giallombardo, C., Papagno, L., and Mariotto, G. (2008). Investigations on Raman bands from carbon linear chains in multiwalled carbon nanotubes, *Diam. Relat. Mater.*, **17**, 1716–1723.

39. Akita, S., Ashihara, H., and Nakayama, Y. (2000). Optical emission spectroscopy of arc flame plasma for generation of carbon nanotubes, *Jpn. J. Appl. Phys.*, **39**, 4939–4944.

40. Farhat, S., Hinkov, I., and Scott, C. D. (2004). Arc process parameters for single-walled carbon nanotube growth and production: experiments and modeling, *J. Nanosci. Nanotechnol.*, **4**, 377–389.

41. Shashurin, A., and Keidar, M. (2008). Factors affecting the size and deposition rate of the cathode deposit in an anodic arc used to produce carbon nanotubes, *Carbon*, **46**, 1826–1828.

42. Zhao, X., Okazaki, T., Kasuya, A., Shimoyama, H., and Ando, Y. (1999). Optical emission spectra during carbon nanotube production by arc discharge in H_2, CH_4, or He gas, *Jpn. J. Appl. Phys.*, **30**, 6014–6016.

43. Liu, Y., and Jones, R. O. (2003). Carbon species confined inside carbon nanotubes: a density functional study, *Phys. Rev. B*, **68**, 125413.

44. Nishio, M., Akita, S., and Nakayama, Y. (2004). Cooling effect on the growth of carbon nanotubes and optical emission spectroscopy in short-period arc-discharge, *Thin Solid Films*, **464**, 304–307.

45. Keidar, M., and Beilis, I. I. (2009). Modeling of atmospheric pressure anodic carbon arc producing carbon nanotubes, *J. Appl. Phys.*, **106**, 103304.

46. Casari, C. S.. and Milani, A. (2018). Carbyne: from the elusive allotrope to stable carbon atom wires, *MRS Commun.*, **8**, 207–219.

Chapter 6

Atomistic Simulations of Formation Processes of Carbon Nanowires

Yi Liu, Fu Liu, and Xinluo Zhao
Materials Genome Institute and Department of Physics, Shanghai University, Shanghai, 200444, China
yiliu@shu.edu.cn

It is intriguing that carbyne now can be synthesized experimentally under the protection of confinement. As the long carbon chains encapsulated in carbon nanotubes (CNTs), the carbon nanowires (CNWs) can be prepared by electrical arc discharge or thermal annealing methods. The formation mechanisms of long carbon chains inside CNTs remain poorly understood at an atomistic level. Understanding the mechanism of carbyne formation in CNTs not only helps to optimize and control the existing experimental synthesis techniques, also guides the development of novel experimental methods to prepare carbynes under various confinements and device applications. This chapter introduces the atomistic simulations of carbyne formation inside CNTs by using newly developed reactive force fields (dubbed $ReaxFF_C$-S22). The

Emerging Carbyne: Truly One-Dimensional Form of Carbon
Edited by Yahachi Saito

ISBN 978-981-5129-15-1 (Hardcover), 978-1-003-51036-9 (eBook)
www.jennystanford.com

insertion and fusion processes of carbon chains are described in terms of configuration evolution associated with the bond breaking and formation. The energy barriers are evaluated for the rate-limiting steps. Several formation mechanisms of CNWs are discussed to guide the development of novel experimental synthesis methods, the essential prerequisite for the success of future applications of carbyne materials in carbon-based electronic devices.

6.1 Introduction

The widely studied nanocarbon materials such as fullerenes, CNTs, and graphene are sp^2 hybridized carbon materials with major distinction in different curvatures. Carbyne is a carbon allotrope composed entirely of *sp* hybridized carbon atoms and exhibited fascinating unique properties. It has the largest effective surface among various carbon allotropes [1]. Carbyne has the specific stiffness ~10^9 Nm/kg, more than twice as much as graphene and carbon nanotubes (CNTs). The Young's modulus and shear modulus are predicted up to 32.7 and 47.2 TPa, respectively, suggesting its remarkable mechanical properties [2]. Combined with its tunable electronic and thermal properties, carbyne has remarkably potential applications in nano-electronic and electromechanical switching devices. Despite the predicted outstanding performance, the experimental synthesis of isolated carbyne is still challenging. Although carbyne-like linear carbon chains (LCCs) can be stabilized by end-capped functional groups [3–7] and graphyne-type materials [8–11], with the increase of length the middle carbon (C) atoms are more likely to react *via* crosslinking reactions and to form more stable sp^2 or sp^3 hybridized nanocarbons, making isolated carbyne elusively hard to prepare. The carbon nanowires (CNWs) can be considered as long LCC (LLCC) encapsulated inside the empty cavity of CNTs. The synthesis of CNWs accelerated the successful experimental syntheses of carbyne (Fig. 6.1) [12]. In 2003, Zhao *et al.* [12] prepared multi-walled CNWs (MWCNWs) in the cathode deposits by hydrogen arc discharge evaporation of pure carbon electrodes. The high-resolution transmission electron microscopy (HRTEM) shows that the innermost CNT is 0.7 nm in diameter and a LLCC composed of more than 100 carbon atoms inside its hollow core. Raman scattering shows that the characteristic Raman peaks of LCCs appeared at around 1850 cm^{-1}. The confinement effect of CNT can prevent LLCCs from crosslinking reactions, providing a reliable approach for the bulk production of LLCCs as well as property investigations [13].

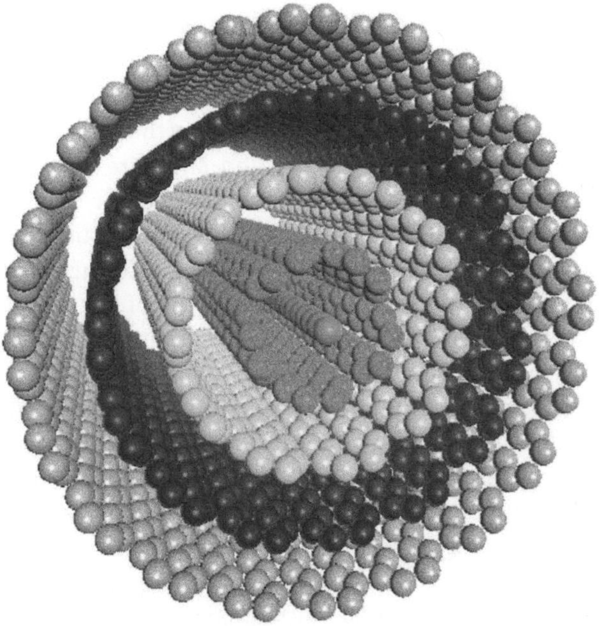

Figure 6.1 Atomic model of carbon nanowire [12].

Since MWCNWs were discovered, the synthesis and preparation of CNWs have achieved significant progresses [14–20]. Theoretical studies on the CNWs have also made considerable progress. Density functional theory (DFT) calculations have shown its ability in the study of carbon and related materials [21–26]. In 2003, Chen *et al.* [27] studied electronic and transport properties of single-walled CNWs (SWCNWs) using the tight-binding calculations. In the same year, Liu *et al.* [28] performed DFT calculations to confirm that carbon chains can be inserted coaxially into 0.7 nm nanotubes with prediction of characteristic Raman peaks, and suggested the existence of carbon nanowires as well as other various carbon species remained inside CNTs. Liu *et al.* predicted that besides carbon chains the CNTs with larger diameters can accommodate other small carbon species such as various types of graphitic sheets, carbon rings, carbon bowls, and carbon cages (Figs. 6.2 and 6.3). Carbon species with dangling bonds can react with the wall atoms of CNTs but the reaction could not occur due to the repulsive interactions if the dangling bonds are saturated. The inserted carbon chains, sheets, and rings inside CNTs could be detected due to the distinct Raman vibration modes.

In 2006, Wang *et al.* [29] using DFT calculations studied the geometric structures and electronic properties of SWCNWs and demonstrated that the encapsulated carbon chains may alter the electronic properties depending on the radii of the CNT and the filling density of carbon atoms. Vadapalli and Mintmire [30] calculated the bond length, energies, and electronic properties of LCCs inside the chiral (7, 3) and (7, 4) CNTs. They reported an equilibrium bond length of about 0.129 nm of cumulenic LCCs and found that the Fermi level of CNWs pinned near the top of the valence band of corresponding CNTs. In addition, DFT calculations have also been extensively used to study the structural and electronic properties [31–34], mechanical properties [31], and optical properties [35] of CNWs. In 2016, Shi *et al.* [13] calculated LCCs inside four small-diameter armchair CNTs using DFT calculations. They found that LLCC@(5, 5) CNT exhibits the lowest interaction energy, further predicted and confirmed the optimum tube diameter is ~0.71 nm for LLCC growth. In 2017, Shi *et al.* [36] predicted the electronic band gaps of LCCs using DFT calculations and explained that the predicted band gaps varied compared with experimental values due to the interactions with the environment.

Compared with DFT calculations, molecular dynamics (MD) uses pairwise potentials to model the interactions between atoms and thus enables the simulations of dynamic evolutions on large systems over long time scales at a lower cost. In 2011, Zhao *et al.* [37] investigated the fusion reactions of LCCs inside the double-walled CNTs (DWCNTs) using tight-binding-based MD (TB-MD) simulations, suggested that C–C bond formation occurs at the end of polyyne molecules and verified the formation of LCCs@DWCNTs. In 2015, Andrade *et al.* [38] investigated the structural behavior of LCCs inside a CNT under high-pressure conditions using reactive force field (ReaxFF) MD simulations and interpreted the irreversible coalescence of different carbon chains. In 2019, Deng *et al.* [39] studied the effects of temperature, confinement, and alignment on the formation of

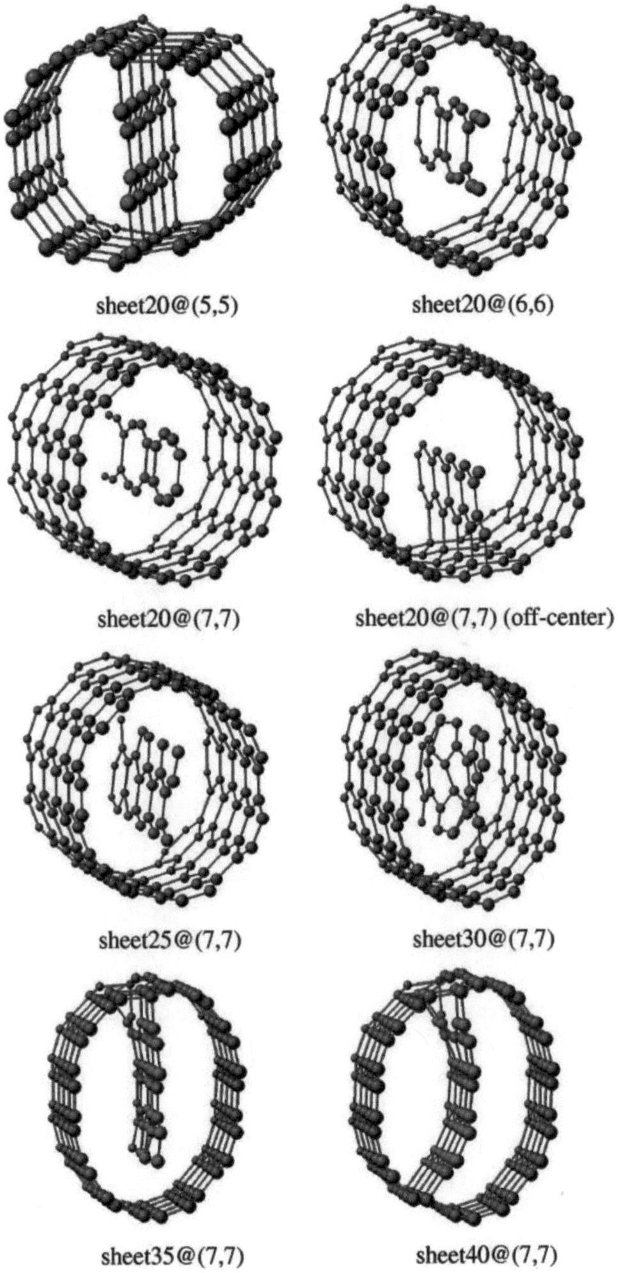

Figure 6.2 Graphitic sheets and carbon rings inside CNTs [28].

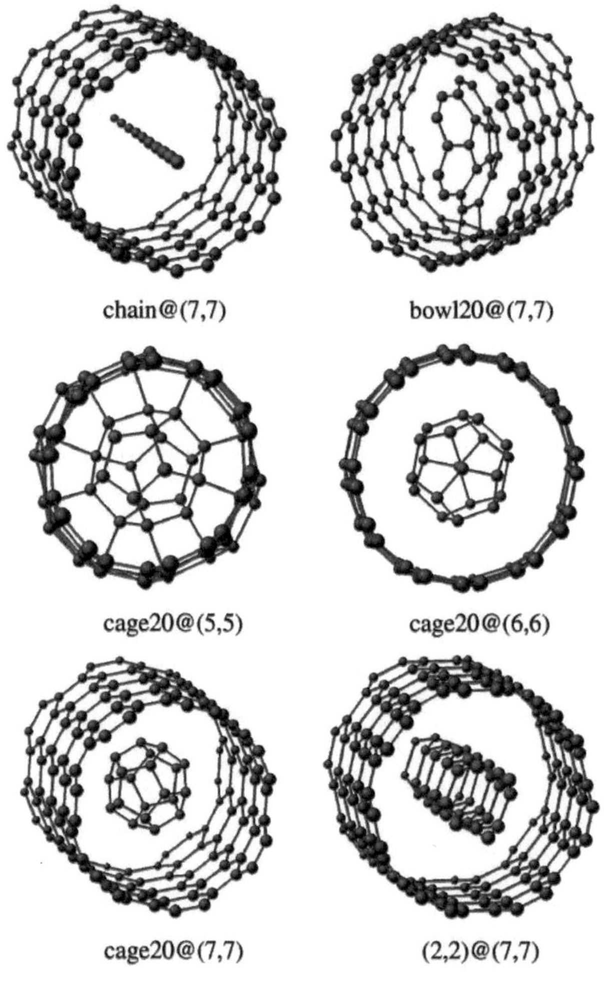

Figure 6.3 Carbon chain, bowl, cage, and tube inside CNTs [28].

carbyne inside CNTs systematically using ReaxFF MD simulations and proposed an empirical/phenomenological relation to predict growth rate based on classical collision theory with a modified Arrhenius relation. In the same year, Khalilov *et al.* [40] studied the nucleation and growth mechanism of LCCs@DWCNTs and compared the role of the Ni catalyst and the hydrogen (H) atoms using ReaxFF MD simulations and DFT calculations. In 2021, Chang *et al.* [20] prepared the LLCCs@SWCNTs structure with high yield by a controllable fusion reaction of polyynes for the first time and reported the ReaxFF MD simulations on the possibilities of the formation of CNWs from the thermodynamics point of view. In the same year, Shi *et al.* [41] performed MD simulations of the formation process of carbon chains inside a (6, 4) and (20, 0) SWCNT using C_2 species as precursors and interpreted experimental findings of controlled growth of confined carbyne inside the SWCNTs.

Simulation algorithm based on the Monte Carlo (MC) method has also been developed by Wong [42] to predict the environmental impacts of carbyne inside carbon nanotubes. Although the growth process of LLCCs inside CNTs has been investigated both experimentally and theoretically, the formation mechanisms especially the detailed dynamics of growth process and refined configurations of carbon chains inside CNTs are still elusive from kinetic point of view. Moreover, the effects of the residual H adatoms on fused LCCs have not yet been fully explored.

Regarding the H effects on nanocarbon materials, most existing simulations focus on the edge H atoms that intrinsically exist to saturate the dangling bonds of the edge carbon atoms. The single C atom [43, 44] or carbon dimer [45, 46] without H atoms have often been used as the precursors and building blocks in the simulations of nanocarbon growth. The ignorance of H atoms is understandable because the naked C atoms without H are more easily to bind and grow to bigger species at short time scale of simulations, plus the simplicity of configuration construction avoiding the introduction of additional degree of freedom. This H-free assumption has been considered valid for years since the final products often do not have internal H adatoms or are not carefully confirmed due to the challenge of locating light H atoms in experiments. Theoretically speaking, it would be energetically favorable that H adatoms turn the C–C orbital hybridization of 1D carbon chain from original *sp* to more stable sp^2 locally. However, what process can lead to the formation of the H adatom configurations and what the consequence of these H adatoms on the configurations, electronic, and magnetic properties are still poorly understood. Specifically, the effects of the residual H adatoms on fused LCCs have not yet been fully explored.

ReaxFF force fields, one of most widely used reactive force field, have been extensively applied to study the chemical reactions of various organic and inorganic systems, especially for hydrocarbon systems that mainly involve C–H bond breaking and formation [47–51]. In this chapter, we show the atomistic simulations of formation processes of carbon nanowires by ReaxFF MD simulations and DFT calculations from the following aspects:

- Physical properties of CNWs
- Insertion process of CNW formation
- Fusion process of CNW formation
- Configurations of carbon chains with hydrogen adatoms
- Diameter effects on the binding energies of carbon nanowires

6.2 Properties of CNWs

As a carbon allotrope composed entirely of *sp* hybridized carbon atoms, carbyne, the LLCCs with infinite length, were predicted to have excellent mechanical properties. In 2013, Liu *et al.* [2] investigated the mechanical response to tension, bending, and torsion deformations of carbyne using first-principles calculations. They predicted the specific strength of carbyne is up to 7.5×10^7 N·m/kg, the torsional stiffness can be zero and can be tuned by end-capped functional groups, the nominal Young's modulus (Y) is 32.7 TPa with an effective mechanical thickness of 0.772 Å, the effective shear modulus (G) is about 47.2 TPa and the formal Poisson's ratio (n) is –0.65, and the specific stiffness is about 10^9 N·m/kg, twice as much as the CNTs and graphene [52–54]. Using MD simulations, Wang *et al.* [55] predicted the thermal transport properties and obtained the thermal conductivity (κ) of carbyne is 80 ± 26 kW/m/K, more than one order of magnitude of monolayer graphene (~5 kW/m/K) [56] and SWCNTs (~3.5 kW/m/K) [57]. Moreover, the structural configurations, electronic properties, magnetic behavior, and thermal properties can be tuned by means of stress tension [2, 55, 58, 59].

The properties of LLCCs confined in CNTs have also been studied. Liu *et al.* [28] calculated the energies, structures, and vibration frequencies of carbon chains, rings, and other species inside the SWCNTs using DFT, obtained the stretching modes of carbon chains and rings are between 1800 and 2160 cm^{-1} and the chain modes are softened when inserted into CNTs. Wang *et al.* [29, 31] calculated the change of geometric structures and electronic properties as well as bonding features of the encapsulated LCCs inside zigzag and armchair CNTs. They found that the interaction between the CNTs and the LCCs becomes more obvious when the diameters of the CNT decreases. The tube size and the filling density of carbon atoms will modulate the structures and electronic properties of corresponding CNWs. The impact of the inserted LCCs on the electronic properties of CNT have also been studied by Tapia *et al.* [32]. They found despite the fact that both LCCs and (8, 0) CNT have a band gap, the composite LCC inside the (8, 0) CNT shows metallic characters. This phenomenon is interpreted as the charge transfer from CNT to the LCC and is provided by the electronic states from the LCC exclusively rather than CNT.

The external pressure has an important effect on the Raman signature of confined LCCs. Yang *et al.* [34] simulated and observed that the bond length alternation (BLA, the bond length difference between single and triple C–C bonds) of LCCs@(5, 5) CNT decreases as pressure increases, resulted in the downshift of the Raman frequency. The reduction of BLA is induced by the tube-chain interaction enhancement under pressure. However, if the external pressure is too high, *sp* to sp^2 structural transformation for LCCs will occur due to the fuse of the chain atoms and the inner wall of the CNT. By adding the second outer tube, Bai *et al.* [33] shows the enhanced stabilities of DWCNWs and found that the charge carrier mobility of encapsulated LCC reaches 10^5–10^6 cm^2/V at room temperature, which is much larger than that of the free LCCs. These findings presented the potential applications of CNWs, offered a way to engineer the structures, electronic, and thermal properties of LLCCs, and contributed to the future usage for the nanoelectronics and magnomechanics of the LLCCs as well as CNWs.

6.3 Insertion Process of CNW Formation

To understand the insertion mechanism of CNWs, we address the insertion reactions of C_8H_2 chains into (6, 5) CNT at 973 K using constrained MD simulations. As shown in Fig. 6.4, the host (6, 5) CNT had a finite length of about 50 Å whose edges were saturated with H atoms. By fixing the distance (d) between the end of C_8H_2 and the edge of the CNT, the constrained MD were performed. We define $d < 0$ when the C_8H_2 is located outside the CNT while $d > 0$ when the C_8H_2 moves into the CNT [Fig. 6.4(a)]. The relative energy profile was plotted as a function of separation d during the insertion process with infinite separation of the CNT and the C_8H_2 as a reference [Fig. 6.4(b)]. The insertion process of carbon chain can be described as follows:

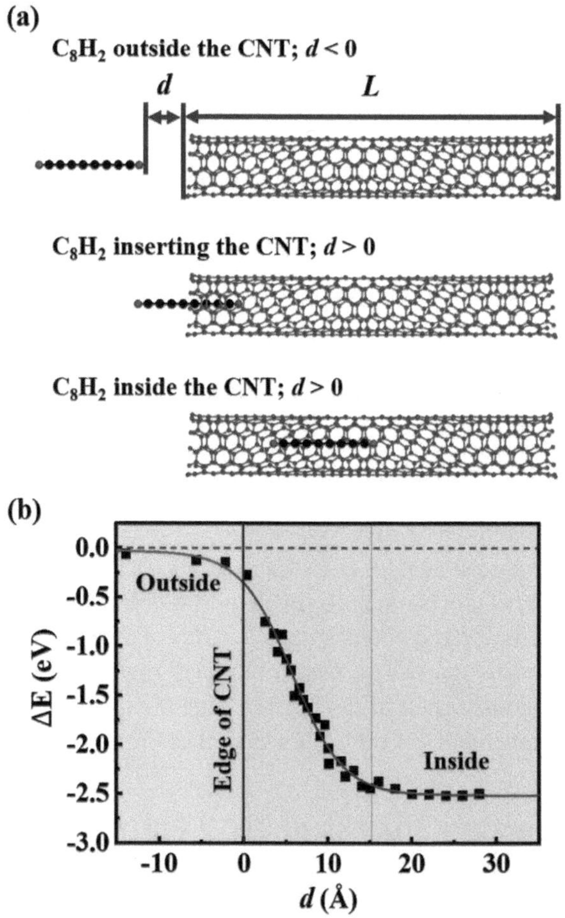

Figure 6.4 (a) Snapshots of a C_8H_2 insertion into the (6, 5) CNT. (b) Relative energy profile during the insertion process with an infinite separation as a reference. The total energy of C_8H_2 inside the CNT is 2.50 eV lower than that outside the CNT.

- When the C_8H_2 was located far from the CNT (d = –13.8 Å), there is little interaction between C_8H_2 and CNT, indicated by the small relative energies close to the reference energy.
- When the C_8H_2 approached to the CNT edge (d = 0), the relative energy of the system began to drop due to the attractive van der Waals (vdW) interactions between C_8H_2 and CNT.
- As the chain was moving into the CNT, the relative energy of the system decreased linearly because of the larger vdW interactions associated with the more inserted carbon atoms. The distance of vdW interactions was ~3.8 Å between the C_8H_2 and the CNT wall, same as the empty CNT before the insertion.
- When the C_8H_2 moved into the CNT completely ($d \approx 15$ Å), the system reached its lowest energy (ΔE = –2.50 eV), meaning that the vdW interaction energy is ~0.3 eV/C atom between the C chain and the CNT.

The energy profile during the insertion process indicates that carbon chains can be inserted into the CNTs with little energy barrier if the diameters of the CNT favor the attractive vdW interactions.

6.4 Fusion Process of CNW Formation

Since the insertion of a carbon chain into CNTs is energetically barrierless, multiple carbon chains may be absorbed into one CNT. We next illustrate the fusion process of two carbon chains inside (6, 5) CNT by constrained MD. Various constraint conditions were considered to examine the fusion process:

- Fixing the distance between two end C atoms of the C_8H_2 chains ($d_{C\text{-}C}$) [Fig. 6.5(a)]
- Fixing the distance between two end-capped H atoms of the C_8H_2 chains ($d_{H\text{-}H}$) [Fig. 6.5(c)]
- Fixing the distance between the end C atom of the C_8H chain and the end-capped H atom of the C_8H_2 chain ($d_{C\text{-}H}$) [Fig. 6.6(a)]
- Fixing the distance between two end C atoms of the C_8H chains ($d_{C\text{-}C}$) [Fig. 6.6(c)]

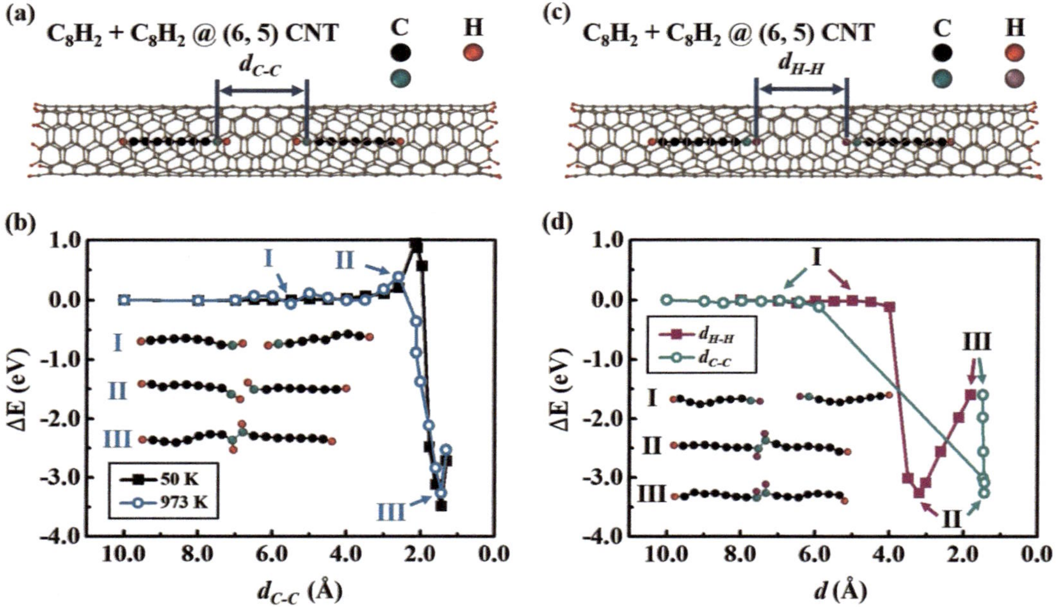

Figure 6.5 (a) Schematic model of the fusion process of two C_8H_2 chains inside (6, 5) CNT. The distance ($d_{C\text{-}C}$) between two end carbon atoms of the C_8H_2 chains was constrained. (b) Relative energies as functions of $d_{C\text{-}C}$ at 50 K and 973 K. (c) The distance ($d_{H\text{-}H}$) between two end-capped H atoms of the C_8H_2 chains was constrained. (d) Relative energies as functions of $d_{H\text{-}H}$ at 973 K. The insets in (b) and (d) are typical configurations during the fusion process, and the CNTs were hidden for clear visual inspection.

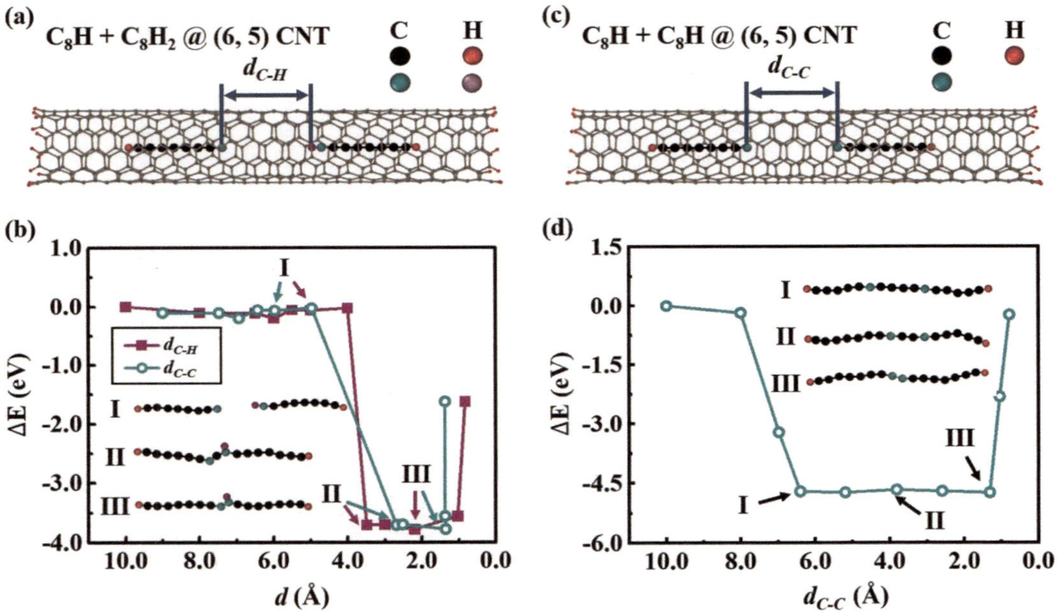

Figure 6.6 (a) Schematic model of the fusion process of C_8H chain and C_8H_2 chain inside (6, 5) CNT. The distance (d_{C-H}) was constrained between the end C atom (in green) of the C_8H chain and the end-capped H atom (in purple) of the C_8H_2 chain. (b) Relative energies as functions of d_{C-H} and d_{C-C} at 973 K. (c) Schematic model of the fusion process of two C_8H chains inside (6, 5) CNT. The distance (d_{C-C}) was constrained between the two end C atoms (in green) of the C_8H chains. (d) Relative energies as functions of d_{C-C} at 973 K. The insets in (b) and (d) are typical configurations during the fusion process and the CNTs were hidden for clear visual inspection.

By fixing the distance ($d_{C\text{-}C}$) between two end C atoms of the C_8H_2 chains [Fig. 6.5(a)], the constrained MD at 973 K were performed. The relative energies were plotted as functions of $d_{C\text{-}C}$ [Fig. 6.5(b)]. When $d_{C\text{-}C}$ is between 10.0–3.5 Å, the energy remains unchanged, indicating that there is little interaction between two C_8H_2 chains far each other [stage I in Fig. 6.5(b)]. As $d_{C\text{-}C}$ decreased further, the repulsive interactions between the two end-capped H atoms of the chains led to the energy increase of the system. The maximum energy barrier was 0.39 eV when $d_{C\text{-}C}$ = 2.6 Å [stage II in Fig. 6.5(b)]. As the $d_{C\text{-}C}$ decreased further, the energy suddenly dropped mainly due to the attraction between two end C atoms of the C_8H_2 chains. The two constrained end C atoms overcame the energy barrier and formed C–C covalent bond, leading to the coalescence of two C_8H_2 chains into a $C_{16}H_4$ chain inside (6, 5) CNT. The minimum energy of the coalesced $C_{16}H_4$@(6, 5) CNT was –3.26 eV when $d_{C\text{-}C}$ = 1.4 Å [stage III in Fig. 6.5(b)], indicating that the fusion energy from C–C bond formation is approximately 10 times larger than the insertion vdW interaction ~0.3 eV/C atom. The two end-capped H atoms still kept their bindings with the end C atoms at the middle of the chain. The system energy increased again if $d_{C\text{-}C}$ decreased further mainly due to the repulsive interaction between the too close end-capped carbon atoms. The fusion process at a low temperature of 50 K was also simulated to examine the temperature effect. The trend of the energy changes at 50 K is similar to that at 973 K except that the energy barrier at 50 K is 0.57 eV higher when $d_{C\text{-}C}$ = 2.2 Å. The higher temperature would facilitate the fusion process starting at farther distance with lower energy barrier. At high temperature the large kinetic energy prompts the rotation of C–H bonds that favors the C–C bond formation in the fusion reactions. The energy barriers in Fig. 6.5(b) were mainly caused by the repulsive interactions between H atoms.

The constrained MD by fixing the distance between two end-capped H atoms of the C_8H_2 chains ($d_{H\text{-}H}$) [Fig. 6.5(c)] was compared to examine the influences of the different reaction coordinates on the minimum energy paths. The relative energies were plotted as functions of the constrained $d_{H\text{-}H}$ and corresponding non-fixed $d_{C\text{-}C}$ in Fig. 6.5(d). Different from the case of C–C constraint, however, there was little fusion energy barriers under H–H constraints.

This comparison implied that the small energy barriers under the C–C constraint might come from the hindrance of H–H repulsion during the C–C fusion. When $d_{H\text{-}H}$ is large (i.e., $d_{H\text{-}H}$ = 10.0–4.0 Å), the energies remained unchanged due to little interactions between the two far separated chains and both the end C–H bonds are parallel to the carbon chains [stage I in Fig. 6.5(d)]. As $d_{H\text{-}H}$ decreased to 3.2 Å, the H–H repulsion and thermal vibration rotated the two C–H bonds to the opposite directions to minimize H–H repulsion. The C–H bond rotation facilitated the C–C fusion and the non-fixed $d_{C\text{-}C}$ decreased suddenly from 5.9 Å to 1.4 Å, reaching the optimal equilibrium C–C bond length of sp^2 hybridization [stage II in Fig. 6.5(d)]. If $d_{H\text{-}H}$ decreased further, the energy increased due to the repulsive interaction between the closed H atoms while $d_{C\text{-}C}$ remained the same since the completion of the C–C bond formation [stage III in Fig. 6.5(d)].

In the fusion simulations we discussed above, the ends of carbon chains were saturated using H atoms that is a reasonable model in low temperature experiments. CNWs also can be synthesized in vacuum and inert gas at high temperatures [37, 60, 61] where carbon chains may also exist without end caps. We next address constrained MD simulations along the other reaction coordinates to examine the fusion reactions without the end-capped H atoms.

As shown in Fig. 6.6(a), the distance ($d_{C\text{-}H}$) was constrained between the end C atom of the C_8H chain where one of end C atoms was not saturated by H and the end-capped H atom of the C_8H_2 chain. The energy remained unchanged when $d_{C\text{-}H}$ = 10.0–4.0 Å [stage I in Fig. 6.6(b)]. The C–C fusion occurred when the energy dropped suddenly to −3.71 eV as the constrained $d_{C\text{-}H}$ decreased to 3.5 Å. Meanwhile, the distance ($d_{C\text{-}C}$) between two end C atoms of the chains decreased from 5.0 Å to 2.7 Å, ~two times length of a common single C–C bond. In fact, an unconstrained C atom jumped over the constrained end-capped C atom and inserted between the two end C atoms as a bridge, leading to the formation of $C_{16}H_3$ chain inside (6, 5) CNT [stage II in Fig. 6.6(b)]. As $d_{C\text{-}H}$ decreased further to the equilibrium distance 1.4 Å, the two end C atoms of two chains form a bond with the minimum energy of −3.78 eV [stage III in Fig. 6.6(b)]. The fusion energy of the unsaturated $C_{16}H_3$ chain was higher than −3.26 eV of saturated $C_{16}H_4$ chain [Fig. 6.5(b)] due to the absence of H–H repulsion at the joint position.

In the case of the fusion of C chains without end-capped H atoms where only C–C contact exists, two C_8H chains were inserted and the distance ($d_{C\text{-}C}$) of two unsaturated end C atoms was constrained as shown in Fig. 6.6(c). As $d_{C\text{-}C}$ decreased to 6.4 Å, thermal vibrations caused more unconstrained C atoms in the middle of the C_8H chains jumping into the fixed gap between the two chains, forming a $C_{16}H_2$ chain with a fusion energy of 4.7 eV [stage I in Fig. 6.6(d)]. Further decrease of the constraint $d_{C\text{-}C}$ could eventually bring the two end C atoms together to form a bond while the fusion energies had little changes [stages II and III in Fig. 6.6(d)]. The C chains with unsaturated end C atoms fused earlier at farther distances with much larger energy drop compared with the C chains with the H saturated ends.

6.5 Configurations of Carbon Chains with Hydrogen Adatoms

Since carbon chains can be inserted into the CNTs and subsequently fused together, the configurations of fused carbon chains inside CNT need to be clarified upon the understanding of the insertion-and-fusion process. In this section, we compared several typical configurations of carbon chains inside a finite (6, 5) CNT [Fig. 6.7(a)]. The isolated carbon chains without CNTs were also calculated for comparison [Fig. 6.7(b)]. In general, the fused carbon chain is thermodynamically more stable than that of two separated C_8H_2 chains inside CNT ($2C_8H_2$@CNT) [Fig. 6.7(a)]. Specifically, the energy of a $C_{16}H_2$ chain inside CNT plus an isolated H_2 ($C_{16}H_2$@CNT + H_2) is 0.67 eV lower than that of $2C_8H_2$@CNT. The energies of $C_{16}H_4$ chains inside the CNT are even lower, despite that the two H adatoms sit at the middle of the chain ($C_{16}H_4$-m) or transfer to the ends of the chain ($C_{16}H_4$-e). Moreover, $C_{16}H_4$-e@CNT is the most thermodynamically stable configuration with the relative energy of −4.34 eV. This indicated that the fusion reactions of carbon chains inside CNT did not necessarily break C–H bonds and produced H_2 molecule as most previous theory and experiments assumed. Instead, the short C chains preferred to form $C_{16}H_4$ chains without C–H bond breaking. The results of the isolated carbon chains had the similar trends [Fig. 6.7(b)] except that the carbon chains would bend without the confinement of the CNT. The atomistic simulations discussed above confirmed that CNTs can provide the radial confinement to prevent polyyne molecules from crosslinking reactions and offer alignment effect along the narrow axial channel, increasing the probability of end-to-end fusion reactions of carbon chains.

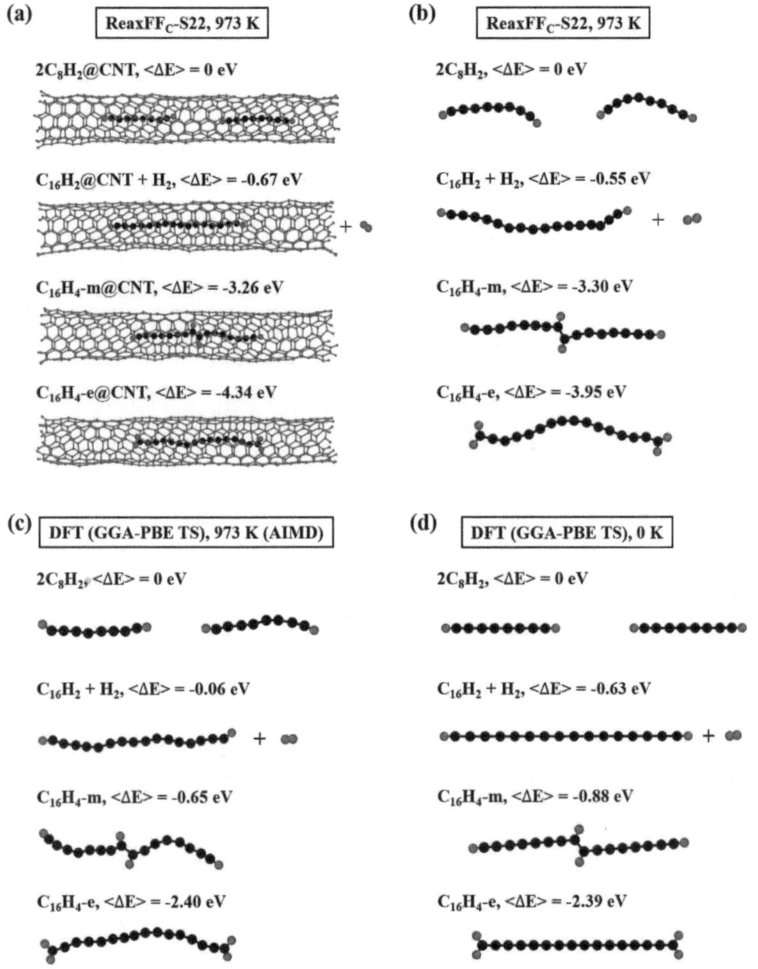

Figure 6.7 ReaxFF MD simulations of various configurations of (a) carbon chains inside a finite (6, 5) CNT and (b) isolated carbon chains at 973 K. (c) AIMD simulations and (d) DFT calculations of various typical configurations of isolated carbon chains at 973 K and 0 K, respectively.

It has been commonly believed that the fusion of C chains involves the C–H bond breaking that produces an isolated H atom. The two isolated H atoms then form a H_2 molecule that moves out of CNT eventually. However, as we discussed in Fig. 6.7(a), the C–H bonds did not necessarily break and H atoms remained on the C chains as H adatoms. As shown in Fig. 6.7(b), the separated $C_{16}H_2$ and H_2 system had lower energies in spite of the existence of CNTs compared with the two separated C_8H_2 chains. However, if the C–H bonds did not break, the $C_{16}H_4$ chains had even lower energies. Moreover, the $C_{16}H_4$ configuration where H adatoms are all located at the chain ends (dubbed $C_{16}H_4$-e) was much more stable than the configuration having H adatoms in the middle (dubbed $C_{16}H_4$-m). The *ab initio* molecular dynamics (AIMD) simulations by DFT at 973 K [Fig. 6.7(c)] and DFT calculations at 0 K [Fig. 6.7(d)] both supported and validated the trends predicted by ReaxFF MD in Fig. 6.7(a) and (b), confirming that the order of stability remained the same despite quantitative difference due to temperature effects. The residual H adatoms were also found in the C_2H_2 addition process during the formation of graphene, the so-called "Add and Dehydrogenate" (ADh) mechanism [62].

According to the ReaxFF MD simulations and experimental observation, the long carbon chains would not fuse with the edge or the wall of carbon nanotubes based on the discussions as follows:

(1) Chain-edge fusion: The insertion of chain into the tube core is energetically barrierless as shown in the energy profile in Fig. 6.4. Also, the chain insertion was driven by the significant energy drop due to the attractive vdW interaction between chain and tube. On the other hand, the chain is hard to fuse with the edge because chain end would have H–H repulsion with the saturated edges of CNT. In CNW the energy barrier would be overcome and collision probability would increase due to the confinement effects inside CNTs plus the activation energy provided by the high temperature heat treatment. Without the geometry confinement of CNT, the collision probability between chain and edge would become extremely low, so the chain-edge fusion could hardly occur from both thermodynamic and kinetic points of view.

(2) Chain-wall fusion: The fusion between chain end and tube wall is energetically unfavorable mainly because the chain would deviate from the optimal parallel alignment along the tube axis that maximizes the attractive vdW interaction. The chain-wall fusion would make the end C atoms of chain changed from *sp* to sp^2 hybridization with downhill energy and the C atom of wall changed from sp^2 to sp^3 hybridization with uphill energy. The fusion-induced energy changes during C_{chain}-C_{wall} bond formation are smaller than the dominated energy penalty of misalignment of long carbon chains off the tube axis, further confirmed by the calculations below.

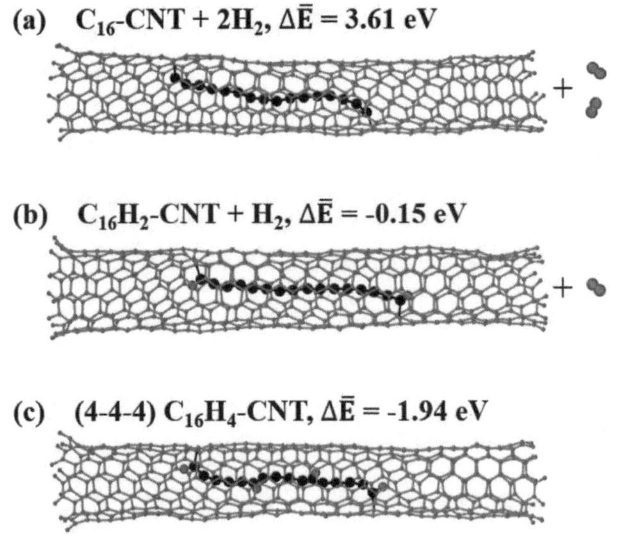

Figure 6.8 ReaxFF MD simulations of various configurations of carbon chains fused with the inner wall of a finite (6, 5) CNT. (a) C_{16}-CNT + $2H_2$, (b) $C_{16}H_2$-CNT + H_2, and (c) (4-4-4) $C_{16}H_4$-CNT.

To examine the possibility of the fusion reactions between carbon chains and CNT tube, we discuss several MD simulations on $C_{16}H_2$ fusion with the (6, 5) CNT wall as shown in Fig. 6.8. The chain-wall fusion configuration (C_{16}-CNT + $2H_2$) with two generated H_2 molecules was unfavorable with very high positive energies, confirming again that the H_2 generation is unlikely [Fig. 6.8(a)]. The chain-wall fusion configuration with one H_2 molecule

[Fig. 6.8(b)] had higher energy than the encapsulated chain configuration ($C_{16}H_2$@CNT + H_2) [Fig. 6.7(a)]. The chain-wall fusion configuration without H_2 generation [Fig. 6.8(c)] was energetically less favorable than the capsulated chain configurations [Fig. 6.7(a)]. It is possible that the very short carbon chain fuse with the CNT wall. Liu *et al.* [28] reported that small carbon species with dangling bonds including carbon chains and graphene sheets could react with the wall atoms of CNT.

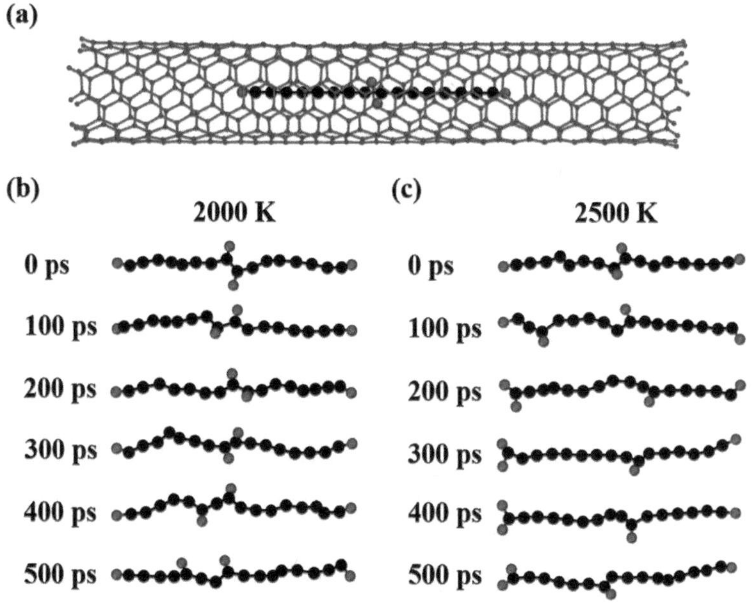

Figure 6.9 (a) Schematic model of a $C_{16}H_4$-m chain inside a finite (6, 5) CNT. The typical configurations during the NVT-MD simulations at (b) 2000 K and (c) 2500 K, respectively. The CNTs were hidden in (b) and (c) for clear visual inspection.

Next, we discuss the effect of higher temperatures on $C_{16}H_4$-m@(6, 5) CNT with the H adatoms located at the joint positions of the two short chains [Fig. 6.9(a)] and examine the stability of H adatoms. The snapshots of the MD trajectories at 2000 K and 2500 K were shown in Fig. 6.9(b) and (c), respectively, where the CNTs were hidden for clear visual inspection. At 2000 K [Fig. 6.9(b)], the H adatoms vibrated at the joint positions at the first 300 ps. Between 300–400 ps, one of the H adatoms transferred to form a bond with the other C atom. Subsequently, the H adatom transferred to another C atom again between 400–500 ps. The H transfer along the chain was driven by the thermal vibration of the system. In the MD simulations at a higher temperature of 2500 K [Fig. 6.9(c)], the C–H bonds broke and formed more frequently and the H adatoms migrated more easily along the fused carbon chain, leading to one H atom moving from the center to the end. The H adatom transferred to and remained at the end of the C chain, while the other H adatom continued to move along the C chain. Statistically speaking, the other H atom may eventually migrate to the end of chain. These high temperature MD simulations demonstrated the thermodynamic limit shown in Fig. 6.7(a). However, only the MD simulations at very high temperatures and very long time could lead to the complete migration of H atoms going from the center to the ends, which may not be easily realized at the experimental conditions. Therefore, we believe the H migrations are most likely incomplete and the H atoms are distributed separately along the carbon chains at the experimental conditions. The thermodynamically stable configuration with two H at each end of carbon chain may be obtained by other experimental approaches with higher energy input in future.

The diffusion dynamics predicted by ReaxFF MD were further validated by DFT calculations at 0 K on a series of H adatom positions on the chains as shown in Fig. 6.10. The H adatom effects on the electronic and magnetic properties of the C chains were also calculated using DFT as shown in Fig. 6.11 and the energy levels of the carbon chains with the typical H adatom configurations are shown in Fig. 6.12. The binding of the H adatoms with C atoms after C–C bond formation led to local sp^2 hybridization at the joint position. Such local hybridization changes from sp to sp^2 would have a long-range domino effect, *e.g.*, inducing the alternative single and triple C–C bonds between the neighboring C atoms that propagate along the C chain. To describe the various configurations with different H atom origins, the H adatoms are denoted as aH type while the terminal H atom as tH type. According to the H origin types and their positions, we can classify various configurations into several categories as follows.

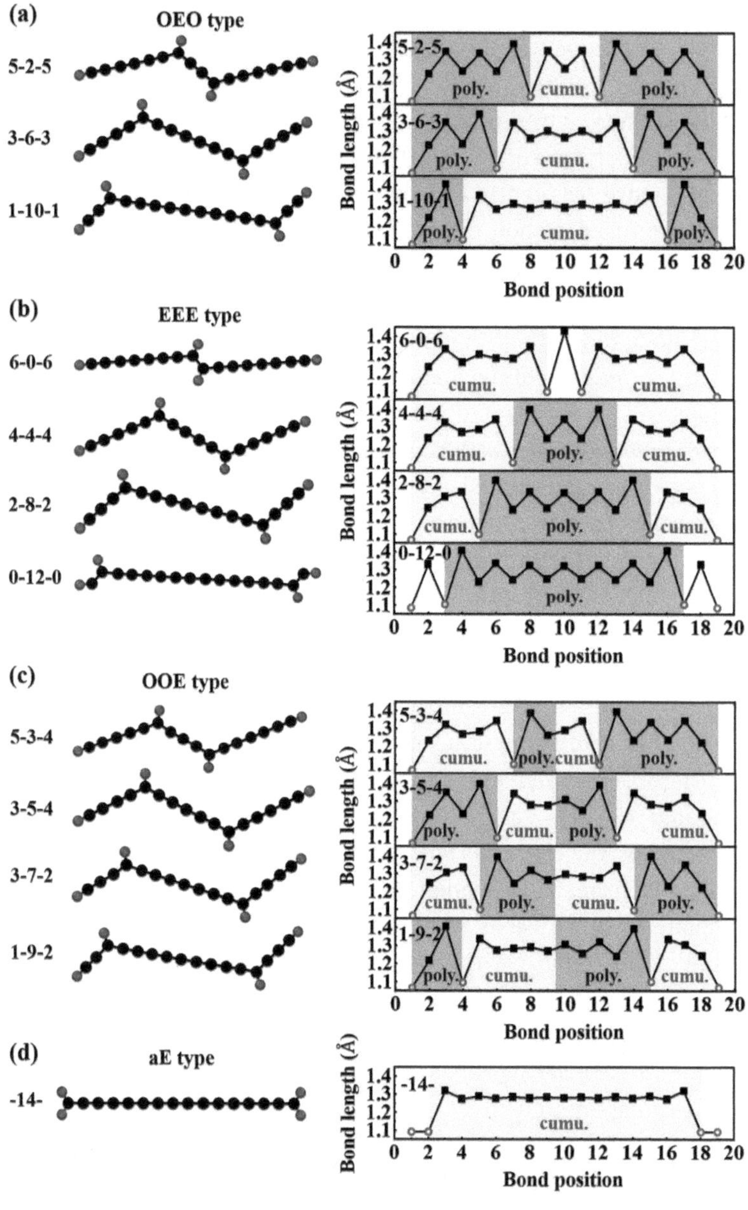

Figure 6.10 Typical configurations of C chains with H adatoms: (a) OEO type, (b) EEE type, (c) OOE type, and (d) aE type of $C_{16}H_4$ chains and corresponding bond length distributions optimized by DFT. The solid squares represent the C–C bond length while the open circles represent the C–H bond length.

(1) Odd-Even-Odd (OEO) type: The Odd-Even-Odd (OEO) type of configurations is defined if the number of C atoms (N_C) between C-tH and C-aH is odd leading to the polyyne type with alternative single and triple C–C bonds, and N_C between the two middle C-aH atoms is even leading to the cumulene type with all double C–C bonds, *e.g.*, 5-2-5, 3-6-3, 1-10-1 where the numbers represent the N_C of different segments excluding the C atoms binding to the tH or aH [Fig. 6.10(a)]. The OEO types of configurations are energetically stable without magnetic moments. The OEO types have the cumulene middle segments with all double C–C bonds responsible for electron conduction. If aH atoms transferred gradually to the ends of C chains, N_C of the cumulene segments increased while those of the polyyne segments decreased [Fig. 6.10(a)]. These changes of bond length alternation (BLA) reduced the HOMO-LUMO gaps [solid circles in Fig. 6.11(a)] that would increase their electron conductivity. On the other hand, the intensity of Raman spectrum may decrease because the Raman spectra are contributed by the segments of polyyne type between the C-aH and C-tH atoms. These OEO types are probably the low energy configurations observed in the experiments.

(2) Even-Even-Even (EEE) type: The Even-Even-Even (EEE) type of configuration is defined if N_C between C-tH and C-aH is even leading to the cumulene type with all double C–C bonds, and N_C between the two middle C-aH atoms is even leading to the polyyne type with alternative single and triple C–C bonds, *e.g.*, 6-0-6, 4-4-4, 2-8-2, 0-12-0 [Fig. 6.10(b)]. The EEE types of configurations are energetically less stable and exhibit magnetic moments. If the aH atoms transfer to the ends, the cumulene segments became shorter while the polyyne segments became longer [Fig. 6.10(b)]. These changes of BLA increased HOMO-LUMO gaps [up and down triangles in Fig. 6.11(a)] that may reduce electron conductivities. The intensity of Raman spectra may be enhanced because Raman spectra are determined by the segments of polyyne type between the two C-aH atoms. The non-magnetic OEO type and magnetic EEE type could be switched by the H transfer *via* temperature control experimentally.

(3) Odd-Odd-Even (OOE) or Even-Odd-Odd (EOO) type: The Odd-Odd-Even (OOE) type of configuration is defined if N_C between C-tH and C-aH is odd on one side while N_C between C-tH and C-aH is even on the other side, and N_C between the two middle C-aH atoms is odd [Fig. 6.10(c)]. In the OOE types of configurations the C–C bonding exhibit alternative cumulene and polyyne types, *e.g.*, 5-3-4,

3-5-4, 3-7-2, 1-9-2 [Fig. 6.10(c)]. The OOE types of configurations are energetically less stable and also exhibit magnetic moments [Fig. 6.11(a) and (b)]. If the aH atoms transfer to the ends, the cumulene and the polyyne segments switched each other that kept HOMO-LUMO gaps having little change [Fig. 6.10(c)]. Accordingly, the electron conductivity and Raman intensity may not change. These alternative cumulene and polyyne segments may play a role of P-N junctions in nanoelectronics and thus be used to control direction of electric current.

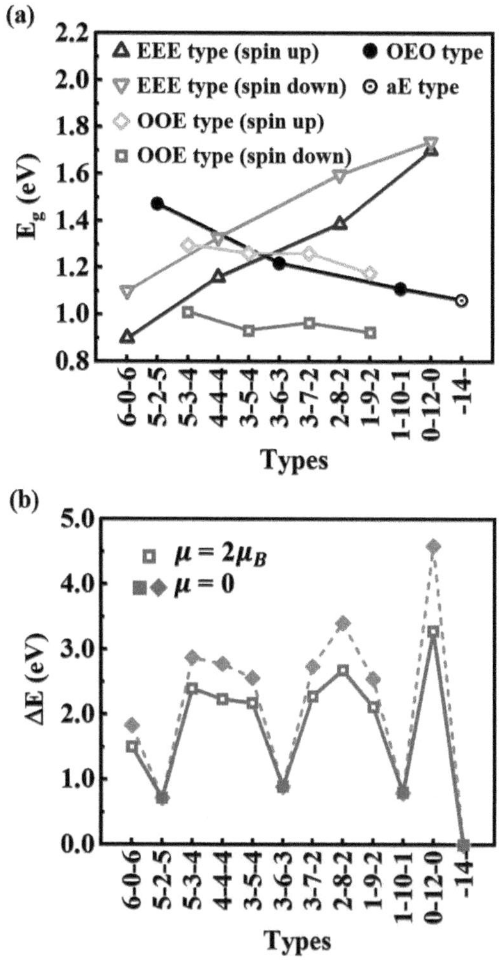

Figure 6.11 (a) HOMO-LUMO gaps and (b) The relative DFT energies of typical types of $C_{16}H_4$ chains including OEO, EEE, OOE, and aE types.

(4) All-Even (aE) type: The All-Even (aE) type of configuration is defined if the tH and aH atoms are bound to the same end C atoms simultaneously and Nc between the C-aH (or C-tH) atoms are even, *e.g.*, dubbed -14- [Fig. 6.10(d)]. The aE type of configurations has the cumulene bonding type with all double C–C bonds that are Raman inactive modes. The aE type of chains is energetically most stable but the H adatom transfer to the end C atoms needs to overcome high energy barriers between 1.28–2.48 eV [Fig. 6.11(b)]. The mild experimental conditions most likely lead to metastable H adatoms sitting in the middle of the carbon chains.

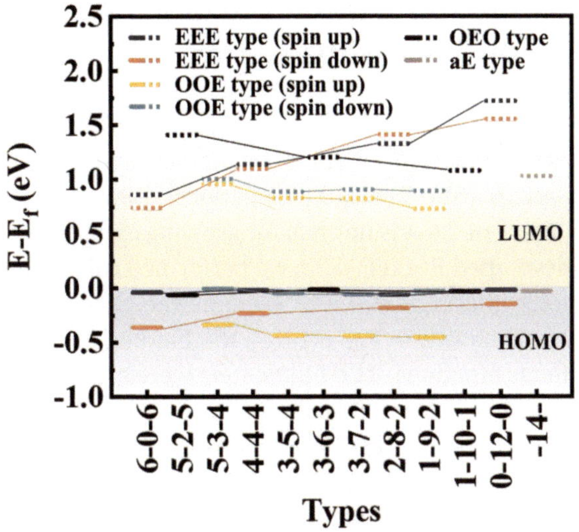

Figure 6.12 Energy level diagrams of the $C_{16}H_4$ chains with the typical types of H adatom configurations, including OEO, EEE, OOE, and aE types.

According to analyses above, every H adatom transfer along the C chain could cause the changes of configurations as well as HOMO-LUMO gaps and magnetic moments. The experimental confirmation on the existence of H on the carbon chain was challenging using conventional electron microscopy or C–H vibration spectrum method. However, the H-induced changes of electronic and magnetic properties may be possible to be measured and verified experimentally. On the other hand, one may induce H transfer *via* temperature control or other methods that in turn tunes the electronic and magnetic properties. Therefore, it is possible to use hydrogen adatoms to tune the properties of carbon-based nanoelectronics devices that can be called carbon-hydrogen (C–H) nanoelectronics. The future sizes of electronic circuits would reduce to below one nm that makes the precise doping control extremely hard if not impossible. The designed H adatoms in C–H nanoelectronics provide a new strategy to tune the electronic properties of nanomaterials without the needs of doping. The hydrogen adatom tuning strategy may be applied to other nanocarbon materials and their associated electronic or magnetic devices.

6.6 Diameter Effects on the Binding Energies of Carbon Nanowires

Chang *et al.* [20] reported the binding energies of $C_{16}H_2$ chains inside finite CNTs with various diameters and chirality using $ReaxFF_{CHO}$-2016 [50]. The MD simulations show that (6, 5) CNT with a diameter of 7.6 Å is optimal for growing carbon chains. Using in-house developed $ReaxFF_C$-S22, we performed calculations on the CNT diameter effects at 300 K to examine the effects of force field parameters. As shown in Fig. 6.13(a), the binding energy curves of $C_{16}H_2$ chains as functions of CNT diameters were qualitatively similar between $ReaxFF_C$-S22 and $ReaxFF_{CHO}$-2016 and both predicted the minimum energy at the CNT diameter of 7.6 Å. The quantitative differences include that $ReaxFF_C$-S22 predicted the larger depth of energy well −4.51 eV and harder repulsion at short distance. The MD results confirmed again that it is thermodynamically stable for carbon chains to be inserted into the CNTs whose diameters are favorable for the vdW interactions despite the chirality and metallic/semiconducting properties, consistent with experimental results [20].

The optimal CNT for insertion predicted in this work was (6, 5) CNT with a diameter of 7.6 Å, larger than the diameter 7.1 Å predicted by Shi *et al.* using DFT [13]. The DFT prediction at 0 K did not consider temperature effects and the distortion of flexible C chains. The H adatoms would induce the local kink distortion of straight chains due to the sp^2 hybridizations. The thermal energy at finite temperatures in MD simulations caused $C_{16}H_2$ chains to vibrate and distort to some extent. There should be a "free" moving

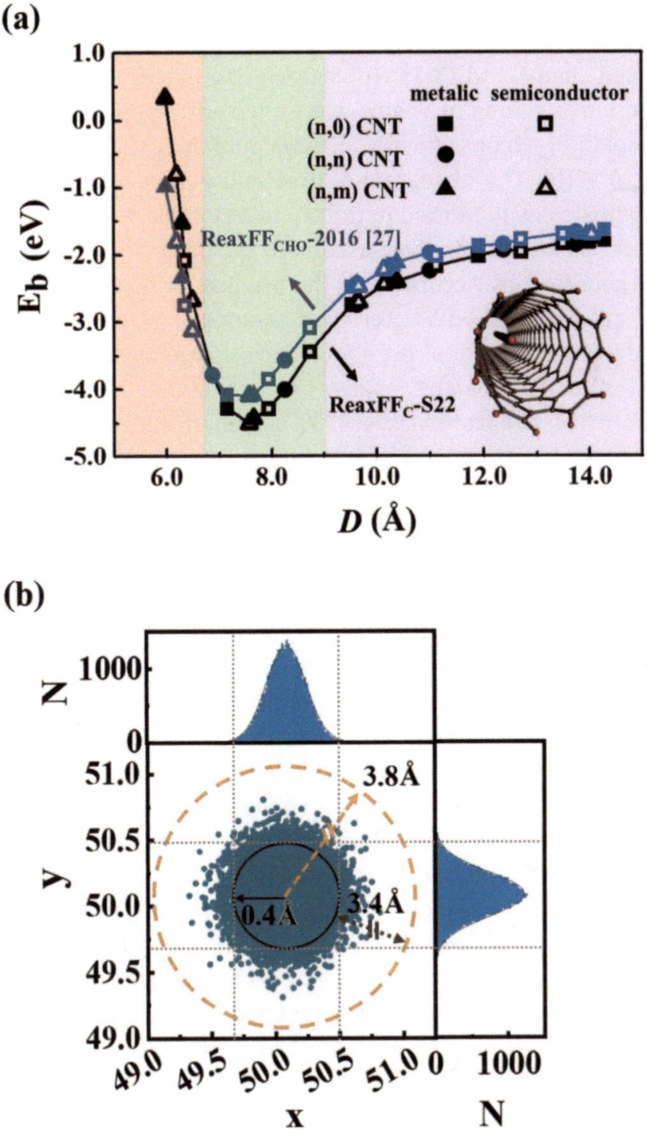

Figure 6.13 (a) Binding energies of $C_{16}H_2$ chain inside finite CNTs with various diameters and chirality calculated by ReaxFF$_{CHO}$-2016 and ReaxFF$_C$-S22, respectively. The inset is the schematic model of a $C_{16}H_2$ chain inside a finite (6, 5) CNT. (b) The projected positions of all C atoms of $C_{16}H_2$ chain at every MD step and the corresponding histogram of probability distribution of x- and y-coordinates.

zone for the easier vibration and distortion of flexible chains in the CNT cores satisfying overall optimal vdW interactions between the distorted chains and CNTs. To characterize quantitatively the size of free moving zone of chains, we projected the positions of all C atoms of $C_{16}H_2$ chains at every MD step onto the x-y plane as plotted in Fig. 6.13(b). The histogram of probability distribution of x- and y-coordinates were plotted in the top and right panels, respectively. The Gaussian functions were used to fit the probability distribution of the coordinates. According to the predicted optimal CNT radius 3.8 Å and the ideal vdW interlayer distance of graphite 3.4 Å, the core of free moving zone core should have an average optimal radius of 0.4 Å. Our MD trajectory analyses show that the free moving zone with R = 0.4 Å encompasses 97% maximum height of Gaussian distribution functions and 98% occurrence probability of C chain atoms [Fig. 6.13(b)]. These mean that the majority of C chain atoms stay within the free moving zone except that a very few atoms move into the repulsive vdW region *via* thermal energy.

The MD simulations show that the CNTs with the radii slightly larger than the ideal vdW interaction distance still can protect and prevent the carbon chains from crosslinking reactions. The free moving zone allows the carbon chains to have enough room for vibration and distortion but not to rearrange themselves to form sp^2 configurations. If the tube diameter is too large, the CNT lose its confinement effect and carbon chains tend to form sp^2 carbon materials *via* crosslinking reactions. On the other hand, if the tube diameter is too small, *e.g.*, the smallest 3.0 Å diameter carbon nanotube [63], the carbon chains are hard to be inserted into the tube due to repulsion. Both ReaxFF_{C}-S22 and $\text{ReaxFF}_{\text{CHO}}$-2016 predicted the similar optimal diameters of CNW, confirming further the main conclusion that the diameter effects on the binding energies of carbon nanowires is less sensitive to the force field parameters. More importantly, we reported two new results: (1) More accurate ReaxFF_{C}-S22 predicted the deeper energy well and harder repulsion at short distance of CNWs compared with $\text{ReaxFF}_{\text{CHO}}$-2016; (2) We proposed the existence of "free" moving zone of carbon chains at the center of SWCNT while most previous studies assumed the single ideal CNT radius with optimal vdW interaction. The CNTs with the radii slightly larger than the ideal vdW interaction distance suggested that the choices of CNT diameters can be broader in the insertion-and-fusion experimental preparation, providing a guidance for the CNT precursor selection in the experimental synthesis of CNWs. Indeed the SWCNWs prepared in the experiments are observed to have a range of diameters between 7.0–9.5 Å [20].

6.7 Summaries and Perspectives

In this chapter, we illustrated the insertion and fusion processes of carbon chains inside CNTs by ReaxFF MD and DFT calculations, providing theoretical evidence of insertion of carbon chains into CNTs to form carbon nanowires, consistent with recent experimental synthesis. The MD simulations show that the thermodynamically optimal CNTs have a free moving zone to encapsulate the distorted carbon chains that leads to a range of candidate CNT diameters for the carbon chain insertion. The constrained MD simulations show that both the insertion and fusion processes are energetically barrierless, indicating that the formation of carbon nanowire is kinetically a spontaneous process. Other carbon chain formation processes are also possible, *e.g.* stretching CNTs to an atomic thin carbon thread or direct growth of carbon chain from small carbon precursors inside CNTs, remaining for further study in future.

The analyses of configurations of fused carbon chains show that the end-capped H of short C chains remained as residual adatoms on the fused C chains rather than the C–H bond breaking and formation of H_2. The H adatoms could transfer along the C chains *via* thermal activation or external electrical field. Depending on the varied positions of H adatoms on the C chains, there are several distribution types of bond length alternation, leading to the different HOMO-LUMO energy gaps and magnetic moments. Therefore, the electronic and magnetic properties of carbon chains can be tuned accordingly upon the H adatom transfer. The H adatoms are predicted to have rich configurations, bonding characteristics, and influences on electronic and magnetic properties different from the terminal H of carbon nanomaterials for the saturation of the dangling bonds, opening up the door for the design of novel carbon-hydrogen nanoelectronics in future.

References

1. Sorokin, P. B., Lee, H., Antipina, L. Y., Singh, A. K., and Yakobson, B. I. (2011). Calcium-decorated carbyne networks as hydrogen storage media, *Nano Letters,* **11**, 2660–2665.
2. Liu, M., Artyukhov, V. I., Lee, H., Xu, F., and Yakobson, B. I. (2013). Carbyne from first principles: chain of C atoms, a nanorod or a nanorope, *ACS Nano,* **7**, 10075–10082.
3. Chalifoux, W. A., and Tykwinski, R. R. (2010). Synthesis of polyynes to model the *sp*-carbon allotrope carbyne, *Nature Chemistry,* **2**, 967–971.
4. Januszewski, J. A., and Tykwinski, R. R. (2014). Synthesis and properties of long [*n*]cumulenes ($n \geq 5$), *Chemical Society Reviews,* **43**, 3184–3203.
5. Zhao, J., Zhang, Y., Fang, Y., Fan, Z., Ma, G., Liu, Y., and Zhao, X. (2017). Synthesis of polyynes by intense femtosecond laser irradiation of SWCNTs suspended in methanol, *Chemical Physics Letters,* **682**, 96–100.
6. Wu, Y., Zhang, Y., Zhu, T., Li, H., Liu, Y., and Zhao, X. (2019). Effects of precursor molecules on polyyne formation by arc discharge between two copper electrodes, *Chemical Physics Letters,* **730**, 64–69.
7. Fang, L., Zhu, T., Chang, W., Liu, Y., and Zhao, X. (2021). Purification of polyynes via carbides, *Carbon,* **179**, 28–32.
8. Baughman, R. H., Eckhardt, H., and Kertesz, M. (1987). Structure-property predictions for new planar forms of carbon: layered phases containing sp^2 and *sp* atoms, *The Journal of Chemical Physics,* **87**, 6687–6699.
9. Chi, B., Liu, Y., Li, X., Xu, J., Qin, X., Sun, C., Bai, C., and Zhao, X. (2015). Energetic stability, atomic and electronic structures of extended γ-graphyne: a density functional study, *Journal of Molecular Modeling,* **21**, 154.
10. Sun, C., Liu, Y., Xu, J., Chi, B., Bai, C., Liu, Y., Li, S., Zhao, X., and Li, X. (2015). Density functional study of α-graphyne derivatives: energetic stability, atomic and electronic structure, *Physica E: Low-dimensional Systems and Nanostructures,* **70**, 190–197.
11. Liu, F., Tang, X., Du, W., Chi, B., Zhao, X., and Liu, Y. (2022). Radial and axial vibration modes of graphyne nanotubes, *Materials Today Communications,* **31**, 103610.
12. Zhao, X., Ando, Y., Liu, Y., Jinno, M., and Suzuki, T. (2003). Carbon nanowire made of a long linear carbon chain inserted inside a multiwalled carbon nanotube, *Physical Review Letters,* **90**, 187401.

13. Shi, L., Rohringer, P., Suenaga, K., Niimi, Y., Kotakoski, J., Meyer, J. C., Peterlik, H., Wanko, M., Cahangirov, S., Rubio, A., Lapin, Z. J., Novotny, L., Ayala, P., and Pichler, T. (2016). Confined linear carbon chains as a route to bulk carbyne, *Nature Materials,* **15**, 634–639.

14. Endo, M., Kim, Y. A., Hayashi, T., Muramatsu, H., Terrones, M., Saito, R., Villalpando-Paez, F., Chou, S. G., and Dresselhaus, M. S. (2006). Nanotube coalescence-inducing mode: a novel vibrational mode in carbon systems, *Small,* **2**, 1031–1036.

15. Cazzanelli, E., Castriota, M., Caputi, L. S., Cupolillo, A., Giallombardo, C., and Papagno, L. (2007). High-temperature evolution of linear carbon chains inside multiwalled nanotubes, *Physical Review B,* **75**, 121405.

16. Shi, L., Sheng, L., Yu, L., An, K., Ando, Y., and Zhao, X. (2011). Ultra-thin double-walled carbon nanotubes: a novel nanocontainer for preparing atomic wires, *Nano Research,* **4**, 759–766.

17. Andrade, N. F., Vasconcelos, T. L., Gouvea, C. P., Archanjo, B. S., Achete, C. A., Kim, Y. A., Endo, M., Fantini, C., Dresselhaus, M. S., and Souza Filho, A. G. (2015). Linear carbon chains encapsulated in multiwall carbon nanotubes: resonance Raman spectroscopy and transmission electron microscopy studies, *Carbon,* **90**, 172–180.

18. Zhang, Y., Zhao, J., Fang, Y., Liu, Y., and Zhao, X. (2018). Preparation of long linear carbon chain inside multi-walled carbon nanotubes by cooling enhanced hydrogen arc discharge method, *Nanoscale,* **10**, 17824–17833.

19. Zhang, Y., Chang, W., Liu, Y., Maruyama, T., and Zhao, X. (2020). High-yield growth of multi-walled carbon nanowires by magnetic field controlled arc technique, *Carbon,* **158**, 672–680.

20. Chang, W., Liu, F., Liu, Y., Zhu, T., Fang, L., Li, Q., Liu, Y., and Zhao, X. (2021). Smallest carbon nanowires made easy: Long linear carbon chains confined inside single-walled carbon nanotubes, *Carbon,* **183**, 571–577.

21. Qin, X., Liu, Y., Li, X., Yang, G., Zhao, D., and Ju, L. (2021). Six-fold symmetry origin of Dirac cone formation in two-dimensional materials, *New Journal of Physics,* **23**, 113033.

22. Qin, X., Liu, Y., Yang, G., and Zhao, D. (2020). Mirror symmetry origin of Dirac cone formation in rectangular two-dimensional materials, *Physical Chemistry Chemical Physics,* **22**, 6619–6625.

23. Chi, B., Liu, Y., Li, X., Qin, X., and Zhao, X. (2017). Exceptional ring topology makes diamond allotropes as light-weight superhard materials, *Diamond and Related Materials,* **80**, 140–146.

24. Qin, X., Wu, Y., Liu, Y., Chi, B., Li, X., Wang, Y., and Zhao, X. (2017). Origins of Dirac cone formation in AB_3 and A_3B (A, B = C, Si, and Ge) binary monolayers, *Scientific Reports,* **7**, 10546.
25. Qin, X., Liu, Y., Chi, B., Zhao, X., and Li, X. (2016). Origins of Dirac cones and parity dependent electronic structures of α-graphyne derivatives and silagraphynes, *Nanoscale,* **8**, 15223–15232.
26. Qin, X., Liu, Y., Li, X., Xu, J., Chi, B., Zhai, D., and Zhao, X. (2015). Origin of dirac cones in SiC silagraphene: a combined density functional and tight-binding study, *The Journal of Physical Chemistry Letters,* **6**, 1333–1339.
27. Chen, J., Yang, L., Yang, H., and Dong, J. (2003). Electronic and transport properties of a carbon-atom chain in the core of semiconducting carbon nanotubes, *Physics Letters A,* **316**, 101–106.
28. Liu, Y., Jones, R. O., Zhao, X., and Ando, Y. (2003). Carbon species confined inside carbon nanotubes: a density functional study, *Physical Review B,* **68**, 125413.
29. Wang, Y., Huang, Y., Yang, B., and Liu, R. (2006). Structural and electronic properties of carbon nanowires made of linear carbon chains enclosed inside zigzag carbon nanotubes, *Carbon,* **44**, 456–462.
30. Vadapalli, R. K., and Mintmire, J. W. (2006). Endohedral carbon chains in chiral single-wall carbon nanotubes, *International Journal of Quantum Chemistry,* **106**, 2324–2330.
31. Wang, Y., Huang, Y., Yang, B., and Liu, R. (2008). Crystal orbital study on carbon chains encapsulated in armchair carbon nanotubes with various diameters, *Carbon,* **46**, 276–284.
32. Tapia, A., Aguilera, L., Cab, C., Medina-Esquivel, R. A., de Coss, R., and Canto, G. (2010). Density functional study of the metallization of a linear carbon chain inside single wall carbon nanotubes, *Carbon,* **48**, 4057–4062.
33. Bai, H., Qiao, W., Zhu, Y., and Huang, Y. (2015). Crystal orbital study on the combined carbon nanowires constructed from linear carbon chains encapsulated in zigzag double-walled carbon nanotubes, *Current Applied Physics,* **15**, 342–351.
34. Yang, X., Lv, C., Yao, Z., Yao, M., Qin, J., Li, X., Shi, L., Du, M., Liu, B., and Shan, C. (2020). Band-gap engineering and structure evolution of confined long linear carbon chains@double-walled carbon nanotubes under pressure, *Carbon,* **159**, 266–272.
35. Bonabi, F., Brun, S. J., and Pedersen, T. G. (2017). Excitonic optical response of carbon chains confined in single-walled carbon nanotubes, *Physical Review B,* **96**, 155419.

36. Shi, L., Rohringer, P., Wanko, M., Rubio, A., Waßerroth, S., Reich, S., Cambré, S., Wenseleers, W., Ayala, P., and Pichler, T. (2017). Electronic band gaps of confined linear carbon chains ranging from polyyne to carbyne, *Physical Review Materials,* **1**, 075601.

37. Zhao, C., Kitaura, R., Hara, H., Irle, S., and Shinohara, H. (2011). Growth of linear carbon chains inside thin double-wall carbon nanotubes, *The Journal of Physical Chemistry C,* **115**, 13166–13170.

38. Andrade, N. F., Aguiar, A. L., Kim, Y. A., Endo, M., Freire, P. T. C., Brunetto, G., Galvão, D. S., Dresselhaus, M. S., and Souza Filho, A. G. (2015). Linear carbon chains under high-pressure conditions, *The Journal of Physical Chemistry C,* **119**, 10669–10676.

39. Deng, Y., and Cranford, S. W. (2019). Mapping temperature and confinement dependence of carbyne formation within carbon nanotubes, *Carbon,* **141**, 209–217.

40. Khalilov, U., Vets, C., and Neyts, E. C. (2019). Catalyzed growth of encapsulated carbyne, *Carbon,* **153**, 1–5.

41. Shi, L., Senga, R., Suenaga, K., Kataura, H., Saito, T., Paz, A. P., Rubio, A., Ayala, P., and Pichler, T. (2021). Toward confined carbyne with tailored properties, *Nano Letters,* **21**, 1096–1101.

42. Wong, C. H. *Routines to Synthesize Carbyne of more than 6000 Atoms.* https://doi.org/10.48550/arXiv.2207.14558

43. Xu, Z., Yan, T., Liu, G., Qiao, G., and Ding, F. (2016). Large scale atomistic simulation of single-layer graphene growth on Ni(111) surface: molecular dynamics simulation based on a new generation of carbon–metal potential, *Nanoscale,* **8**, 921–929.

44. Dong, J., Wang, H., Peng, H., Liu, Z., Zhang, K., and Ding, F. (2017). Formation mechanism of overlapping grain boundaries in graphene chemical vapor deposition growth, *Chemical Science,* **8**, 2209–2214.

45. Xu, L., Jin, Y., Wu, Z., Yuan, Q., Jiang, Z., Ma, Y., and Huang, W. (2013). Transformation of carbon monomers and dimers to graphene islands on Co(0001): thermodynamics and kinetics, *The Journal of Physical Chemistry C,* **117**, 2952–2958.

46. Xu, Z., Yan, T., and Ding, F. (2015). Atomistic simulation of the growth of defect-free carbon nanotubes, *Chemical Science,* **6**, 4704–4711.

47. van Duin, A. C. T., Dasgupta, S., Lorant, F., and Goddard, W. A. (2001). ReaxFF: a reactive force field for hydrocarbons, *The Journal of Physical Chemistry A,* **105**, 9396–9409.

48. Chenoweth, K., van Duin, A. C. T., and Goddard, W. A. (2008). ReaxFF reactive force field for molecular dynamics simulations of hydrocarbon oxidation, *The Journal of Physical Chemistry A,* **112**, 1040–1053.

49. Liu, L., Liu, Y., Zybin, S. V., Sun, H., and Goddard, W. A. (2011). ReaxFF-lg: Correction of the ReaxFF reactive force field for London dispersion, with applications to the equations of state for energetic materials, *The Journal of Physical Chemistry A,* **115**, 11016–11022.
50. Ashraf, C., and van Duin, A. C. T. (2017). Extension of the ReaxFF combustion force field toward syngas combustion and initial oxidation kinetics, *The Journal of Physical Chemistry A,* **121**, 1051–1068.
51. Xu, J., Bian, Y., Liu, Y., and Zhai, D. (2017). Reactive molecular dynamics study of thermal decomposition of nanocarbon energetic composite materials, *Computational Materials Science,* **131**, 126–131.
52. Lee, C., Wei, X., Kysar, J. W., and Hone, J. (2008). Measurement of the elastic properties and intrinsic strength of monolayer graphene, *Science,* **321**, 385.
53. Liu, F., Ming, P., and Li, J. (2007). Ab initio calculation of ideal strength and phonon instability of graphene under tension, *Physical Review B,* **76**, 064120.
54. Sánchez-Portal, D., Artacho, E., Soler, J. M., Rubio, A., and Ordejón, P. (1999). Ab initio structural, elastic, and vibrational properties of carbon nanotubes, *Physical Review B,* **59**, 12678–12688.
55. Wang, M., and Lin, S. (2015). Ballistic thermal transport in carbyne and cumulene with micron-scale spectral acoustic phonon mean free path, *Scientific Reports,* **5**, 18122.
56. Balandin, A. A., Ghosh, S., Bao, W., Calizo, I., Teweldebrhan, D., Miao, F., and Lau, C. N. (2008). Superior thermal conductivity of single-layer graphene, *Nano Letters,* **8**, 902–907.
57. Pop, E., Mann, D., Wang, Q., Goodson, K., and Dai, H. (2006). Thermal conductance of an individual single-wall carbon nanotube above room temperature, *Nano Letters,* **6**, 96–100.
58. Liu, X., Zhang, G., and Zhang, Y.-W. (2015). Tunable mechanical and thermal properties of one-dimensional carbyne chain: phase transition and microscopic dynamics, *The Journal of Physical Chemistry C,* **119**, 24156–24164.
59. Kim, H. S., Kim, T. H., and Kim, Y.-H. (2019). Odd-even phonon transport effects in strained carbon atomic chains bridging graphene nanoribbon electrodes, *Carbon,* **142**, 107–114.
60. Li, X., Zhang, Y., Wu, Y., and Shi, L. (2021). Pressure-tailored synthesis of confined linear carbon chains, *Journal of Applied Physics,* **129**, 064302.
61. Lang, N. D., and Avouris, P. (2000). Carbon-atom wires: charge-transfer doping, voltage drop, and the effect of distortions, *Physical Review Letters,* **84**, 358–361.

62. Chen, R., Liu, F., Tang, Y., Liu, Y., Dong, Z., Deng, Z., Zhao, X., and Liu, Y. (2022). Combined first-principles and machine learning study of the initial growth of carbon nanomaterials on metal surfaces, *Applied Surface Science,* **586**, 152762.

63. Zhao, X., Liu, Y., Inoue, S., Suzuki, T., Jones, R. O., and Ando, Y. (2004). Smallest carbon nanotube is 3 Å in diameter, *Physical Review Letters,* **92**, 125502.

Chapter 7

Synthesis of Confined Carbyne inside Carbon Nanotubes by Thermal Annealing

Weili Cui and Lei Shi
State Key Laboratory of Optoelectronic Materials and Technologies, Nanotechnology Research Center, Guangdong Basic Research Center of Excellence for Functional Molecular Engineering, School of Materials Science and Engineering, Sun Yat-sen University, Guangzhou 510275, P. R. China
shilei26@mail.sysu.edu.cn

The technique of confined synthesis *via* thermal annealing enables the generation of long linear carbon chains within carbon nanotubes. It has been observed that the properties of the resulting long linear carbon chains are independent of length if they exceed a threshold of at most 30 nm, and instead depend primarily on the diameter of the host carbon nanotubes. These long linear carbon chains are henceforth referred to confined carbyne. In this chapter, we provide a comprehensive summary of the most significant advancements achieved thus far in the synthesis of confined carbyne by controlling various parameters such as environmental temperature, pressure, carbon nanotube diameter, and carbon source from a range of

---

Emerging Carbyne: Truly One-Dimensional Form of Carbon
Edited by Yahachi Saito

ISBN 978-981-5129-15-1 (Hardcover), 978-1-003-51036-9 (eBook)
www.jennystanford.com

precursor molecules. We also proposed a growth mechanism with the aid of tailored carbon nanotube diameters and the introduction of isotopic precursor molecules. Furthermore, we demonstrate the effectiveness of Raman spectroscopy in characterizing the properties of confined carbyne.

7.1 Introduction

Linear carbon chains (LCCs), such as polyyne, cumulene, and carbyne, are widely considered to be unstable even under ambient conditions. However, their synthesis requires extreme conditions, such as high temperature that can be achieved through the techniques like laser ablation, arc-discharge, field emission, electron impact, joule heating, and furnace heating. Among these methods, thermal annealing in a furnace is considered as one of the most controllable for conducting repeatable experiments. This approach also enables the introduction of various environments, including vacuum, inert, hydrocarbon, or other gaseous environments, making it a versatile method for synthesizing carbon materials. In particular, this method has been extensively used for synthesizing carbon nanotubes (CNTs) [1], graphene [2], and LCCs [3], including confined carbyne (CC) inside carbon nanotubes [4].

High-temperature annealing has been proven as an effective method for removing metal catalysts, enabling the efficient purification of as-grown carbon nanotubes [5]. Additionally, with high enough temperatures, full or partial coalescence between adjacent carbon nanotubes becomes possible [6, 7], allowing for various processes such as coalescing, merging [8], enlarging [9], and even graphitizing [5] carbon nanotubes. Remarkably, the formations of inner carbon nanotubes [1] and/or carbon chains [3] inside single-walled carbon nanotubes (SWCNTs) provide a pathway for stabilizing pristine carbon nanotubes under high temperature, as the outer tube protects the formed inner carbon nanotubes and/or chains from oxidation [10].

7.2 Formation of Linear Carbon Chains by Thermal Annealing of DWCNTs

Early on, it was proposed that short carbon chains, composed of 3–7 carbon atoms, could covalently bond to the interstitial positions between adjacent double-walled carbon nanotubes (DWCNTs) [11]. High-resolution transmission electron microscopy (HRTEM) has demonstrated that coalescence occurred in the DWCNTs thermally annealed in the presence of boron, as compared to the cross-sections of DWCNTs presented in Figs. 7.1a and b. While the short carbon chains were not visible in the HRTEM images, molecular dynamics calculations have suggested that the presence of these chains accelerated the coalescence process (Fig. 7.1c), with the carbon chains acting as precursors to merge the adjacent DWCNTs.

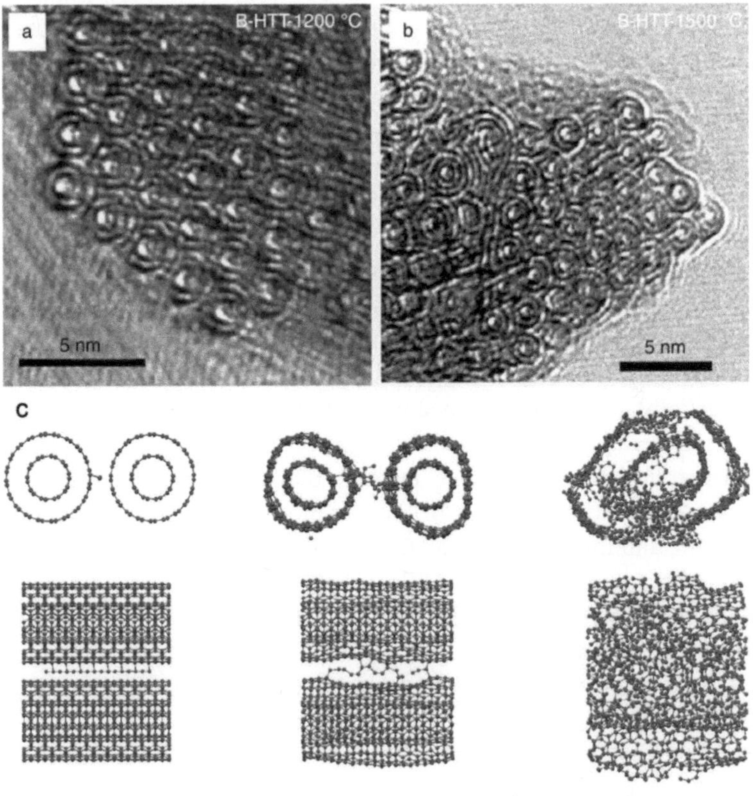

Figure 7.1 HRTEM images of the cross sections of DWCNTs thermally annealed in the presence of boron: (a) at 1200 °C, (b) at 1500 °C. (c) Molecular models show that a carbon chain between two adjacent DWCNTs accelerated the coalescence of nanotubes. Reprinted with permission from Ref. [11].

In contrast to short carbon chains, LCCs have been proposed to form inside DWCNTs with an inner diameter of around 0.6 nm when heated at 1550 °C under anaerobic conditions [12]. However, the LCCs cannot be formed at lower temperatures or inside SWCNTs, and they decompose at higher temperatures (1600 °C) due to the short and defective inner tubes of such DWCNTs that were transformed from C_{60}. DWCNTs prepared by arc-discharge method annealed at lower temperatures gave an even higher yield of LCCs [13]. Raman spectroscopy has shown that the vibrational mode of the LCCs appears between 1780 and 1900 cm^{-1} (Fig. 7.2), which is consistent with the results obtained for LCCs inside multi-walled carbon nanotubes (MWCNTs) prepared by arc-discharge method [14]. However, this signal was not observed for heat-treated SWCNTs, as their average diameter is around 1.3 nm, which is greater than the optimal 0.6 nm. In another study, SWCNTs produced by high pressure carbon monoxide (HiPco) decomposition with a diameter of 0.6–1.2 nm were also unable to synthesize LCCs, suggesting that the diameter of the nanotubes is not the only key factor for the formation of LCCs [3]. Instead, SWCNTs with an average diameter of around 1.7 nm synthesized by the enhanced direct injection pyrolytic synthesis (eDIPS) method can be used for LCC synthesis (Fig. 7.2b), as the inner tubes with an average diameter of around 1.0 nm were synthesized, and the LCCs were formed inside the newly formed inner tubes [3]. This suggests that the optimal diameter of the CNTs for the synthesis of LCCs is most likely between 0.6 and 1.1 nm, and the quality of the CNTs is also important. Interestingly, the Raman mode of the LCCs consists of several components (Fig. 7.2d), which can be clearly seen when the yield of the LCCs is high (Fig. 7.2c) and the laser energy used for the resonance excitation of the Raman signal changes to the one close to the energy gaps of the LCCs [15].

In 2016, a significant milestone was achieved through the synthesis of long LCCs—consisting of more than 6,000 carbon atoms inside DWCNTs *via* a high-temperature annealing process of chemical vapor deposition (CVD)-produced DWCNTs under a high vacuum environment [16]. The high yield of LCCs was obtained by annealing the CVD-DWCNTs at 1460 °C for 30 minutes under a high vacuum with the pressure lower than 2×10^{-6} mbar, enabling the observation of LCC@DWCNT hybrids using HRTEM, as illustrated in Fig. 7.3a. Through the examination of several individual LCC@DWCNT using near-field Raman spectroscopy, the length of carbon chains was measured by overlaying the near-field Raman signal with an atomic force microscopic image (Fig. 7.3b). The particular measured lengths derived from the intensity maps of two LCCs are 290 and 190 nm. The longest chain observed had a length of 797 nm with a corresponding Raman frequency of 1,791 ± 1 cm^{-1}. Additionally, it was practically confirmed that LCCs exist inside the inner tube of DWCNTs. The Raman signal from the LCC is even higher than the G-band of DWCNT (Fig. 7.3c), suggesting a high yield of the LCC inside DWCNT. Consequently, the high-temperature and high-vacuum thermal annealing procedure has been demonstrated as an effective method for achieving the bulk realization of the longest LCCs ever recorded inside DWCNTs.

In LCCs@DWCNTs system, the Raman signals of confined LCCs within the DWCNTs are found to be correlated with specific inner tubes, as illustrated in Figs. 7.3d–f and confirmed by near-field Raman testing on multiple individual LCC@DWCNT (Figs. 7.3g and h). This sheds light on an important question about the relation between the LCCs and the hosted nanotubes, which can be answered through an analysis of the radial breathing modes (RBMs) of the DWCNTs. The peaks of RBMs with indexed chirality from the Kataura plot, which includes a correction term for the inter-tube distance corresponding to the thin inner tubes, are split and enhanced, indicating an interaction with the LCCs. Figure 7.3h displays the Raman frequencies of the encapsulated chains plotted against the encasing inner carbon nanotube's diameter. The observed linear decrease in the encapsulated chain's Raman frequency with decreasing nanotube diameter provides compelling evidence that the properties of the encapsulated carbyne chains are primarily influenced by their interaction with the encasing inner carbon nanotubes. Thinner host tubes interact more strongly with the LCCs, resulting in a lower-frequency Raman response, which tentatively assigns the bulk LCC line shape fine structures (Fig. 7.3e) to different chirality of inner host tubes. For example, (6,4) tube is more likely to host LCCs with lower Raman frequency than the (6,5) tube due to the larger interaction between a (6,4) tube and its encapsulated LCC.

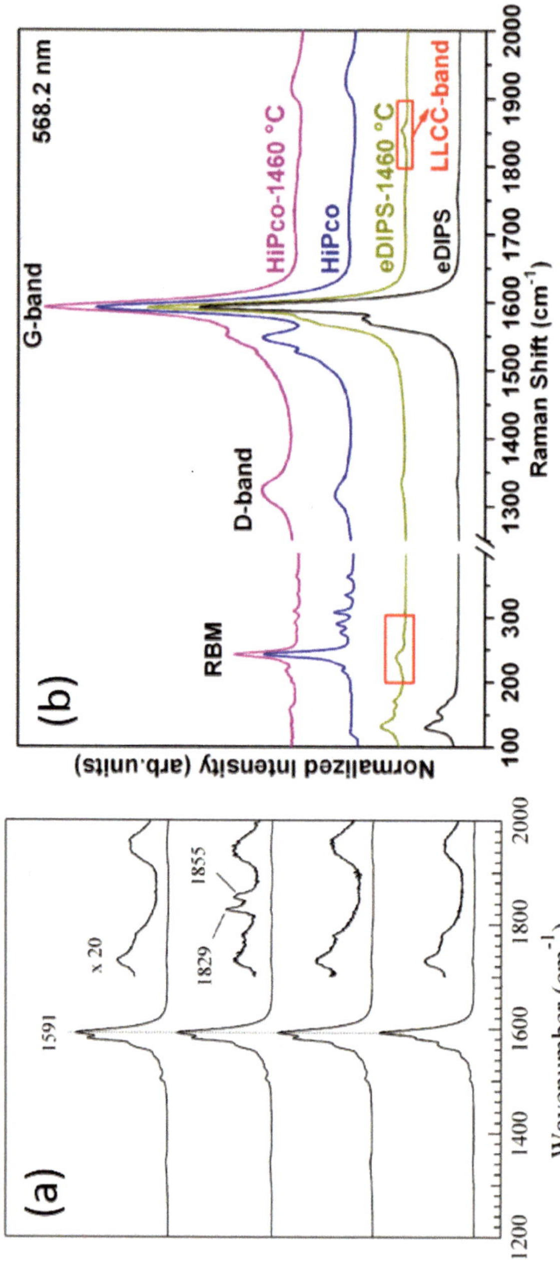
(b)
568.2 nm
G-band
D-band
RBM
HiPco-1460 °C
HiPco
eDIPS-1460 °C
eDIPS
LLCC-band
Normalized Intensity (arb.units)
Raman Shift (cm^{-1})
100 200 300 1300 1400 1500 1600 1700 1800 1900 2000
(a)
1591
x 20
1829
1855
Wavenumber (cm^{-1})
1200 1400 1600 1800 2000

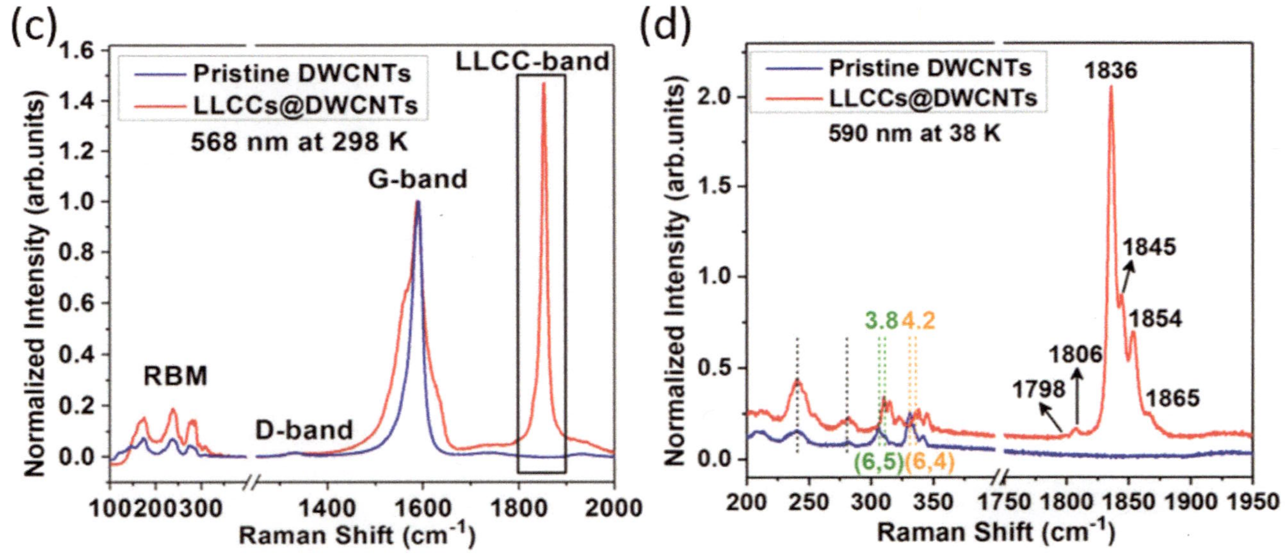

Figure 7.2 Raman spectra of high-temperature annealed (a) DWCNTs transformed from C_{60} peapods, (b) HiPco SWCNTs and eDIPS SWCNTs, (c) DWCNTs produced by chemical vapor deposition. (d) Peak fitting of the Raman signal of LCCs. Reprinted with permission from Refs. [3, 12, 15].

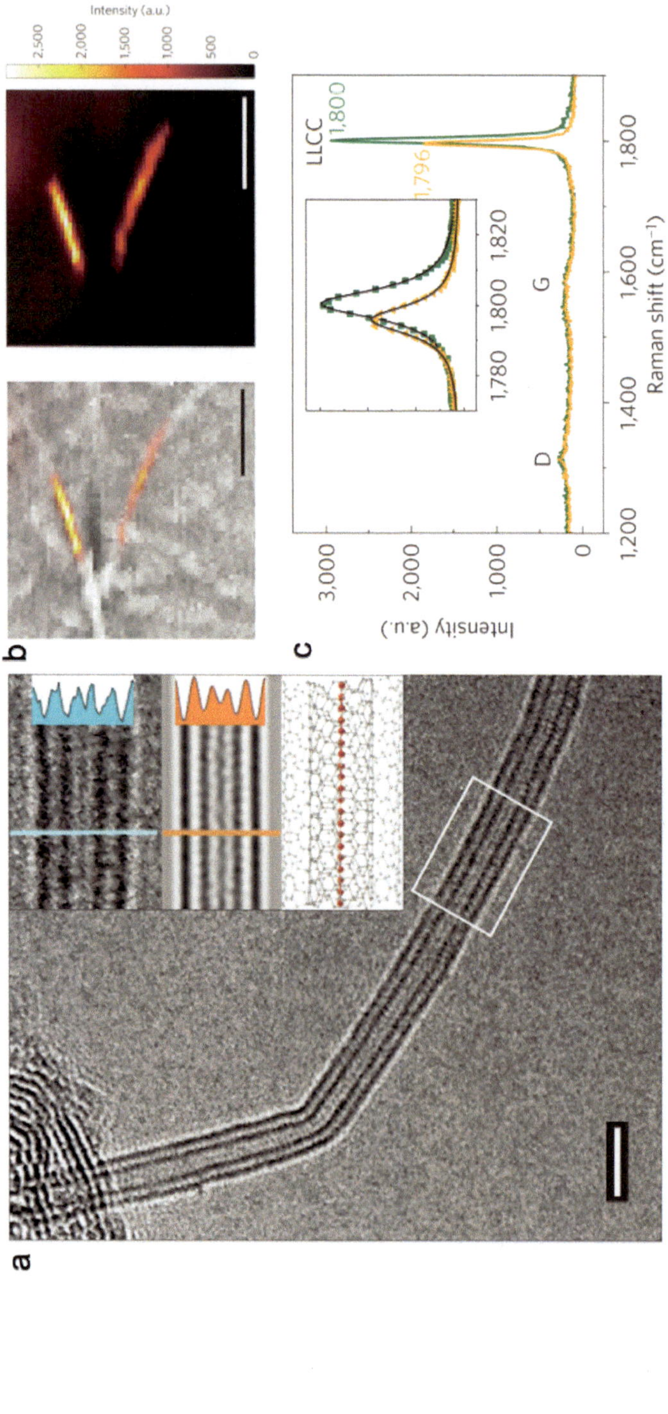
a
b
Intensity (a.u.)
2,500
2,000
1,500
1,000
500
0
c
Intensity (a.u.)
3,000
2,000
1,000
0
LLCC
1,800
1,796
1,780 1,800 1,820
D
G
1,200
1,400
1,600
1,800
Raman shift (cm⁻¹)

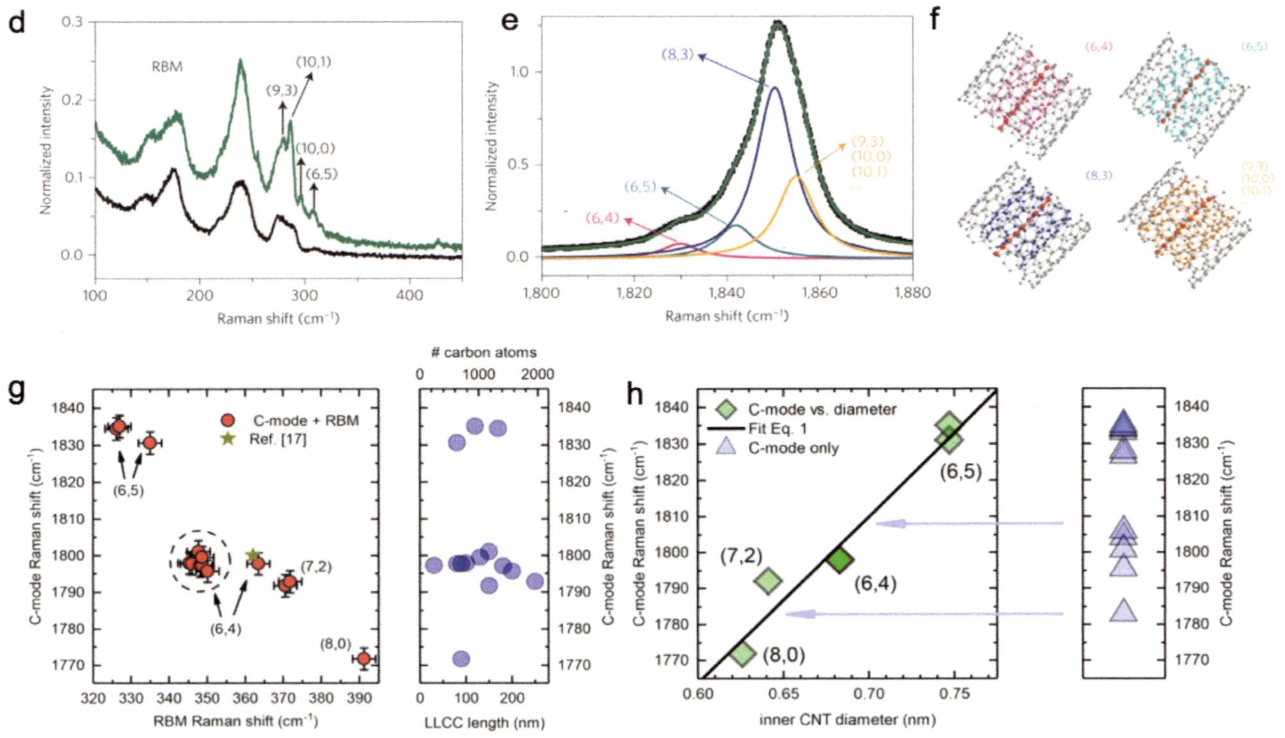

Figure 7.3 (a) HRTEM image of an CC@DWCNT. (b) Overlay of Raman signal of the CC-band (color) on top of the topographic AFM image (greyscale). (c) Raman spectra of the top and bottom LCC@DWCNT in (b). (d) RBM of CCs@DWCNTs (green line) as compared to pristine DWCNTs (black line). (e) Raman peak fitting of Raman signal of the CCs. (f) Schemes of the inner tube chirality and corresponding carbon chain configurations. (g) Raman signal of the CCs as a function of RBM of the inner tubes (left) and the length of the CCs (right). (h) Raman signal of the CCs as a function of the diameter of inner tubes with (left) and without (right) observable RBM. Reprinted with permission from Refs. [16, 17]. Copyright 2023, American Chemical Society.

In addition, by the help of near-field Raman spectroscopy, the Raman frequency of confined LCCs has been found to be independent of their length when the LCCs exceed 30 nm in length [17]. Instead, the Raman frequency is determined by the diameter of the inner tubes, where LCCs with small energy gaps are associated with stronger interactions with thinner host tubes. This demonstrates that long LCCs are indeed carbyne and are referred as confined carbyne (CC) now and later. When plotting the Raman frequencies of the encapsulated chains as a function of the encasing inner carbon nanotube's diameter, a linear decrease of the encapsulated chain's Raman frequency with decreasing nanotube diameter gives strong evidence that the properties of the encapsulated carbyne chains are dominated by interaction with the encasing inner carbon nanotube.

The CC-mode intensity, as illustrated in Fig. 7.3c, is significantly higher than that of the G-band of the nanotubes, which is unexpected given the low carbon atom ratio between the CC and the DWCNTs. Further comparison of the G-band intensity of graphene, measured under the same conditions, indicates that the individual CC inside DWCNT is the strongest Raman scatterer ever reported, exceeding that of any known material by over 2 orders of magnitude [18]. In addition, an anomalously large anti-Stokes/Stokes ratio in the tip-enhanced Raman spectrum of the CC, indicating its potential for use in local temperature monitoring [19, 20]. The superior Raman cross-section and high yield of CCs synthesized inside DWCNTs also enable the use of wavelength-dependent resonant Raman spectroscopy to investigate the optical band gaps of these CCs [21], and it allows to resolve vibronic and electronic fine structures of the CC in different inner tube chirality [22]. Previous theoretical studies have predicted that the band gap of CC can be tuned by the length of the chains, the end groups, and their interactions with the environment, resulting in theoretically predicted band gap values ranging from 0.2 to 8.5 eV for infinitely long carbyne biased by different density functionals [23, 24]. However, in experiment, CCs possess optical energy gaps in the range of 2.253–1.848 eV, which were determined by resonance Raman spectroscopy (Fig. 7.4a). Further investigation using resonance Raman spectroscopy on an individual CC@DWCNT revealed gap values consistent with those obtained from bulk materials (Fig. 7.4c), thus confirming the validity of the above findings [22]. Interestingly, a linear relation between

Raman frequency and optical energy gap of the CC was found as shown in Fig. 7.4d, and the slope is independent of the surrounding environment [25]. This is particularly valuable when analyzing an unknown carbon chain sample, as the optical energy gap can be calculated from the corresponding Raman frequency by applying the linear relation. Different from the environment-independent linear relation, the downshift of the optical band gap of the CCs@DWCNTs compared to the polyynes in gas phase or in solution (Fig. 7.4b) can be attributed to the interaction between the carbon chain and the host tube, i.e., environment-dependent Raman frequency. Therefore, the Raman frequency of CCs is closely related to their lengths and interactions with the environment.

7.3 Tailored Growth of Confined Carbyne by Thermal Annealing of SWCNTs

The host nanotube-encapsulated chain interaction allows for the control of the diameter of the hosted CNTs, enabling the synthesis of desired CCs with defined band gaps [26]. While linear carbon chains have been formed inside MWCNTs by arc-discharge method [14] and inside DWCNTs by high-temperature annealing [16], neither method has produced CCs with controllable properties. However, thin inner carbon nanotubes can be templated grown inside SWCNTs with tailored sizes using the selected size of pristine SWCNTs as outer tubes [1]. This approach serves as a universal route to create hosted tubes with proper sizes for further confined synthesis of desired CCs. A systematic study was conducted to investigate the van der Waals interaction-induced formation of CC by the hosted tubes inside SWCNTs with different diameters. HRTEM images shown in Figs. 7.5a–c revealed an interaction-induced diameter-changed nanotube, further confirming the influence of the interaction between host nanotubes and encapsulated chains. SWCNTs with diameters of 1.0–1.7 nm were used for tailoring the synthesis of CC with energy gaps of 1.82–2.21 eV, as shown in Figs. 7.5d and e. The Raman frequencies of the CCs varied in SWCNTs with different diameters, and the eDIPS-1.3 nm tubes yielded more CCs with the smallest energy gaps. As CCs with small energy gaps are more desirable for practical semiconducting applications, the

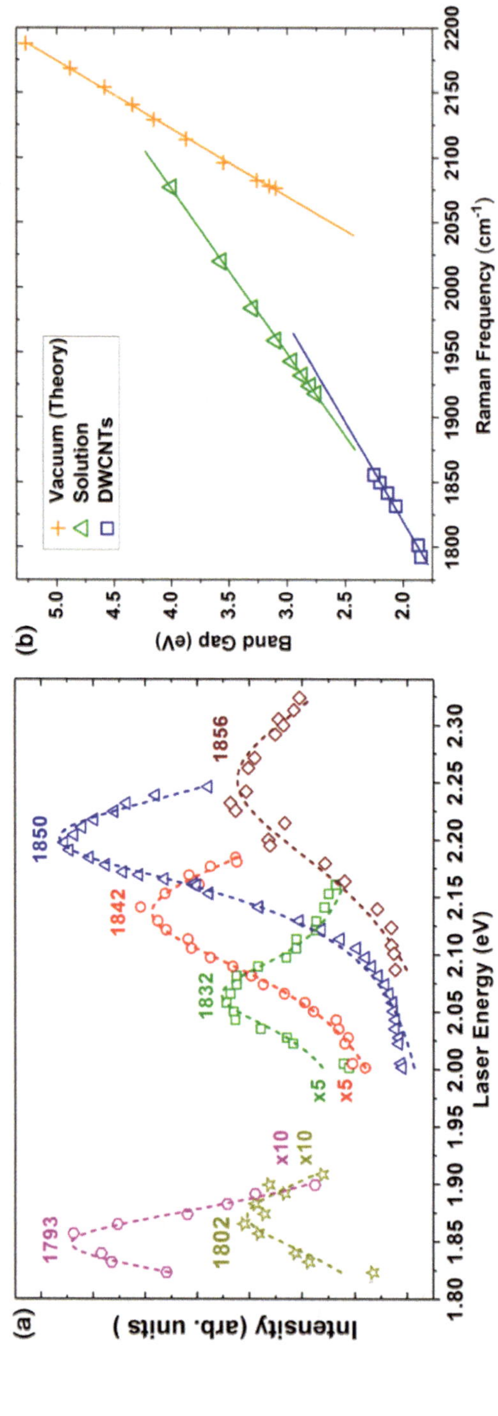
(a)
Intensity (arb. units)
Laser Energy (eV)
1793
1802
x10
x10
1832
x5
x5
1842
1850
1856
(b)
Band Gap (eV)
Raman Frequency (cm⁻¹)
Vacuum (Theory)
Solution
DWCNTs

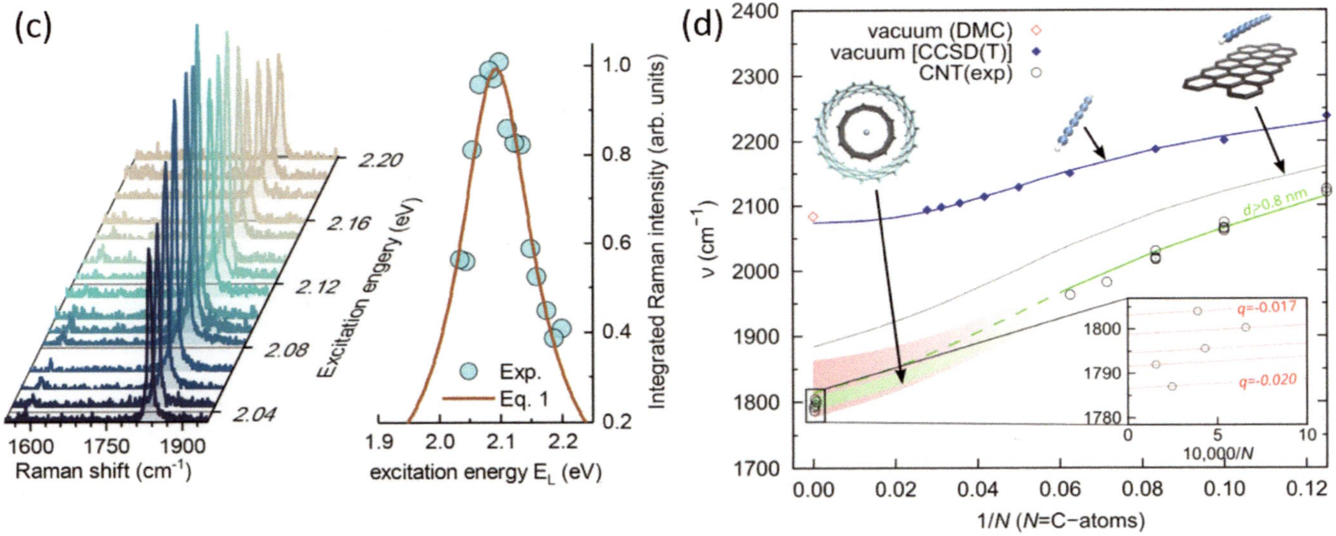

Figure 7.4 (a) Resonance Raman excitation profiles for six LCC-band peaks. (b) Band gap of the LCCs as a function of Raman frequency of the LCCs. (c) Raman resonance profile of an individual long linear carbon chain recorded at different excitation energies (left) and the normalized intensity of chains extracted from the spectra (right). (d) CC-mode in different environments as a function of inverse chain length. Reprinted with permission from Refs. [15, 22, 25].

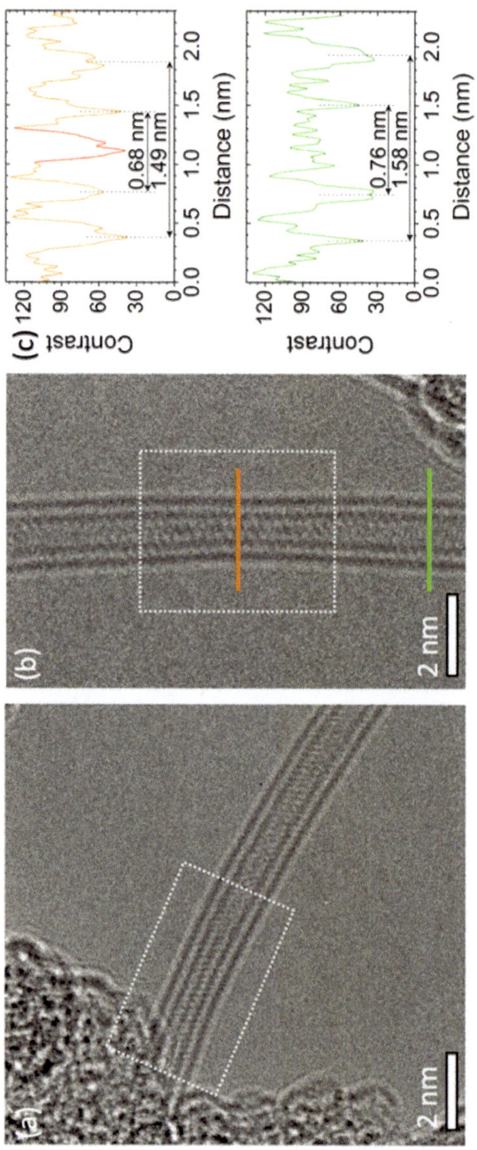
(a)
2 nm
(b)
2 nm
(c)
Contrast
0.68 nm
1.49 nm
Distance (nm)
Contrast
0.76 nm
1.58 nm
Distance (nm)

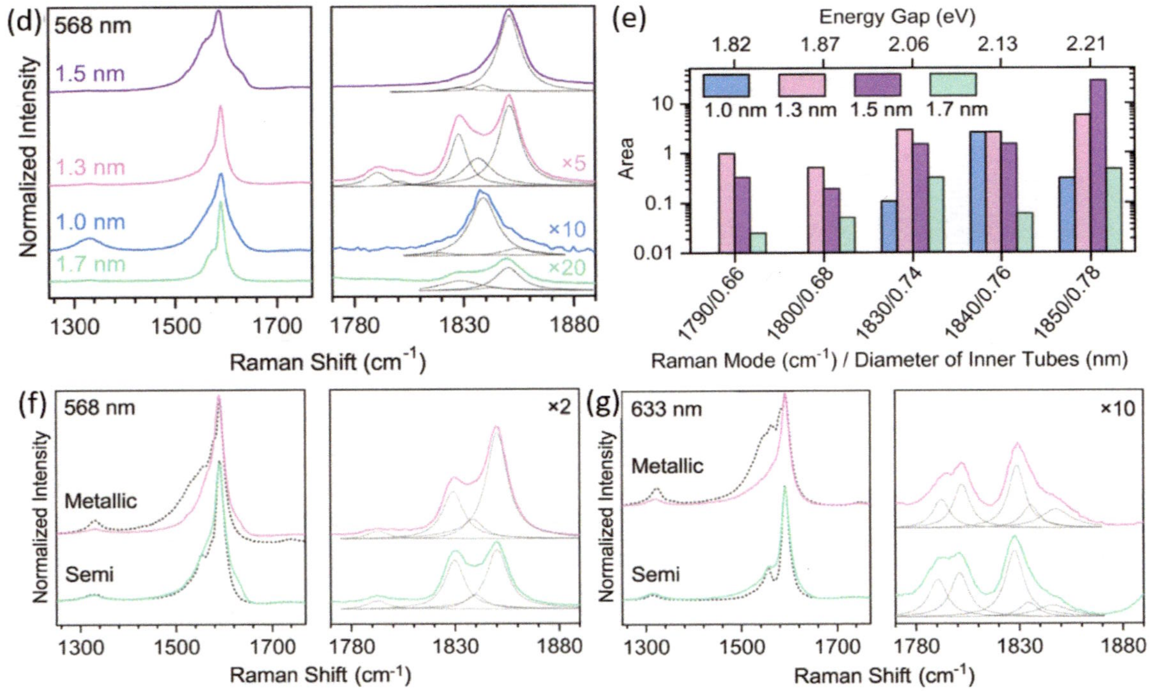

Figure 7.5 TEM images of a CC inside (a) a thin nanotube and (b) an interaction-induced diameter-changed nanotube. (c) The contrast profiles for the marked lines in (b). (d) Raman spectra of annealed SWCNTs with different marked diameters excited at 568 nm. (e) Integrated area of each component in the CC-bands of the annealed samples excited at 568 nm (for peaks at ~1830, 1840, and 1850 cm^{-1}) and 633 nm (for ~1790 and 1800 cm^{-1}). Raman spectra of pristine (dotted lines) and annealed (solid lines) semiconducting and metallic eDIPS-1.3 nm SWCNTs excited at (f) 568 nm and (g) 633 nm, respectively. Reprinted with permission from Ref. [26]. Copyright 2023, American Chemical Society.

eDIPS-1.3 nm SWCNTs were selected for the next metallicity-sorted study. However, the performance of separated semiconducting and metallic eDIPS-1.3 nm SWCNTs presented in Figs. 7.5f and g did not show any significant selectivity for the synthesis of CCs with different energy gaps. Therefore, the desired CCs can be tailored by the diameters of SWCNTs, rather than being greatly affected by their metallicity.

Yield is a crucial factor in the synthesis of CC by high-temperature annealing, in addition to achieving the target structure with tailored properties. The synthesis temperature and pressure can be used to tune the yield of CC [27], while the quality, defects, and purity of the CNTs should also be considered. Temperature-dependent synthesis can be identified by changes in the CC-band in Raman spectra, as demonstrated in Figs. 7.6a–e, regardless of whether DWCNTs produced by chemical vapor deposition or SWCNTs produced by the enhanced direct injection pyrolytic synthesis method. For pressure-dependent synthesis of CC (Figs. 7.6c–e), annealing under 10^{-3} and 10^{1} Pa yields a similar amount of the LCCs, suggesting that low vacuum is sufficient for the synthesis of CCs. However, when the pressure increased to 10^{5} Pa, the yield decreased dramatically.

In addition to annealing conditions, the carbon source used to grow the CCs is also an important factor affecting yield. The initial carbon source for CCs synthesized by high-temperature annealing is considered to be residual carbon from the synthesis of pristine SWCNTs. Therefore, the methods used to produce CNTs can significantly influence their performance in the synthesis of CCs. For example, as shown in Figs. 7.5d and e, 1.3 or 1.7 nm SWCNTs produced by the eDIPS method and 1.5 nm DWCNTs obtained from the CVD method exhibit significant yield differences in synthesizing CCs with different energy gaps. However, 1.4 nm DWCNTs produced by the laser-ablation method hardly produce CC after the same high-temperature annealing process [28]. Therefore, the yield of CCs highly depends on the residual carbon inside the pristine CNT systems if no extra carbon source is introduced.

To further increase the yield of CC, various organic molecules were filled inside SWCNTs to introduce an additional carbon source for growing the CC (Fig. 7.7). The molecules were selected based on their different C to H ratios and molecular sizes, while non-carbon containing H_2O was used as a comparison. It was found that only water molecules prevented the formation of CC, while all other carbon-containing liquid precursors successfully increased the yield of LCCs to some extent, with an optimal yield enhancement up to 2 times (Fig. 7.7a). The water-resulted low yield can be explained as follows: first, the newly formed oxygen and hydrogen molecules etch the CC and the carbon nanotubes at high temperature; second, the hydrogen ions from the decomposed water molecules can saturate the ends of the CC, stopping their growth; and lastly, the water molecules hinder the movement of carbon atoms inside the carbon nanotubes due to steric effects, leading to a decreased yield of CC. Among the effective liquid precursors, ethanol, 2-propanol, and acetone were more suitable for synthesizing CCs with large energy gaps, while methanol was the best for growing chains with small energy gaps, making it desirable for semiconductor applications. Based on this route, high purity ^{13}C-labelled methanol was confined inside SWCNTs with the help of C_{60} molecules to prevent escaping of methanol molecules during the heating (Fig. 7.7b). The results showed that isotope engineered CC with approximately 11.9% of ^{13}C incorporation was achieved experimentally for the first time (Fig. 7.7c), confirming that introducing liquid precursors as an extra carbon source before the high-temperature annealing process not only effectively improves the yield of target CC but also allows for engineering the property of CC *via* isotopic labelling, in addition to controlling the chain length or interaction with hosted CNTs.

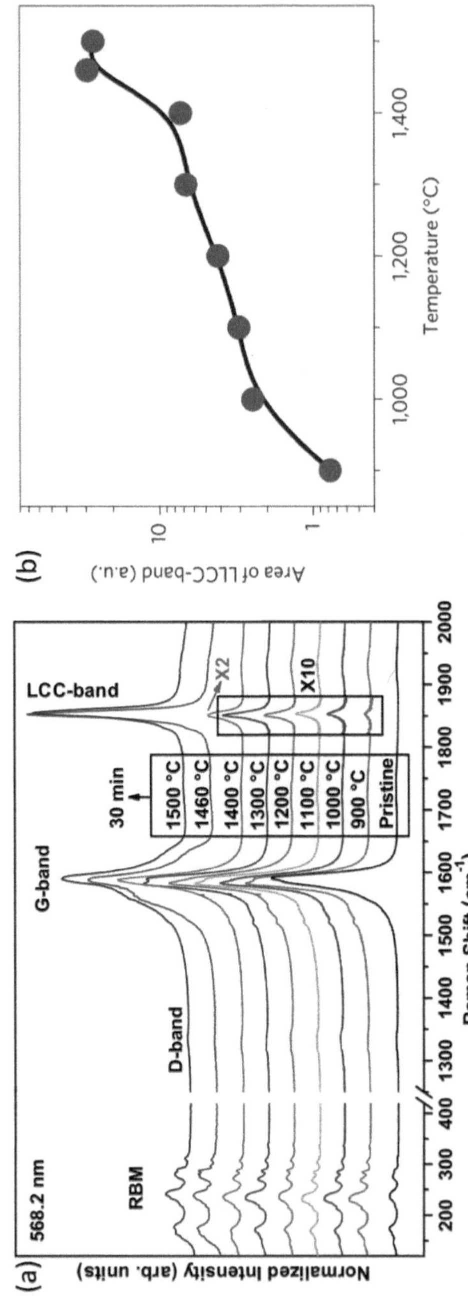
(a)
568.2 nm
RBM
D-band
G-band
LCC-band
30 min
1500 °C
1460 °C
1400 °C
1300 °C
1200 °C
1100 °C
1000 °C
900 °C
Pristine
X2
X10
Normalized Intensity (arb. units)
Raman Shift (cm^{-1})
(b)
Area of LLCC-band (a.u.)
Temperature (°C)

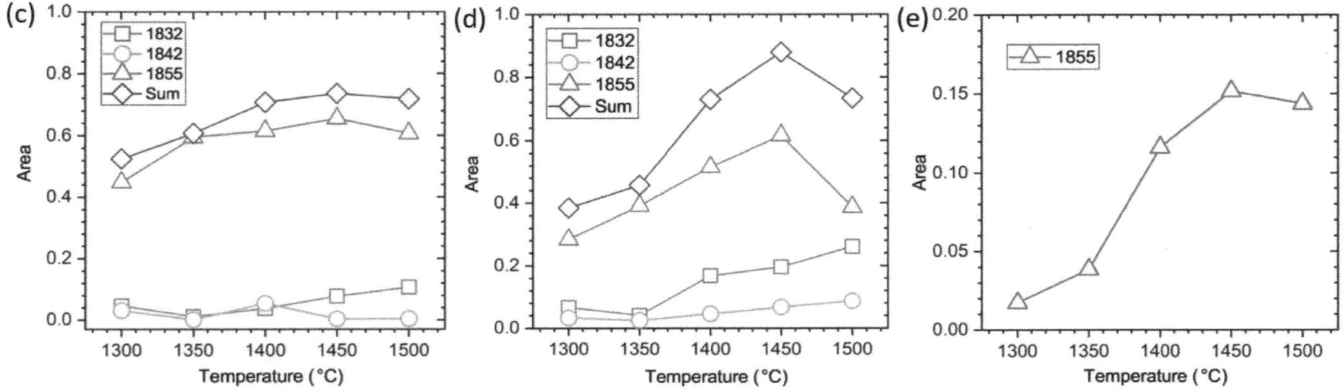

Figure 7.6 (a) Raman spectra and (b) Area of the LCC-band as a function of temperatures of the LCCs growth *via* annealing CVD-DWCNTs for 30 min. (c) Temperature-dependent growth of LCC *via* annealing eDIPS-1.3 nm SWCNTs under (c) 10^{-3} Pa, (d) 10^{-1} Pa and (e) 10^{5} Pa, respectively. Reprinted with permission from Refs. [16, 27].

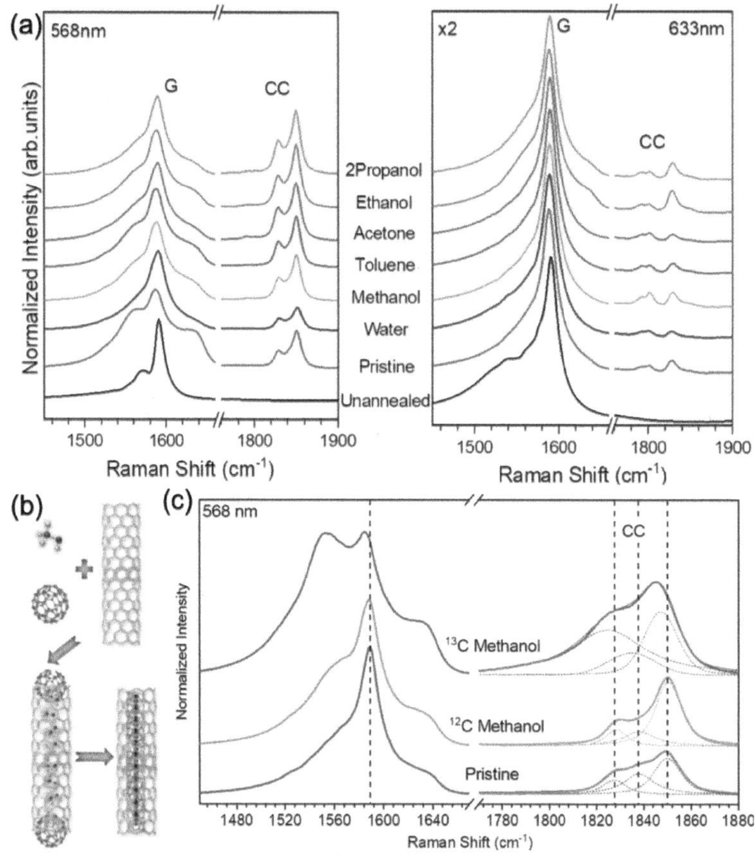

Figure 7.7 (a) Raman spectra of unannealed and optimally annealed empty/liquid precursor filled SWCNTs excited by lasers at 568 and 633 nm. (b) Scheme of the synthesis process for isotope labelling of the CC. (c) Raman spectra of the annealed pristine, ^{12}C methanol, and the ^{13}C-labelled methanol filled SWCNTs. Reprinted with permission from Ref. [4].

7.4 The Role of CNTs in the Formation of LCCs

With the assistance of isotope engineering of CC, it is possible to gain a deeper understanding of the behaviors of CNTs under high-temperature annealing and the role of CNTs in the formation process of CC. Both residual carbon and introduced precursors inside pristine CNTs have been shown to be effective carbon sources for the synthesis of CC. However, the role of CNTs during the high-temperature annealing process is still not well understood, which prevents our understanding of the formation mechanism of CC by thermal annealing. To address this, ultra-clean DWCNTs are needed to exclude the impact of carbon sources other than carbon nanotubes.

SWCNTs with natural isotope content (1.1% ^{13}C) produced by arc-discharge method were filled with isotopic C_{60} (81% ^{13}C) inside and the C_{60} were then converted into the inner tubes, thus DWCNTs with isotopic inner tubes were obtained, called as ^{Iso}DWCNTs. The resulting ^{Iso}DWCNTs, consisting of outer walls with natural abundant ^{13}C and inner walls with ~80% ^{13}C content, were thermally annealed to grow CC. Note that the direct thermal annealing of ^{Iso}DWCNTs cannot produce CC (Fig. 7.8a), indicating that the starting ^{Iso}DWCNTs were ultra-clean without precursor carbon atoms for the formation of CC. However, during annealing, additional carbon atom exchange between the walls of DWCNTs took place, as revealed by the changes in the isotope content distributions in the inner and outer walls (Table in Fig. 7.8). To grow CC, ^{Iso}DWCNTs were oxidized in air before thermal annealing (Fig. 7.8b). This two-step annealing yielded the growth of isotope-labelled CC with the highest ^{13}C concentration of 28.8% to date (Fig. 7.8c), concomitant with the atom exchange. The oxidation step not only introduced defects but also provided additional isotope-labelled carbonaceous precursors from the ^{Iso}DWCNTs themselves. Under the high-temperature annealing environment, not only did atom exchange occur between the inner and outer walls of DWCNTs, but also the carbonaceous precursors released by oxidized DWCNTs were utilized as a carbon source for formation of CC.

In summary, in the process of synthesizing CCs through thermal annealing, achieving a high selectivity and yield of desired CCs is of paramount importance. Annealing pressure and temperature are crucial factors that determine the overall selectivity and yield of target CCs. Additionally, methods such as increasing residual carbons left behind pristine CNTs, incorporating additional carbon sources from external precursors, and oxidizing the CNTs can greatly enhance the CC yield. Finally, tuning the host tube diameter allows to tailor the CC properties through interaction, resulting in the growth of specific CCs with distinct energy gaps [29].

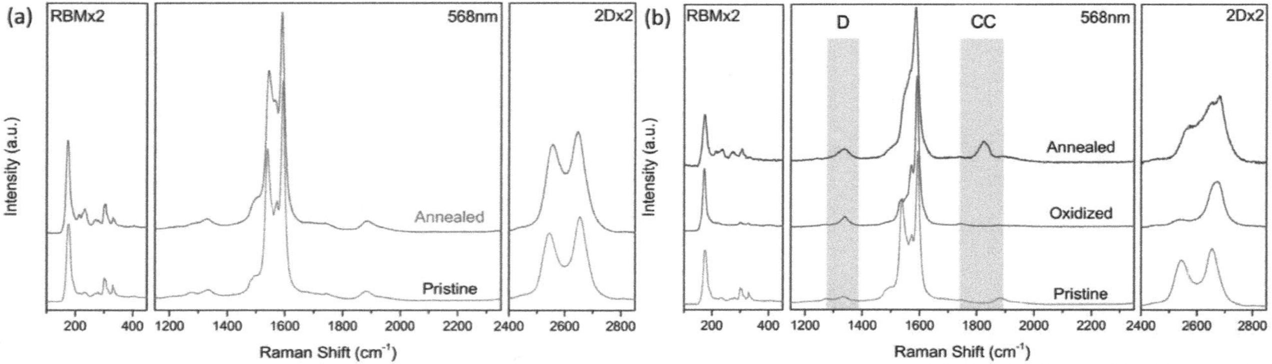

^{13}C concentration	Outer wall	Inner wall	CC
Pristine ^{Iso}DWCNTs	1.1%[30]	80.4 ± 0.8%	–
One-step annealed ^{Iso}DWCNTs	4.1 ± 0.4%	68.5 ± 2.2%	–
Stepwise oxidized and annealed ^{Iso}DWCNTs	6.5 ± 0.8%	57.4 ± 3.6%	28.8 ± 1.4%

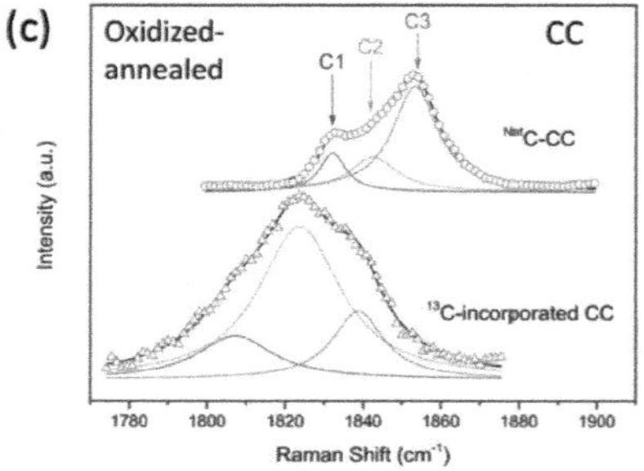

Figure 7.8 (a) The one-step annealed ^{Iso}DWCNTs and (b) the stepwise oxidized and annealed ^{Iso}DWCNTs. The ^{13}C concentrations of the samples in (a) and (b) are calculated by line-shape analysis of their 2D-mode and listed in the table. (c) The CC-mode of the stepwise oxidized and annealed samples in (b) were fitted with three Lorentzian peaks corresponding to the CCs with different energy gaps. Reprinted with permission from Ref. [28].

Acknowledgments

This work was supported by National Natural Science Foundation of China (51902353), Characteristic Innovation Project of Guangdong Provincial Department of Education (2022KTSCX001), and Fundamental Research Funds for the Central Universities, Sun Yat-sen University (22lgqb03).

References

1. L. Shi, J. Wei, K. Yanagi, T. Saito, K. Cao, U. Kaiser, P. Ayala, T. Pichler. Templated direct growth of ultra-thin double-walled carbon nanotubes. *Nanoscale* **10** (2018), 21254–21261. https://doi.org/10.1039/C8NR06925E.
2. K. Yan, L. Fu, H. Peng, Z. Liu. Designed CVD growth of graphene via process engineering. *Acc. Chem. Res.* **46** (2013), 2263–2274. https://doi.org/10.1021/ar400057n.
3. L. Shi, P. Rohringer, P. Ayala, T. Saito, T. Pichler. Carbon nanotubes from enhanced direct injection pyrolytic synthesis as templates for long linear carbon chain formation. *Phys. Status Solidi* **250** (2013), 2611–2615. https://doi.org/10.1002/pssb.201300148.
4. W. Cui, L. Shi, K. Cao, U. Kaiser, T. Saito, P. Ayala, T. Pichler. Isotopic labelling of confined carbyne. *Angew. Chem. Int. Ed.* **60** (2021), 9897–9901. https://doi.org/10.1002/anie.202017356.
5. R. Andrews, D. Jacques, D. Qian, E. C. Dickey. Purification and structural annealing of multiwalled carbon nanotubes at graphitization temperatures. *Carbon* **39** (2001), 1681–1687. https://doi.org/10.1016/S0008-6223(00)00301-8.
6. K. Méténier, S. Bonnamy, F. Béguin, C. Journet, P. Bernier, M. Lamy de La Chapelle, O. Chauvet, S. Lefrant. Coalescence of single-walled carbon nanotubes and formation of multi-walled carbon nanotubes under high-temperature treatments. *Carbon* **40** (2002), 1765–1773. https://doi.org/10.1016/S0008-6223(02)00044-1.
7. M. Endo, T. Hayashi, H. Muramatsu, Y.-A. Kim, H. Terrones, M. Terrones, M. S. Dresselhaus. Coalescence of double-walled carbon nanotubes: formation of novel carbon bicables. *Nano Lett.* **4** (2004), 1451–1454. https://doi.org/10.1021/nl0492483.

8. M. Terrones, F. Banhart, N. Grobert, J.-C. Charlier, H. Terrones, P.M. Ajayan. Molecular junctions by joining single-walled carbon nanotubes. *Phys. Rev. Lett.* **89** (2002), 075505. https://doi.org/10.1103/PhysRevLett.89.075505.
9. M. Yudasaka, H. Kataura, T. Ichihashi, L.-C. Qin, S. Kar, S. Iijima. Diameter enlargement of HiPco single-wall carbon nanotubes by heat treatment. *Nano Lett.* **1** (2001), 487–489. https://doi.org/10.1021/nl010037x.
10. W. Cui, T. Saito, P. Ayala, T. Pichler, L. Shi. Oxidation stability of confined linear carbon chains, carbon nanotubes, and graphene nanoribbons as 1D nanocarbons. *Nanoscale* **11** (2019), 15253–15258. https://doi.org/10.1039/C9NR04924J.
11. M. Endo, Y. A. Kim, T. Hayashi, H. Muramatsu, M. Terrenes, R. Saito, F. Villalpando-Paez, S. G. Chou, M. S. Dresselhaus. Nanotube coalescence-inducing mode: a novel vibrational mode in carbon systems. *Small* **2** (2006), 1031–1036. https://doi.org/10.1002/smll.200600087.
12. M. Jinno, Y. Ando, S. Bandow, J. Fan, M. Yudasaka, S. Iijima. Raman scattering study for heat-treated carbon nanotubes: the origin of ≈1855 cm^{-1} Raman band. *Chem. Phys. Lett.* **418** (2006), 109–114. https://doi.org/10.1016/j.cplett.2005.10.089.
13. L. Shi, L. Sheng, L. Yu, K. An, Y. Ando, X. Zhao. Ultra-thin double-walled carbon nanotubes: a novel nanocontainer for preparing atomic wires. *Nano Res.* **4** (2011), 759–766. https://doi.org/10.1007/s12274-011-0132-y.
14. X. Zhao, Y. Ando, Y. Liu, M. Jinno, T. Suzuki. Carbon nanowire made of a long linear carbon chain inserted inside a multiwalled carbon nanotube. *Phys. Rev. Lett.* **90** (2003), 4. https://doi.org/10.1103/PhysRevLett.90.187401.
15. L. Shi, P. Rohringer, M. Wanko, A. Rubio, S. Waßerroth, S. Reich, S. Cambré, W. Wenseleers, P. Ayala, T. Pichler. Electronic band gaps of confined linear carbon chains ranging from polyyne to carbyne. *Phys. Rev. Mater.* **1** (2017), 075601. https://doi.org/10.1103/PhysRevMaterials.1.075601.
16. L. Shi, P. Rohringer, K. Suenaga, Y. Niimi, J. Kotakoski, J. C. Meyer, H. Peterlik, M. Wanko, S. Cahangirov, A. Rubio, Z. J. Lapin, L. Novotny, P. Ayala, T. Pichler. Confined linear carbon chains as a route to bulk carbyne. *Nat. Mater.* **15** (2016), 634–639. https://doi.org/10.1038/nmat4617.

17. S. Heeg, L. Shi, L. V. Poulikakos, T. Pichler, L. Novotny. Carbon nanotube chirality determines properties of encapsulated linear carbon chain. *Nano Lett.* **18** (2018), 5426–5431. https://doi.org/10.1021/acs.nanolett.8b01681.

18. C. D. Tschannen, G. Gordeev, S. Reich, L. Shi, T. Pichler, M. Frimmer, L. Novotny, S. Heeg. Raman scattering cross section of confined carbyne. *Nano Lett.* **20** (2020), 6750–6755. https://doi.org/10.1021/acs.nanolett.0c02632.

19. C. D. Tschannen, M. Frimmer, G. Gordeev, T. L. Vasconcelos, L. Shi, T. Pichler, S. Reich, S. Heeg, L. Novotny. Anti-stokes Raman scattering of single carbyne chains. *ACS Nano* **15** (2021), 12249–12255. https://doi.org/10.1021/acsnano.1c03893.

20. C. D. Tschannen, M. Frimmer, T. L. Vasconcelos, L. Shi, T. Pichler, L. Novotny. Tip-enhanced stokes–anti-stokes scattering from carbyne. *Nano Lett.* **22** (2022), 3260–3265. https://doi.org/10.1021/acs.nanolett.2c00154.

21. M. Martinati, W. Wenseleers, L. Shi, S. M. Pratik, P. Rohringer, W. Cui, T. Pichler, V. Coropceanu, J.-L. Brédas, S. Cambré. Electronic structure of confined carbyne from joint wavelength-dependent resonant Raman spectroscopy and density functional theory investigations. *Carbon* **189** (2022), 276–283. https://doi.org/10.1016/j.carbon.2021.12.059.

22. S. Heeg, L. Shi, T. Pichler, L. Novotny. Raman resonance profile of an individual confined long linear carbon chain. *Carbon* **139** (2018), 581–585. https://doi.org/10.1016/j.carbon.2018.07.007.

23. S. Yang, M. Kertesz. Bond length alternation and energy band gap of polyyne. *J. Phys. Chem. A* **110** (2006), 9771–9774. https://doi.org/10.1021/jp062701+.

24. E. Mostaani, B. Monserrat, N. D. Drummond, C. J. Lambert. Quasiparticle and excitonic gaps of one-dimensional carbon chains. *Phys. Chem. Chem. Phys.* **18** (2016), 14810–14821. https://doi.org/10.1039/C5CP07891A.

25. M. Wanko, S. Cahangirov, L. Shi, P. Rohringer, Z. J. Lapin, L. Novotny, P. Ayala, T. Pichler, A. Rubio. Polyyne electronic and vibrational properties under environmental interactions. *Phys. Rev. B* **94** (2016), 195422. https://doi.org/10.1103/PhysRevB.94.195422.

26. L. Shi, R. Senga, K. Suenaga, H. Kataura, T. Saito, A. P. Paz, A. Rubio, P. Ayala, T. Pichler. Toward confined carbyne with tailored properties. *Nano Lett.* **21** (2021), 1096–1101. https://doi.org/10.1021/acs.nanolett.0c04482.

27. X. Li, Y. Zhang, Y. Wu, L. Shi. Pressure-tailored synthesis of confined linear carbon chains. *J. Appl. Phys.* **129** (2021), 064302. https://doi.org/10.1063/5.0035854.

28. W. Cui, F. Simon, Y. Zhang, L. Shi, P. Ayala, T. Pichler. Ultra-clean isotope engineered double-walled carbon nanotubes as tailored hosts to trace the growth of carbyne. *Adv. Funct. Mater.* **32** (2022), 2206491. https://doi.org/10.1002/adfm.202206491.

29. W. Cui, P. Ayala, T. Pichler, L. Shi. Unraveling the governing properties of confined carbyne through the interaction with its carbon nanotube host. *Carbon,* **219** (2024), 118784. https://doi.org/10.1016/j.carbon.2024.118784

Chapter 8

Linear Carbon Chains Formed by Electric Discharge of Carbon Nanotube Cathode

Yahachi Saito
Toyota Physical and Chemical Research Institute,
Nagakute, Aichi 480-1192, Japan
yysaito@sky.plala.or.jp

A novel method to produce long linear carbon-chains (LCCs) in a sheet of carbon nanotubes (CNTs) which was used as a cathode of field emission is presented in this chapter. Raman scattering spectroscopy and transmission electron microscopy revealed that LCCs, being confined stably inside CNTs, were formed in the CNT cathode subjected to severe electric discharges. Long LCCs were encapsulated in single-wall CNTs as well as double-wall and triple-wall CNTs. The length of LCCs is longer than 30 nm (i.e., more than 230 carbon atoms). Not only a single LCC but also two or three LCCs in parallel inside a CNT were found. Raman spectra showed prominent peaks, called "LCC bands", in a wavenumber region from

Emerging Carbyne: Truly One-Dimensional Form of Carbon
Edited by Yahachi Saito

ISBN 978-981-5129-15-1 (Hardcover), 978-1-003-51036-9 (eBook)
www.jennystanford.com

1790 to 1860 cm^{-1} originating from longitudinal optical phonons of LCCs. The observation of strong LCC bands is due to resonance Raman scattering, i.e., the excitation photon energies between 1.85–2.54 eV match with energy gaps of the long LCCs inside CNTs.

8.1 Introduction

One-dimensional monoatomic chains of carbon, so called carbyne, is highly reactive and unstable, but in 2003 Zhao *et al.* discovered that carbyne is stably formed and conserved in CNTs [1]. During transmission electron microscopy (TEM) study on multiwall carbon nanotubes (MWCNT) prepared by arc discharge between graphite electrodes in hydrogen gas, they found a linear carbon chain (LCC) consisting of about 100 carbon atoms encapsulated in the hollow core of MWCNTs. Later, the presence of LCC-encapsulated MWCNTs was shown by arcing in liquid nitrogen [2, 3], argon gas [4], and helium gas [5]. Thermochemically highly stable CNTs generate and protect LCCs in their inner space.

In addition, pre-synthesized $C_{10}H_2$ molecules were encapsulated in methanol into double-walled carbon nanotubes (DWCNTs), and then heated (800 to 1000 °C) in a high vacuum to fuse $C_{10}H_2$ molecules inside CNTs to lengthen the molecules [6]. Furthermore, in 2011, Shi *et al.* succeeded in generating LCCs with about 100-6000 carbon atoms encapsulated in DWCNTs by heat-treating CNTs in vacuum at high-temperature [7].

Long linear chains of carbon are very interesting as a model material for one-dimensional (1D) materials, but they are unstable and their physical properties are almost unknown. In order to explore the science of 1D materials, it is desirable to develop new methods for stably synthesizing a large amount of LCCs. In the course of our research of CNT-based field emission electron sources, LCCs encapsulated inside CNTs were found to be formed in the CNT-cathode suffered electric discharge [8, 9]. In this chapter, the experiment of field emission (FE) from CNT films is first presented, and atomic and electronic structures of LCCs formed after discharge, being evaluated by Raman scattering spectroscopy and TEM, are shown.

8.2 Field Emission and Discharge of CNT Film Cathode

The FE cathode used for the formation of LCCs is a thick film of CNTs that were synthesized by the method called "enhanced Direct Injection Pyrolytic Synthesis (eDIPS)" [10]. eDIPS-CNTs were processed into a sheet having a thickness of about 60 μm by Meijo Nano Carbon Co., Ltd. CNTs contained in the eDIPS sheet are mainly single-wall carbon nanotubes (SWCNTs) and DWCNTs, and a small amount of triple-wall CNTs (TWCNTs) is also included. The eDIPS-CNTs is featured by high purity and high crystallinity, with few impurity carbon and defects. The carbon purity is >90%, the diameter of SWCNTs is widely distributed from 1 to 3 nm, and the diameter of the inner tube of DWCNTs is distributed in the range from 0.8 to 1.8 nm [11]. Hereafter, the CNT sheets produced by eDIPS method are simply referred to as CNT sheets.

Figure 8.1 shows a schematic diagram of the electrode structure for FE experiments. A CNT sheet, 2 mm square or 5 mm square in size, was glued on upper surfaces of cylinders (diameter 10 mm or 5 mm) of metal (aluminum or stainless steel) with conductive paste to form FE cathodes. Corners of the metal cylinders were rounded to prevent concentrating of electric field to the edges. The CNT sheet on the metal cylinder (cathode) was faced to an anode with a small distance on the order of 1 mm. Figure 8.2 shows scanning electron microscope (SEM) images of a CNT sheet attached to a metal cylinder before FE measurement. A low magnification image (Fig. 8.2(a)) shows that the surface as a whole is flat though some unevenness is seen. An enlarged SEM image (Fig. 8.2(b)) shows that the bundles of CNTs are interwound to form a film. A tungsten grid was used for the anode, and a high voltage (up to 3 kV) was applied between electrodes with a gap of about 1.6 mm in vacuum (~10^{-4} Pa) until discharge occurred. Figure 8.3(a) is an example of the current-voltage characteristics of FE from a CNT sheet, in which discharge occurred at the cross mark. A picture taken at the moment when the discharge occurred is shown in Fig. 8.3(b). When a discharge occurs, a sharp light emission is observed between the cathode and the anode.

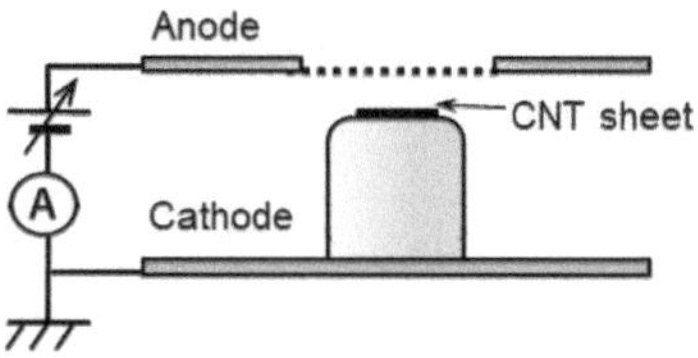

Figure 8.1 Schematic of a diode-type field emission setup with a CNT sheet cathode.

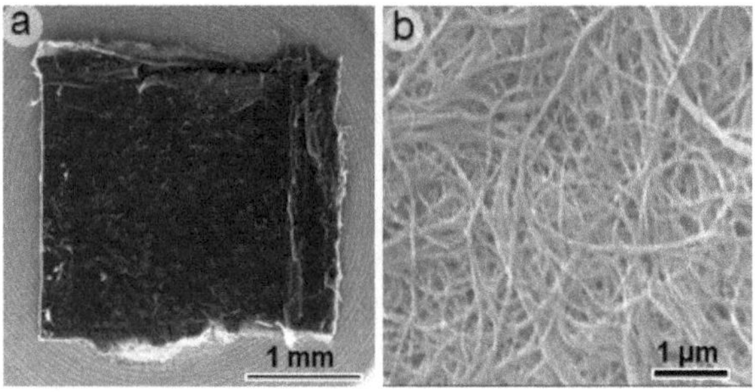

Figure 8.2 SEM images of a pristine CNT sheet with (a) low and (b) high magnifications before field emission experiment.

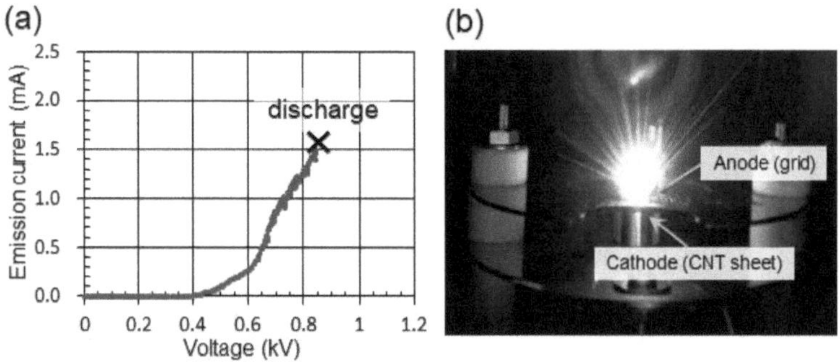

Figure 8.3 (a) Current-voltage characteristic of field emitted electrons from a CNT sheet. (b) Photo of a discharge event between the CNT cathode and anode.

8.3 Raman Spectroscopy of SWCNT Film After Discharge

Figure 8.4(a) shows a SEM image of a surface of the CNT sheet after discharge. Some areas are intact, but damaged areas where the film is turned up and heavily rugged are observed. Raman scattering spectra (excitation wavelength 532 nm) from the intact area (Fig. 8.4 (upper region)) and the damaged area (Fig. 8.4 (lower region)) are shown in Figs. 8.5(a) and (b), respectively [9]. Raman measurements were carried out by using the confocal Raman microscopes (Renishaw inVia, and Nanophoton RAMANview). The spectrum from the intact region represents typical Raman spectral features from SWCNTs/DWCNTs; a prominent G band at about 1585 cm^{-1}, radial breathing modes (RBM) at low-wavenumber region (130–270 cm^{-1}), D at ~1340 cm^{-1}, and 2D (G') bands (~2680 cm^{-1}) are observed. RBM is characteristic of slender CNTs, such as SWCNTs and DWCNTs. The D band is induced by disorders in the six-membered ring structure, and the 2D band (also called G' band, not related to defects) is due to double resonance Raman scattering of phonon modes with a high density of states near the Brillouin zone boundary [12]. On the other hand, spectra from the damaged region (Fig. 8.5(b)), a new peak is observed near 1850 cm^{-1}. In carbon materials, the vibration mode near from 1800 to 2200 cm^{-1} corresponds to the frequency of the triple bond of carbon in *sp* coordination, and is a longitudinal optical phonon mode caused by the expansion and contraction oscillation of the single and triple bonds of the one-dimensional carbon chain (insert in Fig. 8.5(b)). We will call this Raman band the "LCC" band. Figure 8.4(b) shows the Raman mapping of the LCC band on the surface of the CNT cathode after discharge. The brighter the red, the stronger the strength of the LCC band. As can be seen in comparison with the SEM image of the same area (Fig. 8.4(a)), the LCC band appears strongly in the area damaged by the discharge, and the LCC band does not appear in the intact part. The regions containing LCCs are not uniform, but are rugged in fairly fine sizes (on the order of 10 to 100 μm). In some samples, scars of discharge with an interesting shape are found as shown in Fig. 8.6. The "volcano" like shape suggests an explosive release of thermal energy from a localized hot spot within the CNT film.

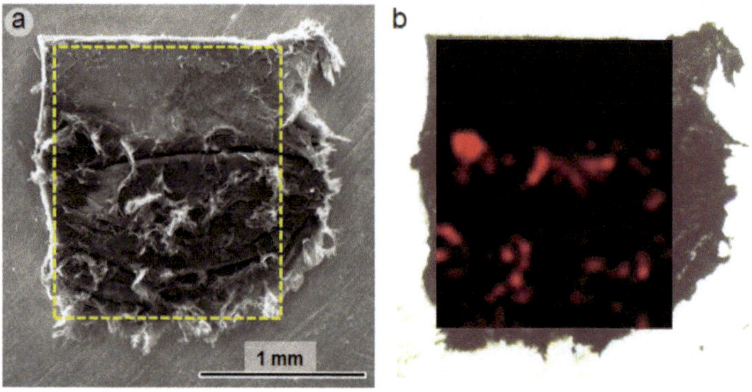

Figure 8.4 (a) SEM image of a CNT sheet after electric discharge. The region of Raman mapping is outlined with the yellow dotted rectangular. (b) Raman mapping showing a distribution of the LCC band on the surface of a CNT sheet after discharge. Brighter red color represents stronger LCC band.

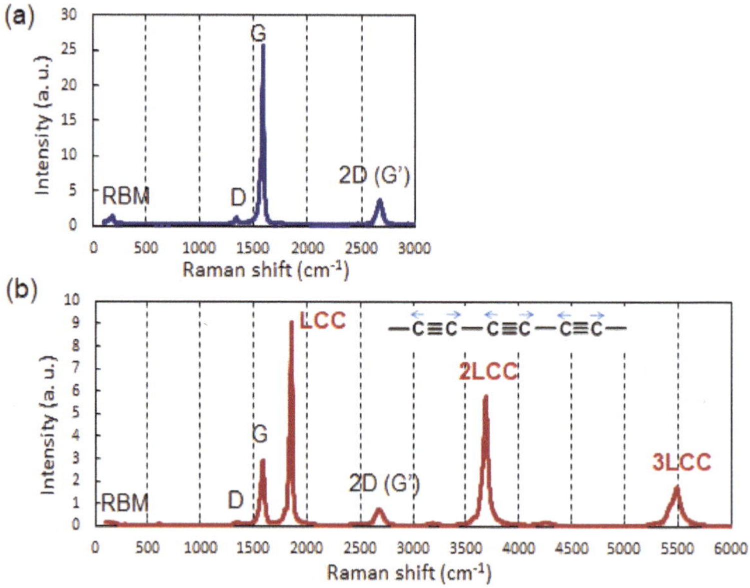

Figure 8.5 Raman scattering spectra of (a) an intact and (b) a damaged region of the CNT sheet, measured with 532 nm laser excitation at room temperature. Figure reprinted from Ref. [9].

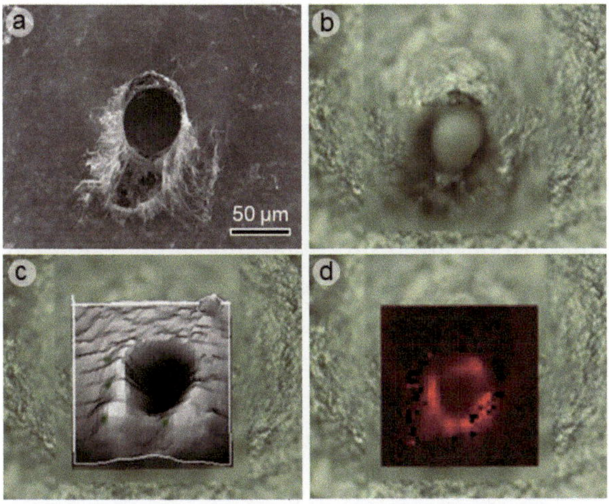

Figure 8.6 Discharge scar with "volcano" like shape. (a) SEM image, (b) optical micrograph, (c) Raman track, and (d) Raman mapping of LCC band (532 nm excitation).

In the spectrum of Fig. 8.5(b), the intensity of the LCC band is stronger than that of the G band, and also higher harmonics of the LCC band are clearly observed, showing that resonant Raman scattering occurs, i.e., the energy of the incident photon coincides with that of the electronic transition. By using resonant Raman, information on the carbon chain length and the energy band gap can be obtained.

To investigate the Raman shift and the intensity of the LCC band depending on the excitation photon energy, the Raman scattering spectra were measured using a range of incident lasers with wavelengths 355, 488, 532, 633, 671 and 785 nm. The results are shown in Fig. 8.7 [9]. LCC bands are observed at excitation wavelengths 488, 532, 633 and 671 nm, though they are not distinct at 355 and 785 nm. The intensity and peak shape of the LCC band depend on the excitation wavelength, and the 532 nm excitation gives the highest intensity. The width of the LCC band is significantly wide at 633 and 671 nm, and splitting of the band into peaks and shoulder is clearly observed. An example of peak separation of the LCC band is shown in Fig. 8.8. Raman shifts and relative intensity of LCC band depending on the excitation photon energy are shown in Fig. 8.9. In this figure, the intensity of the LCC band is represented by the size of the circles. The Raman shift of the LCC band shows a tendency to move to higher wavenumber with increasing the photon energy. For example, at 2.54 eV (488 nm) excitation, the median of the LCC band is close to 1855 cm^{-1}, and at 2.33 eV (532 nm) excitation, the median of the LCC band is close to 1850 cm^{-1}. At the lower excitation energies of 1.96 eV (633 nm) and 1.85 eV (671 nm), the low-frequency end of the band drops to about 1790 cm^{-1}. The frequency of LCC mode depends on the length of the carbyne, and it has been reported that it changes to a lower frequency with the increase of the length, and saturates around 1790 cm^{-1} when the number of carbons exceeds 1000 [13].

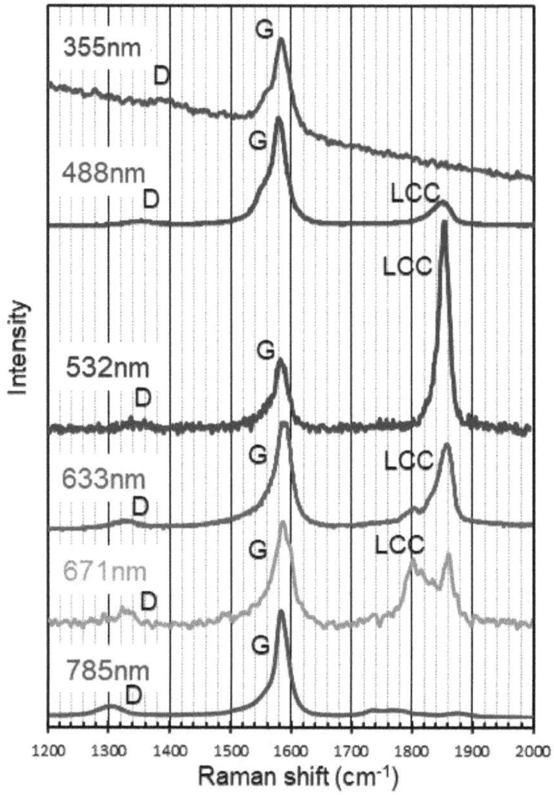

Figure 8.7 Raman spectra depending on excitation wavelength. Figure reprinted from Ref. [9].

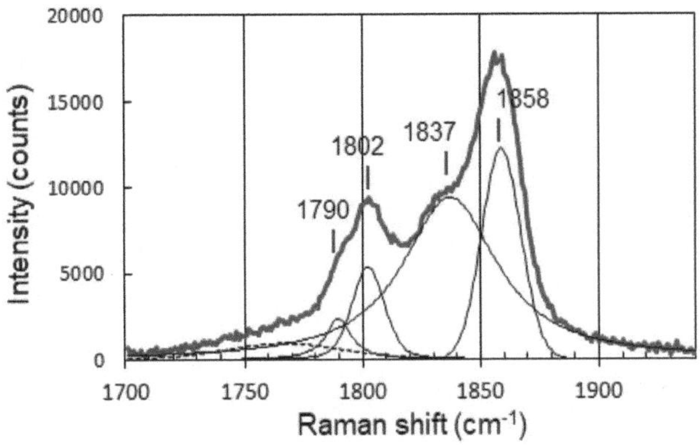

Figure 8.8 Peak separation of a LCC band observed with 633 nm excitation.

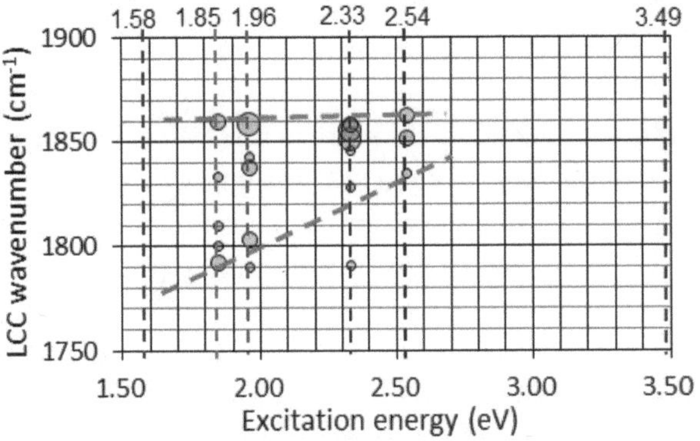

Figure 8.9 LCC peak positions and intensity versus excitation energy. Size of circles represents the intensity of LCC peaks; larger circles show higher intensity.

The intensity of the LCC bands is strongly affected by the resonance between the incident photon and the energy gap. The observed multi-peaks and shoulders in the LCC bands (*e.g.*, Fig. 8.8) suggest the presence of distributed lengths of LCCs and different environments surrounding them. In fact, in a well dispersed, diluted sample in which only one or few CNTs are shined by the excitation laser beam (~1 μm in diameter), a single sharp LCC signal is obtained as shown in Fig. 8.10.

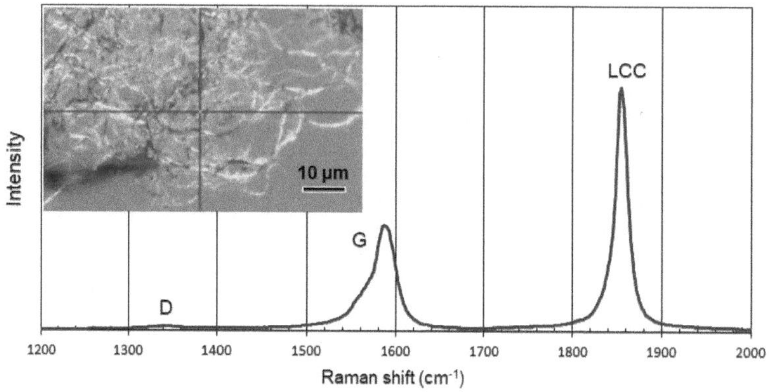

Figure 8.10 Raman spectrum from a well dispersed, diluted sample. The inset is an optical micrograph, showing the location where the spectrum was obtained.

Since the LCC bands are observed by incident photon energies in a range of 1.85–2.54 eV, it is suggested that the electron energy gaps of the LCCs formed in the CNT sheet are in this range. In addition to the distribution of the length of the LCC, the energy gaps are presumed to be influenced by differences in the environment surrounding the LCC, such as the diameter of CNTs containing LCCs and the presence or absence of adjacent LCCs, as described in the next section.

The G-band was peaked at 1589 cm^{-1} in a pristine CNT sheet, but it was observed at 1584 cm^{-1} on average in the damaged region of CNTs sheet after discharge, shifting to the lower wavenumber by about 5 cm^{-1}, as shown in Fig. 8.11. It is known that the shift of the G peak occurs due to lattice strain or charge doping. When CNTs are strained, the G-band shifts to the low frequency [14], while when the charge is doped, it shifts to the high frequency regardless of the sign of charge [15], suggesting that the observed redshift of the G-band is caused by CNT distortion due to LCC inclusion and/or discharge damage. The G/D intensity ratio of the Raman spectrum (the intensity ratio of the G to the D bands) was 38 for the pristine CNT sheet and 16 for the damaged CNT sheet (containing the LCC). CNT damage due to discharge is also reflected in the G/D intensity ratio of the Raman spectra.

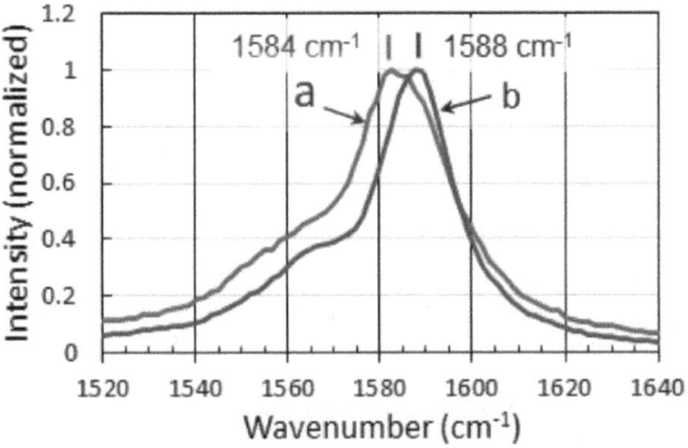

Figure 8.11 G-band spectra from (a) a CNT sample with LCC signals, and (b) a pristine CNT film. Figure reprinted from Ref. [9].

8.4 Transmission Electron Microscopy of LCC inside CNT

Small clumps of CNTs peeled off from the surface of CNT sheets after discharge were ultrasonically dispersed in isopropanol, and the dispersion was dripped onto TEM grids to obtain samples for TEM observation. TEM observations revealed that carbyne is contained in SWCNT, DWCNT, and TWCNT. Figure 8.12 shows typical TEM images of respective CNTs containing LCCs and cross-sectional profiles of TEM contrast. The inner diameters of SWCNTs, DWCNTs, and TWCNTs containing the single carbyne shown here are 1.5, 1.1 and 0.8 nm, respectively. Even in SWCNTs with large inner diameters, the carbyne is located almost in the center of the cavity. Interestingly, it was also observed that two or three carbynes were encapsulated in parallel in a single CNT (Fig. 8.13).

The inner diameter of the DWCNT in Fig. 8.13(a) containing two carbynes is 1.1 nm, and the spacing between the inner carbynes is 0.35 nm, which corresponds to the spread of the electron cloud of π orbits. On the other hand, the inner diameter of the DWCNT containing three carbynes (Fig. 8.13(b)) is 1.2 nm, which is about the same as that containing two inner carbynes, but the distance between the internal carbynes is 0.24 nm being unusually narrow. Since TEM pictures are projected images, information on depth is not known; i.e., the locations of each carbyne along the direction of depth (perpendicular to the image plane) are unclear. Presumably, the three carbynes are not aligned in the same plane, but are displaced vertically from each other to form a triangular cross section.

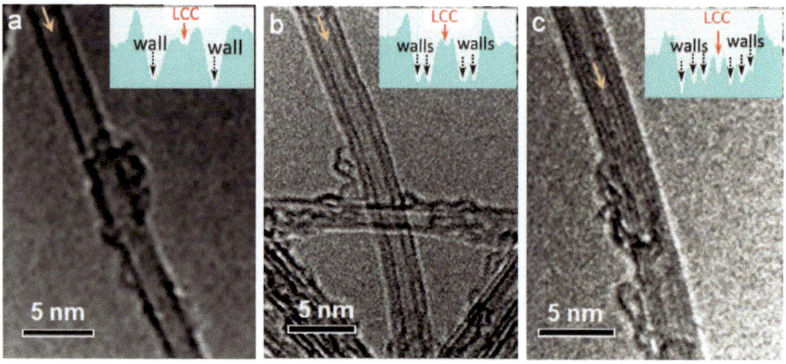

Figure 8.12 TEM images of an LCC encapsulated in (a) SWCNT, (b) DWCNT and (c) TWCNT. An arrow in each TEM image indicates the position of an LCC. Inset in each TEM image is a contrast profile across the CNT axis. Dips in the profiles correspond to dark places; red arrow indicates an LCC, and dotted arrows, CNT walls.

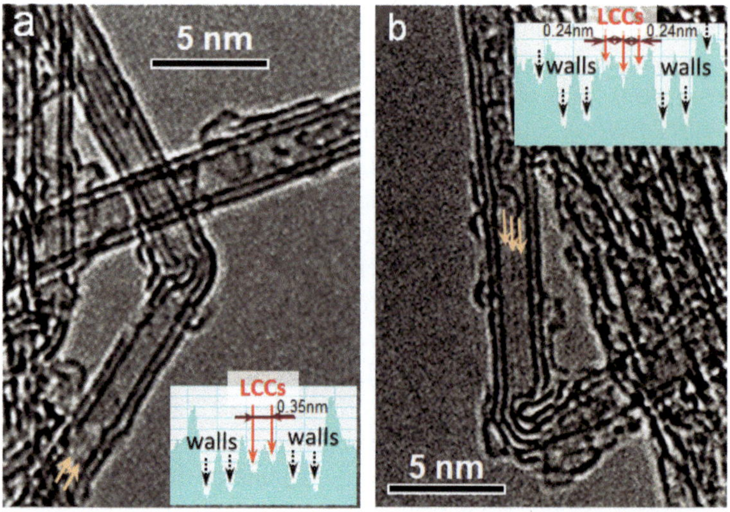

Figure 8.13 TEM images of (a) two and (b) three LCCs in parallel with each other inside CNT. Arrows in TEM images indicate the positions of LCCs. Inset in each TEM image is a contrast profile; red arrows indicate LCCs, and dotted arrows, CNT walls.

As for the length of the carbyne contained in CNT, it is difficult to capture the total length from an end to the other end by TEM, but the length of the carbyne is at least 30 nm (~230 in terms of the number of carbon atoms). Raman scattering spectroscopy suggests that the number of carbon atoms constituting carbyne is distributed in the range of tens to 1,000 or more from the observed range of LCC modes (1860 to 1790 cm^{-1}) [16].

8.5 Features of LCC@CNT Formed by Discharge Method

Although it is difficult to accurately measure the yield of LCCs (the proportion of CNTs containing LCCs), we will estimate them roughly from the intensity ratio (I_{LCC}/I_G) of the LCC band and the G band in the Raman spectra. The maximum I_{LCC}/I_G ratio was about 2.2. On the other hand, Shi *et al.* reported that Raman scattering from a single DWCNT being completely filled with an LCC shows the I_{LCC}/I_G ratio of 6.1 [7]. If this value is the I_{LCC}/I_G ratio when all CNTs contain LCCs (i.e., 100% yield), the yield of LCC@CNT formed by the discharge method is estimated roughly 36% at best. On the other hand, in TEM observations, LCC@CNT is found only about 1 in 10 to 20 CNTs. In the sample preparation for TEM, tiny lumps of CNTs of about 1 mm in size are peeled off from a film of CNT cathode, so CNTs are also captured from regions with low LCC@CNT density (or non-existent). Since the confocal Raman microscope probes with a resolution of ~1 μm, it can detect localized regions with high density LCC@CNT, but in TEM, the yield is assumed to be much lower than Raman's evaluation because the high and low densities are averaged.

In previously reported LCCs encapsulated in CNTs, the CNTs that stably contained LCCs were limited to DWCNTs and MWCNTs [1, 17]. Shi *et al.* report that LCC inclusions are limited to DWCNTs after high-temperature annealing [7]. Zhao *et al.* first discovered LCCs confined inside MWCNTs formed by arc discharge between graphite electrodes [1]. The TEM results presented here shows that LCCs can be included in SWCNTs as well as DWCNTs and TWCNTs. The diameter of SWCNTs containing a single LCC is in the range from 1.22 to 1.9 nm, which is wider than the diameter (0.7–0.85 nm) of the internal cavities of DWCNTs and MWCNTs previously reported

[1, 17], respectively. The number of LCCs contained in a single CNT was also limited to one in previous reports, but in the FE discharge method, two or three LCCs were encapsulated side by side parallel to each other. The internal cavity of a CNT containing multiple LCCs in parallel is expected to be wider than in the case of a single LCC. In fact, while the diameter of the internal cavities of DWCNTs and TWCNTs containing one LCC was 0.83–1.3 nm, that of CNTs containing two LCCs was 1.1–1.2 nm, and that of CNTs containing three LCCs was 1.4 nm. The diameter of the internal cavity is not much different between CNTs containing one LCC and two LCCs, but the internal cavity of a CNT containing three LCCs is clearly larger.

8.6 Hypothetical Formation Mechanism of LCC@CNT

The mechanism of inclusion of LCCs in CNTs *via* FE discharge is not clear, the discussion here is speculative. Besides FE discharge of CNT cathode described in this chapter, carbon arc discharge for MWCNT production [1, 18, 19], heat treatment of DWCNTs (1400–1450 °C) [7, 17], and heat treatment of X-ray irradiated SWCNTs (450 -500 °C) [20] are known. All the physical methods of generating LCC@CNT utilize radical processes involving high temperatures and high-energy particles. This suggests that the destruction of carbon six-membered rings and the following reconstruction of CNTs are involved in LCC@CNT formation. In the case of carbon arc for MWCNT synthesis, LCCs were formed inside MWCNTs when the diameter of the central cavity was thin (about 0.7 nm). A hypothetical mechanism explaining how LCCs are trapped at the center of a MWCNT has been proposed by Kang [18]; carbon that was initially molten on the cathode of the arc discharge begins to solidify from the outside during the process of forming MWCNTs, and finally the carbon remaining at the center forms a one-dimensional chain in a long, narrow cavity.

In the FE discharge, DWCNT heat treatment, and X-ray irradiation, in which pre-formed CNTs are used as the start materials for LCC formation, the temperature is not high enough to melt the CNT, so it is considered that other mechanisms are working. Carbon-containing molecules or carbon clusters, which may have been left in the cavity of the CNT during CNT synthesis/purification, are fused with each other to form linear chains in the narrow one-dimensional cavity *via* migration of carbon clusters activated by FE discharge, heat treatment and X-ray irradiation. Alternatively, a catastrophic process, collapse and reconfiguration of CNT structures, maybe considered to work in the carbyne formation, especially for the FE discharge method, i.e., CNTs are collapsed into tiny flakes of carbon clusters by high density of energy injected to CNTs in a short time, and the fragments are subsequently reconstructed into CNTs and encapsulated LCC. The destruction/reconstruction hypothesis does not contradict the historic report claiming the carbyne formation in graphite gneiss by a meteorite impact [21] and previous studies aiming at carbyne formation by using a high-energy process, laser irradiation onto a graphite [22], ion sputtering of graphite in combination with a bombardment of the growing carbon film by Ar^+ ions [23] and so on.

8.7 Summary and Perspectives

Preparation of LCCs longer than 100 carbon atoms stably encapsulated inside CNTs have been demonstrated so far by carbon arc-discharge in hydrogen [1, 24], by argon arc in atmosphere [18, 19], and by heat treatment of DWCNTs [7, 17]. In this chapter, the new method utilizing CNTs as starting materials and electric discharge for the structure modification is described. The LCCs formed by this method have the following characteristics; (1) LCCs longer than 30 nm are encapsulated mainly inside SWCNTs and DWCNTs, and sometimes encapsulated in TWCNTs, and (2) two or three LCCs parallel to each other are encapsulated in a single CNT.

Vibrational spectroscopy using Raman scattering is a powerful measurement method for investigating the structural features, chemical bonds, and atomic coordination of materials, but the fact that carbyne is contained in CNTs suggests that environmental effects are reflected in the carbyne's Raman spectra. In order to observe the electronic and vibration structures inherent in carbyne and to understand the effects of hosting CNTs, it is desirable to extract carbynes from CNTs and isolate them. For this purpose, Shi *et al.* attempted to extract LCCs from LCC@DWCNT by applying a combined tip-ultrasonic and density gradient ultracentrifugation (DGU) process, but it seems to be difficult to extract LCCs alone though LCCs are extracted together with inner tubes of DWCNTs [24]. The extracted LCCs inside the inner tubes (LCC@SWCNT) were significantly shortened due to the ultrasonication treatment; since DWCNTs are quite robust, hard ultrasonic irradiation might be required. Therefore, if the carbyne contained in SWCNTs obtained by the present method is used, it may be possible to destroy the outer SWCNT and extract the carbyne while suppressing the damage of the carbyne.

Considering the methods of physical generation of LCCs reported so far, ion bombardment on CNT sheets/subsequent heat treatment and explosive heating (burnout) of a thread of CNT yarn may also be new methods for forming carbyne in CNTs.

References

1. Zhao, X., Ando, Y., Liu, Y., Jinno, M., and Suzuki, T. (2003). Carbon nanowire made of a long linear carbon chain inserted inside a multiwalled carbon nanotube, *Phys. Rev. Lett.* **90**, 187401.

2. Scalese, S., Scuderi, V., Bagiante, S., Simone, F., Russo, P., D'Urso, L., Compagnini, G., and Privitera, V. (2010). *J. Appl. Phys.* **107**, 014304.

3. Scuderi, V., Scalese, S., Bagiante, S., Compagnini, G., D'Urso, L., and Privitera, V. (2009). Direct observation of the formation of linear C chain/carbon nanotube hybrid systems, *Carbon* **47**, 2134–2137.

4. Andrade, N. F., Vasconcelos, T. L., Gouvea, C. P., Archanjo, B. S., Achete, C. A., Kim, Y. A., Endo, M., Fantini, C., Dresselhaus, M. S. and Filho, A. G. S. (2015). Linear carbon chains encapsulated in multiwall carbon nanotubes: resonance Raman spectroscopy and transmission electron microscopy studies, *Carbon* **90**, 172–180.

5. Cazzanelli, E., Castriota, M., Caputi, L., Cupolillo, A., Giallombardo, C., and Papagno, L. (2007). High-temperature evolution of linear carbon chains inside multiwalled nanotubes, *Phys. Rev. B* **75**, 121405R.

6. Zhao, C., Ksitaura, R., Hara, H., Irle, S., and Shinohara, H. (2011). Growth of linear carbon chains inside thin double-wall carbon nanotubes, *J. Phys. Chem. C* **115**, 13166–13170.

7. Shi, L., Shen, L., Yu, L., Am, K., Ando, Y., and Zhao, X. (2011). Ultra-thin double-walled carbon nanotubes: a novel nanocontainer for preparing atomic wires, *Nano Res.* **4**, 759–766.

8. Toma, S., Asaka, K., Irita, M., and Saito, Y. (2019). Bulk synthesis of linear carbon chains confined inside single-wall carbon nanotubes by vacuum discharge, *Surf. Interface Anal.* **51**, 131–135.

9. Asaka, K., Toma, S., and Saito, Y. (2020). Raman spectroscopy and TEM of long linear carbon chain formed in CNT field emission cathode, *e-J. Surf. Sci. Nanotech.* **18**, 159–163.

10. Saito, T., Ohshima, S., Okazaki, T., Ohmori, S., Yumura, M., and Iijima, S. (2008). Selective diameter control of single-walled carbon nanotubes in the gas-phase synthesis, *J. Nanosci. Nanotechnol.* **8**, 6153–6157.

11. Toma, S. (2019). *Production of One-Dimensional Carbon Chains in Carbon Nanotubes By Vacuum Discharge*, Master's Degree thesis, Nagoya University.

12. Saito, R., Jorio, A., Filho, A. G. S., Dresselhaus, G., Dresselhaus, M. S., and Pimenta, M. A. (2002). Probing phonon dispersion relations of graphite by double resonance Raman scattering, *Phys. Rev. Lett.* **88**, 027401.

13. Yang, S., Kertesz, M., Zolyomi, V., and Kürti, J. (2007). Application of a novel linear/exponential hybrid force field scaling scheme to the longitudinal Raman active mode of polyyne, *J. Phys. Chem. A* **111**, 2434–2441.

14. Cronin, S. B., Swan, A. K., Ünlü, M. S., Goldberg, B. B., Dresselhaus, M. S., and Tinkham, M. (2004). Measuring the uniaxial strain of individual single-wall carbon nanotubes: resonance Raman spectra of atomic-force-microscope modified single-wall nanotubes, *Phys. Rev. Lett.* **93**, 167401.

15. Tsang, J. C., Freitag, M., Perebeinos, V., Liu, J., and Avouris, P. (2007). Doping and phonon renormalization in carbon nanotubes, *Nat. Nanotechnol.* **2**, 725–730.

16. Wanko, M., Cahangirov, S., Shi, L., Rohringer, P., Lapin, Z. J., Novotny, L., Ayala, P., Pichler, T., and Rubio, A. (2016). Polyyne electronic and vibrational properties under environmental interactions, *Phys. Rev. B* **94**, 195422.

17. Shi, L., Rohringer, P., Ayala, P., Saito, T., and Pichler, T. (2013). Carbon nanotubes from enhanced direct injection pyrolytic synthesis as templates for long linear carbon chain formation, *Phys. Status Solidi B* **250**, 611–2615.

18. Kang, C.-S., Fujisawa, K., Ko, Y.-I., Muramatsu, H., Hayashi, T., Endo, M., Kim, H. J., Lim, D., Kim, J. H., Jung, Y. C., Terrones, M., Dresselhaus, M. S., and Kim, Y. A. (2016). *Carbon* **107**, 217–224.

19. Saito, Y., and Asaka, K. (2021). Raman features of linear-carbon-chain and multiwall carbon nanotube composites, *IntechOpen*, DOI: 10.5772/intechopen.99465.

20. Murakami, T., Matsuda, M., Kisoda, K., and Itoh, C. (2016). Origin of a Raman scattering peak generated in single-walled carbon nanotubes by X-ray irradiation and subsequent thermal annealing, *AIP Advances* **6**, 085303.

21. Goresy, A. E., and Donnay, G. (1968). A new allotropic form of carbon from the ries crater, *Science* **161**, 363–364.

22. Whittaker, A. G., and Waites, G. M. (1972). Carbon: a suggested new hexagonal crystal form, *Science* **178**, 54–56.

23. Kudryavtsev, Y. P., Esyukov, S. E., Babaev, V. G., Guscva, M. B., Khvostov, V. V., Krechko, L. M. (1992). Oriented carbyne layers, *Carbon* **30**, 213–221.

24. Shi, L., Yanagi, K., Cao, K., Kaiser, U., Ayala, P., and Pichler, T. (2018). Extraction of linear carbon chains unravels the role of the carbon nanotube host, *ACS Nano* **12**, 8477–8484.

Chapter 9

Structural and Electrical Characterization of Atomic Carbon Chains in the Electron Microscope

Florian Banhart
University of Strasbourg, CNRS, Institut de Physique et Chimie des Matériaux, UMR 7504, 67034 Strasbourg, France
florian.banhart@ipcms.unistra.fr

Individual chains of carbon atoms can be created and characterized by *in situ* experiments in the transmission electron microscope. The imaging of the chains as well as measurements of their electrical properties are getting possible. Fundamental properties of carbon chains as perfectly one-dimensional conductors are revealed such as the existence of the two phases cumulene and polyyne and the applicability of Peierls theorem of linear metals.

Emerging Carbyne: Truly One-Dimensional Form of Carbon
Edited by Yahachi Saito

ISBN 978-981-5129-15-1 (Hardcover), 978-1-003-51036-9 (eBook)
www.jennystanford.com

9.1 Introduction

9.1.1 The Structure and Properties of Individual Carbon Chains

One-dimensional chains of carbon atoms are considered as the elementary building blocks of carbyne, the bulk phase of carbon in the sp^1 hybridization. In analogy to graphene as the two-dimensional unit of the sp^2 carbon phase graphite, we may assume that carbyne consists of a regularly packed arrangement of parallel carbon chains. While bulk carbyne has been reported to be chemically unstable against relaxation to sp^2-bonded carbon, in particular through reactions between neighboring chains, isolated chains, where only the ends are attached to end groups or bulk contacts, appear to be stable in a suitable environment. Indeed, carbon chains have been observed in several electron microscopy studies, and their signatures appear in Raman spectra.

The fundamental interest in carbon chains originates not only from their chemical behavior but in particular from their perfectly one-dimensional nature [1]. We can consider a chain of carbon atoms as the thinnest possible wire. Carbon chains have delocalized electrons and can behave either as metallic conductors or as semiconductors. This allows us to take the carbon chain as a model system to study the electrical properties of perfectly one-dimensional conductors. Unlike carbon nanotubes, which have a certain extension normal to their axis and whose electrical properties depend on their chirality and diameter, carbon chains have monoatomic thickness so that the electrical properties depend solely on the atomic periodicity along the chain. Therefore, carbon chains as one-dimensional conductors can be used to test Peierls' theory of linear metals [2].

According to Peierls, a linear metal with uniform distance between neighboring atoms is unstable against relaxation to a linear semiconductor where dimerization leads to the opening of a bandgap. The associated lowering of the total electronic energy of the system outweighs the increase in elastic energy. Such a relaxation is also expected in linear carbon chains when the lattice parameter a of a chain with uniform distance between neighboring atoms is doubled to $2a$ by the formation of dimers [3, 4].

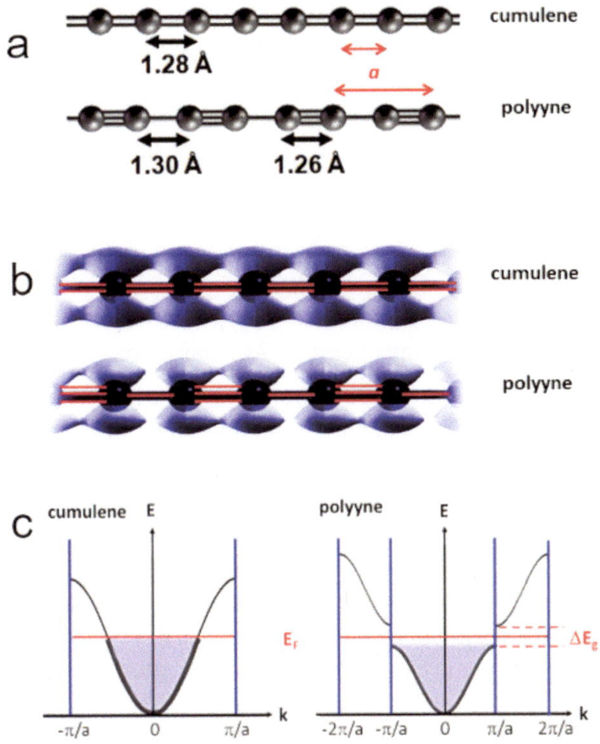

Figure 9.1 (a) Schematic bonding configuration of cumulene and polyyne. The respective lattice parameter *a* is shown in red. (b) Electron distribution in cumulene and polyyne. The bonds are indicated in red. Courtesy: A. Botello-Mendez and J.-C. Charlier. (c) Band structure of cumulene and polyyne. The edges of the Brillouin zones are marked with blue vertical lines at multiples of π/a. The Fermi energy is marked with a red line.

Two electronic configurations of carbon chains are possible (Fig. 9.1):

(1) In *cumulene*, all carbon atoms are distributed with equal interatomic distance *a* so that the uniform overlap of the *p*-orbitals delocalizes the electrons along the whole chain. This is usually symbolized by drawing double bonds between all neighboring carbon atoms (=C=C=C=C=C=). Every carbon atom contributes with two *p*-orbitals (p_x and p_y) normal to the alignment of the chain in the *z*-direction. The delocalization of the electrons along the chain leads to metallic conductivity.

(2) In *polyyne*, dimerization occurs so that the electron density over the bonds alternates between low and high which is depicted by alternating single and triple bonds (–C≡C–C≡C–C≡). Since charge in the *p*-orbitals is mainly localized between the two atoms of a dimer; i.e., over the shorter triple bonds, electron transport along the chain needs to involve hopping over the longer single bonds, which is associated with a certain activation energy and leads to semiconducting behavior. The energy gain of polyyne relative to cumulene is only 70 meV per atom, according to calculations [4], which is small compared to the total bonding energy of 7 eV per atom.

Carbon chains are an ideal system for studying the effects of Peierls distortion in a chain of monoatomic thickness. Simple electrical conductivity measurements allow to decide whether the chain is metallic or semiconducting, i.e., whether Peierls distortion is effective. However, geometric considerations and density functional calculations [5] indicate that long-range interactions in carbon chains are important and that the length of the chains and the bonding character at the terminating atoms play a certain role. While in very long chains (>50 atoms), the Peierls effect should stabilize polyyne over cumulene, in short chains, the bonding of the end atoms determines the configuration. As an example, if an end atom is saturated by a hydrogen atom, the C–H bond is always a single bond. This imposes a triple bond between the first and the second carbon atom, a single bond between the second and third atom, and so on. In reality, however, single, double or triple bonds cannot be clearly defined because the charge distribution may change in a continuous way, and the important parameter is the degree of dimerization.

In experiments by transmission electron microscopy, carbon chains are often attached to carbon particles where single or double bonds can occur at the ends, depending on whether the chain is attached to an open edge of a graphenic sheet or to the surface of the sheet. This is shown schematically in Fig. 9.2. The atom attached to the graphene edge (left hand side) is sp^2-hybridized and can form a double bond with a continuous π-electron system, allowing electrical transport over the contact. An atom attached to the surface of a graphenic structure, such as the cap of a nanotube (right hand side), is sp^3-hybridized with a single bond at the contact, leading to a gap in the π-electron system. Less electron transmission is expected over such a contact.

Figure 9.2 Distribution of π-electrons (in pink colour) over sp^2 (left) and sp^3 (right) contacts. Courtesy: A. Botello-Mendez and J.-C. Charlier.

Another restriction comes into play when the end atoms of the chain are saturated by a defined bond. If, for example, both end atoms are saturated by a single bond, the polyyne configuration can only be realized if the number of atoms in the chain is even. For an odd number of atoms, the charge has to be redistributed, and the conduction properties are changed. Therefore, the bond length alteration in short chains is controlled by the saturation of the end atoms (by either single or double bonds, depending on the attached species) and the number of atoms in the chain.

It is evident that carbon chains have to be stabilized against relaxation or reaction due to their high reactivity. For isolated chains spanning over free space, this is achieved by saturation of the highly reactive end atoms with end groups or metallic contacts, as will be described below. Alternatively, encapsulation of carbon chains in carbon nanotubes has shown to stabilize very long chains. The longest isolated carbon chains that have been reported so far had a length of only 40 atoms [6] whereas carbon chains encapsulated inside carbon nanotubes reached lengths up to 6000 atoms [7].

9.1.2 The Characterization of Individual Carbon Chains

The experimental characterization of individual carbon chains needs tools with high spatial resolution such as transmission electron microscopy (TEM), scanning tunneling microscopy (STM) or tip-enhanced Raman scattering (TERS). In this chapter, we will focus on TEM studies that allowed imaging the structure of carbon chains and on a combination of TEM and STM to study the electrical properties [8].

Carbon nanotubes wouldn't have been discovered without TEM, and TEM has also played a fundamental role in the characterization of graphene, multi-shell fullerenes and carbon composite particles. Modern TEMs have a spatial resolution of 0.5–1 Å. The spacing between two carbon atoms in a chain is of the order 1.2 Å, so that atom-by-atom imaging should be easily feasible with modern instruments. However, a detrimental effect in TEM has to be considered which is the momentum transfer to the atoms by collisions with the energetic electrons. For low-momentum transfers, the atoms remain in their positions, but the chain starts vibrating which makes atomic resolution in the TEM difficult. However, it has been demonstrated that atomic imaging of very short chains is possible nevertheless [9]. High momentum transfers lead to the rupture of the bonds or the complete knock-on displacement of carbon atoms so that the chain breaks. When the end atoms of the chains are attached to flexible supports such as small graphitic aggregates, the supports relax when a strained chain breaks so that qualitative information about the tensile strength of a chain can be obtained. An ultimate strength of 10 nN has been predicted for monoatomic carbon chains [10, 11] which could possibly be measured by using an AFM tip in a TEM specimen holder. Such systems are already available, but the precision of force measurements is still above 10 nN.

To measure the electrical properties of individual carbon chains, STM would actually be the method of choice. However, the imaging of the chain would be necessary at the same time as the electrical measurement to ensure that it is indeed a carbon chain that is spanning between the electrodes. This can be achieved in special TEM specimen stages where an STM tip is integrated and allows electrical measurements while the object is under observation in the TEM [12].

The chains can be created in such an experiment by contacting a graphitic (or graphenic) nanoparticle with the STM tip, passing an electrical current to weld the tip to the particle, and then slowly retracting the piezo-driven tip until (normally after several unsuccessful attempts) a carbon chain can be unraveled from the graphitic particle (Fig. 9.3). Then, the chain spans between the STM tip and the particle so that the electrical properties of the chain can be measured [13]. This can be done by taking current-voltage curves that allow to decide whether the chain is metallic or semiconducting.

Furthermore, the maximum current carrying capacity can be determined by raising the voltage and measuring the current until the chain fails.

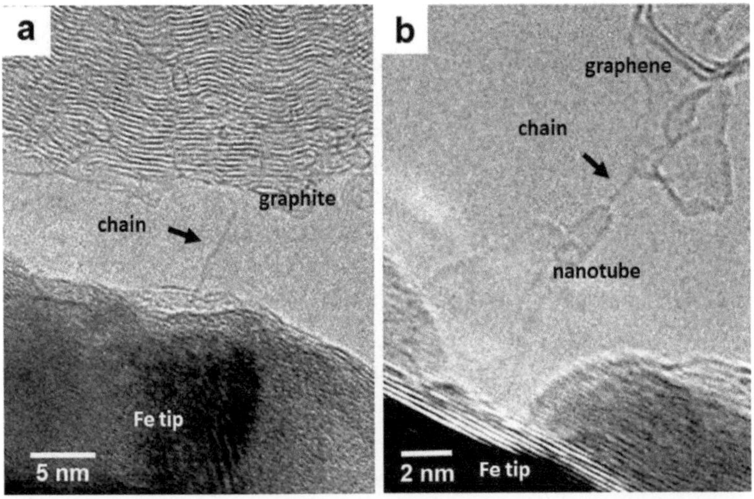

Figure 9.3 Electron microscopy images of carbon chains created *in situ* by "welding" an iron tip to a graphitic substrate (a) or to an aggregate of graphenic material (b). Retraction of the tip leads to the unraveling of carbon chains from the graphitic/graphenic substrate [13].

9.2 Imaging the Structure of Carbon Chains by Electron Microscopy

By imaging of the atomic structure, there are several structural characteristics of the chains that can be revealed, as we have seen above. We have to see if the chain is attached at one or both ends to a support, the nature of the support has to be known to understand the type of bonding to the chain, the length of the chain (i.e., the number of atoms) has to be measured, and the atom positions in the chain have to be determined to decide whether the chain is in the cumulene or in the polyyne configuration.

However, the first task is to prove the existence of isolated carbon chains. This was indeed achieved for the first time by TEM studies of carbon nanomaterials such as graphene or nanotubes [14–19].

By irradiating graphene layers or few-layer carbon nanotubes, it was seen that dark lines appeared where the graphene layer was removed or the nanotube was thinned by sputtering in the electron beam. Although some doubt was raised about imaging artifacts such as graphene ribbons in side-view projection, it became clear that most of the observed lines were indeed monoatomic chains of carbon atoms. Due to inevitable atom displacements by knock-on collisions in the electron beam, the lifetimes of the chains were short. But it is remarkable that the chains are more stable under the electron beam than narrow graphene ribbons. This is due to the large number of dangling bonds all along the edges of graphene, making narrow ribbons energetically unfavorable against sp^1 chains that only have open bonds, if any, at their ends. And this is also believed to be the reason why carbon chains can be unraveled from graphene edges if a metal tip is contacted to the edge (see the section below).

Now, that it is possible to identify carbon chains in TEM images, their length can at least be estimated (they might appear shorter than they are due to projection effects if they are inclined relative to the image plane). However, the more difficult task is to determine their atomic configuration and to find out whether they are in the cumulene or polyyne modification. This is due to the small distance between the atoms of the order 1.3 Å, to the low contrast of single carbon atoms in TEM images and to a possible vibration of the chain under the electron beam. Figure 9.3 shows two typical TEM images of carbon chains where the chains can be identified although an atomic resolution is not possible. By high-resolution scanning TEM (STEM), the imaging of single carbon atoms within chains became finally possible [9]. Surprisingly, the observed dimerization was much stronger than expected, showing bond lengths from 0.93 to 1.58 Å in carbon chains spanning between graphene edges. Since the calculated lengths of single, double and triple bonds in carbon chains are assumed to be 1.265, 1.282 and 1.301 Å, respectively [5], an explanation for the large interval of the measured atom spacings still has to be found.

Carbon chains have also been seen in the hollow channel of carbon nanotubes [7, 20]. They are more difficult to identify than isolated chains due to possible Fresnel contrast fringes along the nanotubes. However, TEM combined with Raman spectroscopy has shown evidence for the existence of such encapsulated chains.

9.3 Measuring the Electrical Properties of Carbon Chains

As described above, electrical measurements on atomic carbon chains have been carried out by contacting graphitic material with an STM tip and successively unraveling the chains while the tip is slowly retracted. To establish a covalent bond between the tip and the chain, tungsten and iron tips were used in these experiments [21]. The initial welding of the tip to the graphitic materials needed currents of the order 100 μA; then the current was lowered to 10 μA while the tip was retracted. The lifetime of the chains is typically some tens of seconds and limited by mechanical stress between the tip and the substrate, by unavoidable electron irradiation or by too high currents that lead to failure.

As long as the chains are stable, their current-voltage characteristics can be measured during a voltage sweep, which was typically in the range of ±500–1000 mV. Figure 9.4 shows two examples of current-voltage curves. Most of the current-voltage measurements taken under such conditions show the S-type curves which are characteristic for a bandgap, indicating a semiconducting behavior [13, 22, 23]. This, in turn, shows that the chains are of the polyyne type which is in accordance with imaging studies [9]. In an asymmetric contact configuration (i.e., metal on one side and graphitic carbon on the other side, a rectifying behavior can be observed [22], and the chain-metal bond behaves like a Schottky contact. In a few cases, however, when the chains were visibly curved and therefore under no stress, linear current-voltage curves were obtained, as expected for metallic conductors [22].

The transition between metallic and semiconducting behavior is described by the influence of the stress state of the chains on the electrical conductivity, as studied by quantum conductivity calculations in the Landauer-Büttiker approach. If an unstrained chain is in the cumulene phase, increasing tensile stress leads to increasing dimerization and so to an opening of the bandgap which increases with increasing strain [4, 13]. Qualitatively, this can be understood by the fact that single bonds between carbon atoms are less strong than double or triple bonds, so that the elongation mainly occurs by stretching the single bonds. DFT calculations of a 10-atom chain in the polyyne phase show that the bandgap increases from 0.4 eV at zero strain up to almost 2 eV at a relative strain of $\Delta l/l = 0.12$ [13].

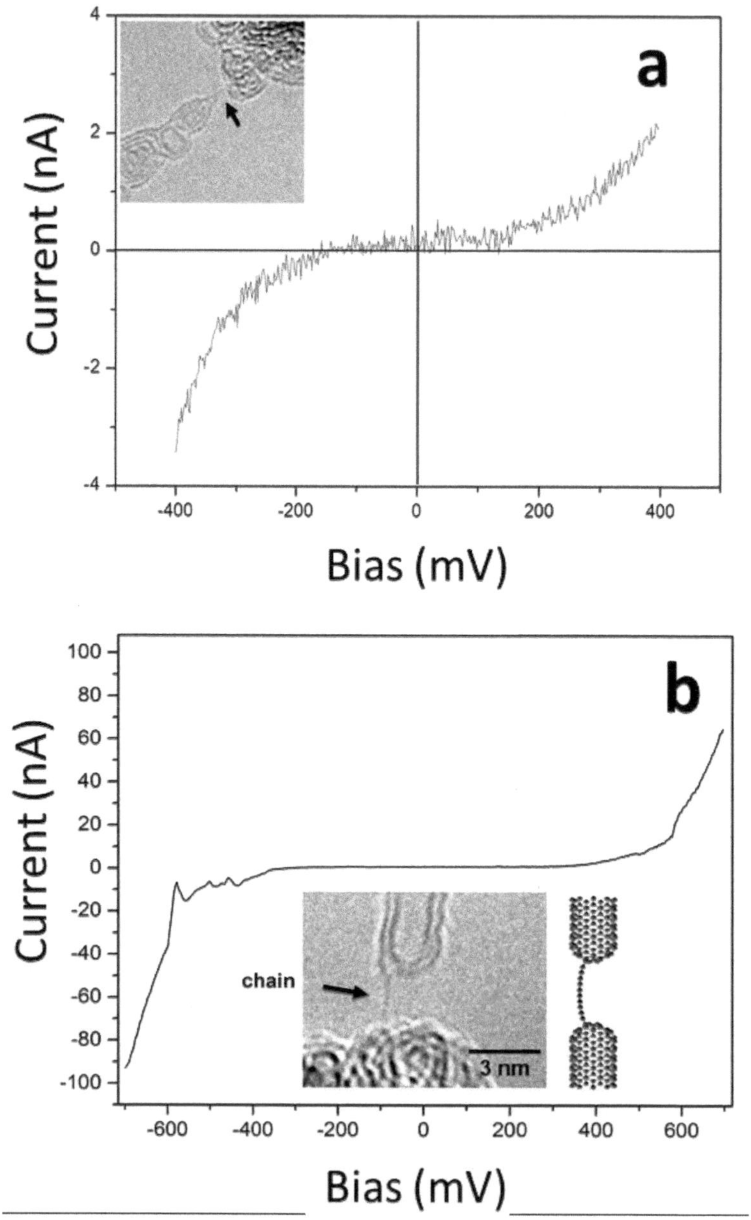

Figure 9.4 Current-voltage curves of two typical carbon chains [22, 23]. In the example in (a), the chain is spanning between two graphitic aggregates, whereas in (b) the chain connects a double-wall carbon nanotube and graphitic material. The width of the plateau regions is different in (a) and (b), presumably due to the different contacts.

An important feature of carbon chains is their current carrying capacity. This was studied by increasing the voltage and simultaneously measuring the current up to the failure of the chain. The maximum current that was measured was 6.5 μA at an applied voltage of 1.8 V [23]. This is a remarkably high current in view of the fact that the whole current flows through a monatomic chain. The current could easily be measured with a simple analog amperemeter without amplifier.

It is interesting to compare these values with the expectation from quantum conductivity as described by the Landauer-Büttiker theory. Since quantized conductance has already been observed earlier in rows of metal atoms [24], a similar behavior was also expected for carbon chains. In contrast to chains of metal atoms, metallic carbon chains in the cumulene configuration have two elementary conduction channels that cross the Fermi level and both half-filled degenerate *p*-bands (p_x and p_y) contribute to conductivity. The quantum conductivity with ideal contacts (transmission factor $T = 1$) is therefore $G = 2\ G_0 = 4e^2/h$ (h is Planck's constant and e the elementary charge) which would result in a current of 300 μA at 1.8 V. This is almost a factor of 50 higher than measured. The discrepancy can be explained by the contacts that reduce the transmission probability so that ballistic conductivity is only achieved in the chain but not over the contacts which limit the overall transmission. Transport through the contacts is determined by narrow resonant states due to reflections at the contacts which reduce the transmission [25]. Furthermore, the contacts at the two ends of the chain might be different materials (metal and graphitic), so that an integration between the different Fermi energies has to be carried out. Like graphene, carbon chains are also expected to change their electrical properties when other species (atoms, molecules) are attached [26].

Besides the electrical characterization, a qualitative mechanical study of the chains can also be carried out in the STM-TEM specimen stage. Once a carbon chain has been established between a graphitic aggregate and the STM tip (Fig. 9.3), the slow retraction of the tip allows to observe the relaxation of the graphitic aggregate when the chain breaks. A remarkable relaxation has already been observed [22], showing that these monoatomic chains can indeed withstand considerable tensile forces.

9.4 Future Experimental Challenges and Outlook

Due to the difficulty of making and characterizing individual chains of carbon atoms, the research on these interesting species is still in its beginnings and some of the results obtained so far have to be considered as preliminary. To measure the electrical and mechanical properties of carbon chains, STM and AFM would be the ideal techniques because they have the necessary precision in the measurements of electrical currents or mechanical forces to reveal the characteristics of individual chains. However, the difficulty is that the chains attached to the STM or AFM tips cannot be seen during the measurements so that it remains uncertain on which object the measurement has been taken. It is therefore important to use STM or AFM stages in TEMs where the chains can be identified. The disadvantage of such combined measurements is the influence of radiation damage by the electron beam. Future studies with lower acceleration voltage of the TEM (20–60 kV) could avoid radiation damage, yet at the expense of somewhat lower image resolution. The high magnetic field in the objective lens of TEMs might also have an influence on the measurements and could be reduced in future experiments either by turning the lenses off during the electrical measurements (which is more difficult than it seems to be) or by experiments in field-free lenses.

More precise measurements of the electrical characteristics of different types of chains are desirable. They could already be done with existing equipment. A reliable correlation between the number of atoms in the chain (odd/even) and the electrical characteristics would be helpful to test theoretical predictions [5]. Of particular importance is the type of the contact at the end atoms of the chain. A better design of the contacts is therefore necessary to find out of which type the bond to the contact could be (single vs. double carbon-carbon bonds or carbon-metal bonds). In this context, the characterization by electron energy-loss spectroscopy (EELS), that has already been shown to be applicable [9], could give more insight, in particular in combination with electrical measurements.

Although TEM stages with integrated AFM tips already exist, no measurements of the elastic properties or a determination of the fracture strength have been reported yet. This is most likely due to the low sensitivity of these integrated stages that cannot measure forces much below the predicted ultimate strength of carbon chains of the order 10 nN [11]. Improved specimen stages could provide reliable data in the future.

In view of our present knowledge about carbon chains, it is not possible to conclude on their applicability in future devices. The obstacle is the high instability of the chains against reaction with their environment and against relaxation to more stable graphitic material. Furthermore, a reliable synthesis of individual chains doesn't exist yet. If we consider the persisting difficulty of making carbon nanotubes with defined chirality in bulk quantities, even after more than 30 years of intense worldwide research, we shouldn't be too optimistic about possible applications of the much more elusive carbon chains. However, carbon chains remain highly interesting model systems for testing fundamental principles of perfectly one-dimensional materials. The mechanisms of electron transport and the mechanical properties, that depend here on single chemical bonds, are unique in atomic one-dimensional systems. And sp^1 carbon is so far the only material that allows us building electrically conductive monoatomic chains of lengths of more than a few atoms.

Acknowledgments

The author is indebted to O. Cretu, A. La Torre, F. Ben Romdhane, A. Botello-Mendez, J.-J. Adjizian, and J.-C. Charlier for fruitful collaboration in the field of carbon chains.

References

1. Banhart, F. (2015). Chains of carbon atoms: a vision or a new nanomaterial? *Beilstein J. Nanotechnol.*, **6**, 559–569.
2. Peierls, R. E. (1964). *Quantum Theory of Solids,* Clarendon, Oxford, UK.
3. Kertesz, M., Koller, J., and Azman, A. (1978). Ab initio Hartree–Fock crystal orbital studies. II. Energy bands of an infinite carbon chain. *J. Chem. Phys.*, **68**, 2779–2782.
4. Artyukhov, V. I., Liu, M., and Yakobson, B. I. (2014). Mechanically induced metal-insulator transition in carbyne. *Nano Lett.,* **14**, 4224–4229.

5. Cahangirov, S., Topsakal, M., and Ciraci, S. (2010). Long-range interactions in carbon chains. *Phys. Rev. B*, **82**, 195444.

6. Chalifoux, W. A., and Tykwinski, R. R. (2010). Synthesis of polyynes to model the sp-carbon allotrope carbyne. *Nature Chem.*, **2**, 967–971.

7. Shi, L., Rohringer, P., Suenaga, K., Niimi, Y., Kotakoski, J., Meyer, J. C., Peterlik, H., Wanko, M., Cahangirov, S., Rubio, A., Lapin, Z. L., Novotny, L., Ayala, P., and Pichler, T. (2016). Confined linear carbon chains as a route to bulk carbyne. *Nature Mater.*, **15**, 634–639.

8. Banhart, F., La Torre, A., Ben Romdhane, F., and Cretu, O. (2017). The potentials and challenges of electron microscopy in the study of atomic chains. *Eur. Phys. J. Appl. Phys.*, **78**, 20701.

9. Lin, Y.-C., Morishita, S., Koshino, M., Yeh, C.-H., Teng, P.-Y., Chiu, P.-W., Sawada, H., and Suenaga, K. (2017). Unexpected huge dimerization ratio in one-dimensional carbon atomic chains, *Nano Lett.*, **17**, 494–500.

10. Castelli, I. E., Salvestrini, P., and Manini, N. (2012). Mechanical properties of carbynes investigated by ab-initio total-energy calculations. *Phys. Rev. B*, **85**, 214110.

11. Liu, M., Artyukhov, V. I., Lee, H., Xu, F., and Yakobson, B. I. (2013). Carbyne from first principles: chain of C atoms, a nanorod or a nanorope. *ACS Nano*, **7**, 10075–10082.

12. Svenson, K., Jompol, Y., Olin, H., and Olsson, E. (2003). Compact design of a transmission electron microscope-scanning tunneling microscope holder with three-dimensional coarse motion. *Rev. Sci. Instrum.*, **74**, 4945–4947.

13. Cretu, O., Botello-Mendez, A. R., Janowska, I., Pham-Huu, C., Charlier, J.-C., and Banhart, F. (2013). Electrical transport measured in atomic carbon chains. *Nano Lett.*, **13**, 3487–3493.

14. Troiani, H. E., Miki-Yoshida, M., Camacho-Bragado, G. A., Marques, M. A. L., Rubio, A., Ascencio, J. A., and Jose-Yacaman, M. (2003). Direct observation of the mechanical properties of single-walled carbon nanotubes and their junctions at the atomic level. *Nano Lett.*, **3**, 751–755.

15. Marques, M. A., Troiani, H. E., Miki-Yoshida, M., Jose-Yacaman, M., and Rubio, A. (2004). On the breaking of carbon nanotubes under tension. *Nano Lett.*, **4**, 811–815.

16. Jin, C., Lan, H., Peng, L., Suenaga, K., and Iijima, S. (2009). Deriving carbon atomic chains from graphene. *Phys. Rev. Lett.*, **102**, 205501.

17. Chuvilin, A., Meyer, J. C., Algara-Siller, G., and Kaiser, U. (2009). From graphene constrictions to single carbon chains. *New J. Phys.*, **11**, 083019.

18. Song, B., Schneider, G. F., Xu, Q., Pandraud, G., Dekker, C., and Zandbergen, H. (2011). Atomic-scale electron-beam sculpting of near-defect-free graphene nanostructures. *Nano Lett.*, **11**, 2247–2250.

19. Robertson, A. W., and Warner, J. H. (2013). Atomic resolution imaging of graphene by transmission electron microscopy. *Nanoscale*, **5**, 4079–4093.

20. Asaka, K., Toma, S., and Saito, Y. (2020). Raman spectroscopy and TEM of long linear carbon chain formed in CNT field emission cathode. *e-J. Surf. Sci. Nanotechnol.*, **18**, 159–163.

21. La Torre, A., Ben Romdhane, F., Baaziz, W., Janowska, I., Pham-Huu, C., Begin-Colin, S., Pourroy, G., and Banhart, F. (2014). Formation and characterization of carbon–metal nano-contacts. *Carbon*, **77**, 906–911.

22. La Torre, A., Botello-Mendez, A., Baaziz, W., Charlier, J.-C., and Banhart, F. (2015). Strain-induced metal-semiconductor transition observed in atomic carbon chains. *Nature Comm.*, **6**, 6636.

23. Ben Romdhane, F., Adjizian, J.-J., Charlier, J.-C., and Banhart, F. (2017). Electrical transport through atomic carbon chains: the role of contacts. *Carbon*, **122**, 92–07.

24. Ohnishi, H., Kondo, Y., and Takayanagi, K. (1998). Quantized conductance through individual rows of suspended gold atoms. *Nature*, **395**, 780–783.

25. Chen, W., Andreev, A. V., and Bertsch, G. F. (2009). Conductance of a single-atom carbon chain with graphene leads. *Phys. Rev. B*, **80**, 085410.

26. Shen, L., Zeng, M., Yang, S. W., Zhang, C., Wang, X., and Feng, Y. (2010). Electron transport properties of atomic carbon nanowires between graphene electrodes. *J. Am. Chem. Soc.*, **132**, 11481–11486.

Chapter 10

Carbon Chain Stripped from Graphene and Carbon Nanotube

Koji Asaka
Department of Mechanical Engineering, Faculty of Engineering, Fukui University of Technology, 3-6-1 Gakuen, Fukui-city, Fukui 910-8505, Japan
asaka@fukui-ut.ac.jp

A one-dimensional monoatomic chain composed of sp-hybridized carbon atoms is known as carbyne, which is one of carbon allotropes including fullerene, carbon nanotube, graphene, and so on. Theoretical studies on carbyne have predicted outstanding mechanical and electrical properties and promising applications of carbyne have been expected, while the existence of carbyne induced controversy for many decades. Much effort dedicated to the synthesis of carbyne resulted in successful achievements and the existence of carbyne was substantiated by Raman spectroscopy and high-resolution transmission electron microscopy. Transmission electron microscopy particularly possesses the advantage that carbyne is directly observed at higher spatial resolution in real time. Moreover, the use of *in situ* transmission electron microscopy has

Emerging Carbyne: Truly One-Dimensional Form of Carbon
Edited by Yahachi Saito

ISBN 978-981-5129-15-1 (Hardcover), 978-1-003-51036-9 (eBook)
www.jennystanford.com

brought about a means to produce carbon chains and investigate their properties. Here, the carbon chains produced from graphene and carbon nanotube in a transmission electron microscope are described.

10.1 Introduction

Carbyne is one of carbon allotropes including diamond, graphite, fullerene, carbon nanotube (CNT), and graphene, which is composed of an infinite linear monoatomic carbon chain [1]. For the ultimate one-dimensional structure of carbyne, two possible configurations are predicted; cumulene with repeating double bonds and polyyne with alternating single and triple bonds. Cumulene may spontaneously transform into polyyne due to a Peierls distortion. Many theoretical studies on carbyne have reported outstanding mechanical, thermal, and electrical properties far superior to other carbon allotropes [2–7] and promising applications including switching nanodevices, spintronic nanodevices, and hydrogen storage [8–11], while the existence of carbyne was debated for many decades till the successful synthesis of a linear acetylenic carbon chain with end-capping groups [12, 13]. Furthermore, long linear carbon chains encapsulated in CNTs were also produced by the arc discharge method, high-temperature fusion reaction of molecules, high-temperature heat treatment, and electric discharge [14–19]. Due to these syntheses of the carbon chains, the existence was substantiated by Raman spectroscopy and high-resolution transmission electron microscopy (TEM). In addition to direct observation at atomic resolution in real time, the use of a high-resolution transmission electron microscope also realized a means to produce short isolated carbon chains from graphene [20–26] and CNT [27–30] under electron-beam irradiation and/or electrical current application. Furthermore, *in situ* TEM combined with the functions of scanning probe microscopy, using a nanomanipulator tip in the transmission electron microscope, allowed measurements of the electric current passing through the carbon chain [23–25, 29] and tensile force at fracture of the carbon chain [28]. Herein, TEM studies on the carbon chains produced from graphene and CNT in the transmission electron microscope are focused on.

10.2 Carbon Chains Produced from Graphene

Isolated carbon chains produced from graphene in a transmission electron microscope have been reported by Jin *et al.* [20] and Chuvilin *et al.* [21] around in the same period. Figures 10.1 and 10.2 show a time-sequential series of high-resolution TEM images of the formation process of a carbon chain. In the formation process, holes are sculptured by focusing electron beam on a graphene membrane. The growth of the holes generates a graphene nanoribbon between

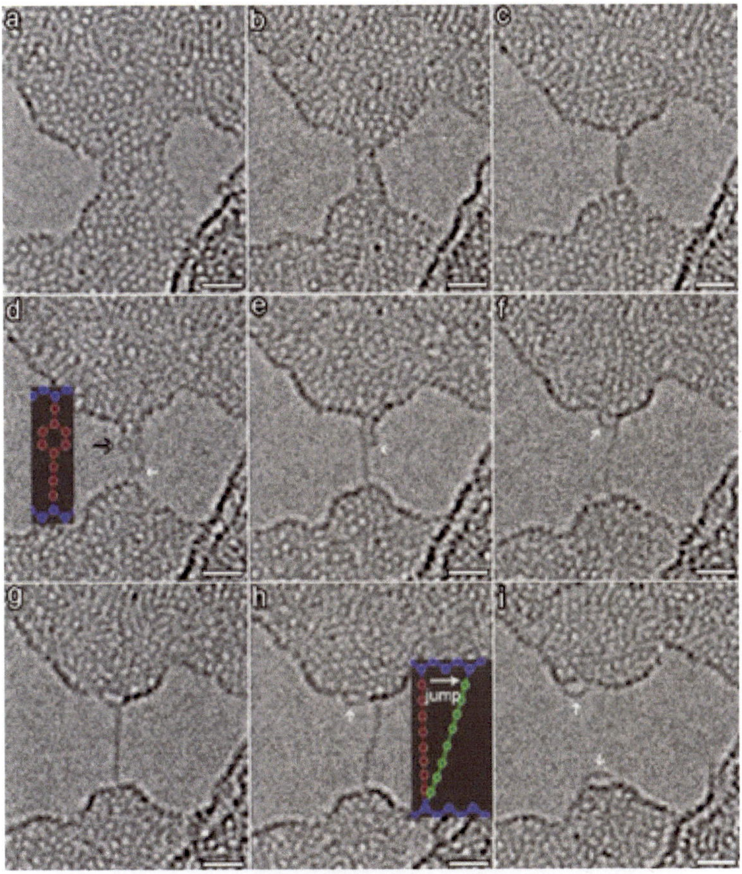

Figure 10.1 Time-sequential series of high-resolution TEM images of the formation process of a carbon chain by focusing electron beam on the graphene membrane. Scale bars are 1 nm. Reprinted with permission from Ref. [20]. Copyright 2009 The American Physical Society.

two neighboring holes. During electron beam irradiation, the nanoribbon is thinned and finally transformed into an isolated double or single carbon chain. According to Lin *et al.* [26], the carbon chains sculptured from graphene can be preserved even at a high temperature of 773 K. The bond lengths in the carbon chains are directly measured by annular dark field imaging in scanning TEM (Fig. 10.3) and high-resolution TEM imaging. The measurement results suggest that dimerization occurs in the carbon chains, i.e., the carbon chains prefer a polyyne-like structure at elevated temperatures.

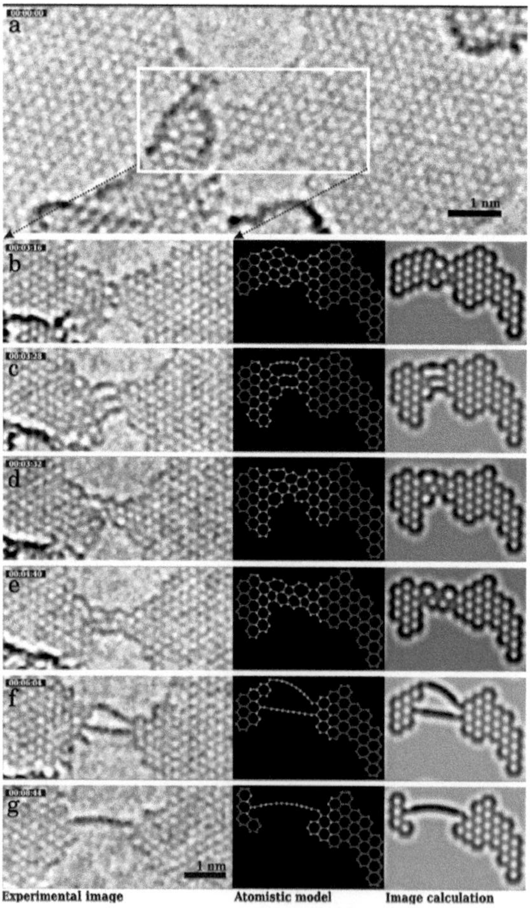

Figure 10.2 Time-sequential series of high-resolution TEM images of the formation process of a carbon chain by focusing electron beam on the graphene membrane, along with atomic models and image simulations. Reprinted with permission from Ref. [21]. licensed under CC BY-NC-SA 3.0.

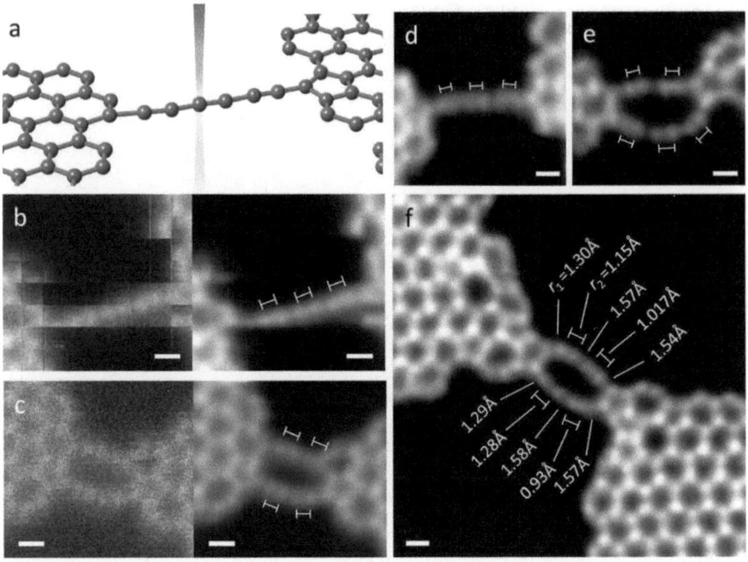

Figure 10.3 Annular dark field images by scanning TEM of dimerized carbon chains at 773 K and their filtered images. Scale bars are 0.2 nm. Reprinted with permission from Ref. [26]. Copyright 2016 American Chemical Society.

The carbon chains have also been obtained by extraction from few-layer graphene nanoribbons using a nanomanipulator tip in a transmission electron microscope [23–25]. Figure 10.4 shows a time-sequential series of high-resolution TEM images of the formation process of a carbon chain. After the tip is brought into contact with an edge of a few-layer graphene and retracted with passing an electric current through the contact region, a graphene nanoribbon suspended between the tip and few-layer graphene is generated, as seen in Fig. 10.4(a). The subsequent retracting of the tip leads to the unraveling of a carbon chain from the graphene nanoribbon, as indicated by an arrow in Fig. 10.4(b). The formed carbon chain is stable for over ten seconds as shown in Fig. 10.4. The stability of the carbon chain formed in the transmission electron microscope allows *in situ* experiments on current-voltage measurements [23–25].

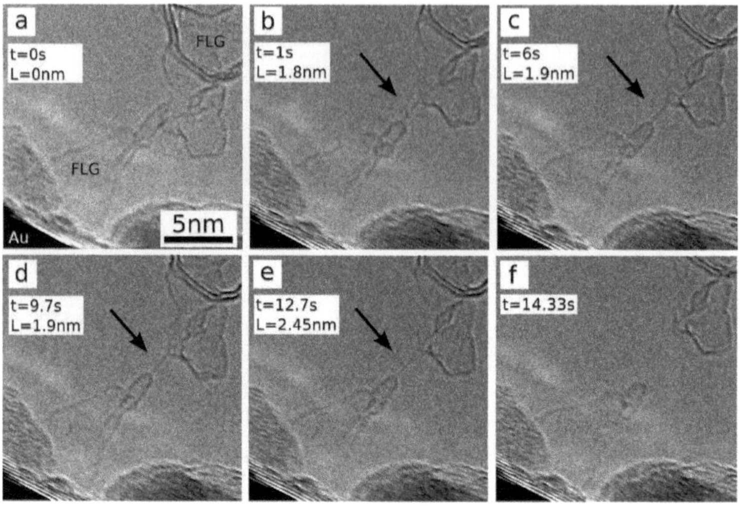

Figure 10.4 Time-sequential series of high-resolution TEM images of a carbon chain formation by extraction from few-layer graphene (FLG) nanoribbons using a nanomanipulator tip. Reprinted with permission from Ref. [23]. Copyright 2013 American Chemical Society.

10.3 Carbon Chains Produced from Carbon Nanotube

The electron beam irradiation into an amorphous carbon or a few-layer graphene have also provided the formation of carbon chains through the transformation into a CNT [27, 30]. Figure 10.5 shows a time-sequential series of high-resolution TEM images of the formation process of a carbon chain *via* a single-walled CNT. During electron beam irradiation, a formed multi-walled CNT continues reducing its diameter and the number of the walls, and is transformed into a single-walled CNT, as shown in Figs. 10.5(a)–(e). Finally, the successive thinning process leads to the fracture of the single-walled CNT. Before the complete separation of the single-walled CNT, a carbon chain suspended between the ends of the two capped CNT fragments appears (see Fig. 10.5(f)). Meanwhile, in the middle of the thinning process, a carbon chain encapsulated in the CNT is also *in situ* observed. The *in situ* observation of the formation process may throw a light on an elucidation as to the formation mechanism of a long linear carbon chain encapsulated in a CNT [14–19].

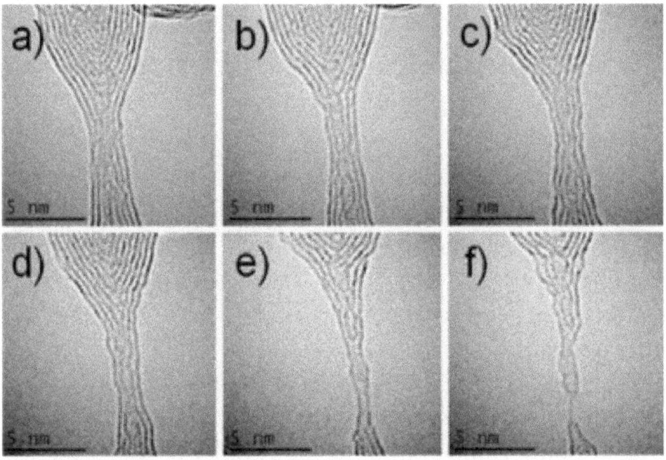

Figure 10.5 Time-sequential series of high-resolution TEM images of the formation process of a carbon chain *via* a single-walled CNT by electron beam irradiation. Reprinted with permission from Ref. [30]. Copyright 2013, Elsevier Ltd.

Manipulation of CNTs with an application of electric current in a transmission electron microscope has also allowed the formation of a carbon chain [29]. Figure 10.6 shows a time-sequential series of high-resolution TEM images of a carbon chain formation by manipulating CNTs. In Fig. 10.6(a), the end of a multi-walled CNT comes into contact with the end of another multi-walled CNT. After the multi-walled CNT on the left-hand side is retracted with applying a bias voltage, a carbon chain is formed between the multi-walled CNTs, as shown in Fig. 10.6(b). An electric current of 1.4 μA at 0.8 V flowed through the carbon chain at the state observed in Fig. 10.6(b), suggesting that the carbon chain can sustain an electric current density of approximately 8×10^{13} A/m^2 at least.

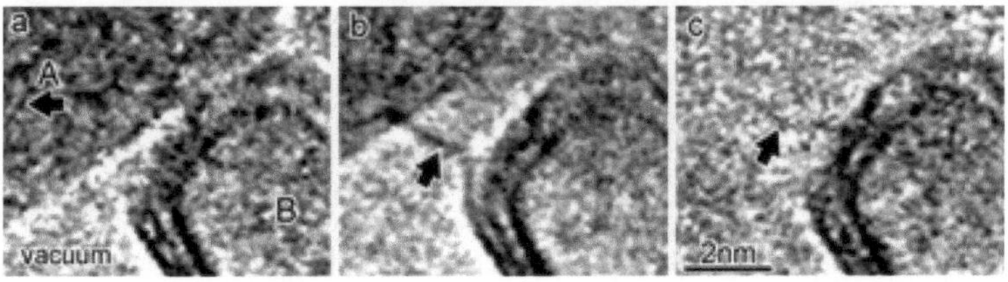

Figure 10.6 Time-sequential series of high-resolution TEM images of a carbon chain formation by manipulating CNTs. Reprinted with permission from Ref. [29]. Copyright 2011, Elsevier B.V.

10.4 Experimental Measurement of Electrical and Mechanical Characteristics of a Carbon Chain

Since precise manipulation of a carbon chain in a transmission electron microscope is extremely difficult, measurements of electrical and mechanical properties have been hardly reported; there are only several reports on the electrical and mechanical characteristics of the carbon chain studied by *in situ* TEM using a nanomanipulator tip.

Experimental and theoretical studies on the electrical properties of the carbon chain have shown that the strain in the carbon chain and the bonding configuration of the contact between the carbon chain and graphene nanoribbon influence the conductivity of the carbon chain [23–25]. The details are described in Chapter 9.

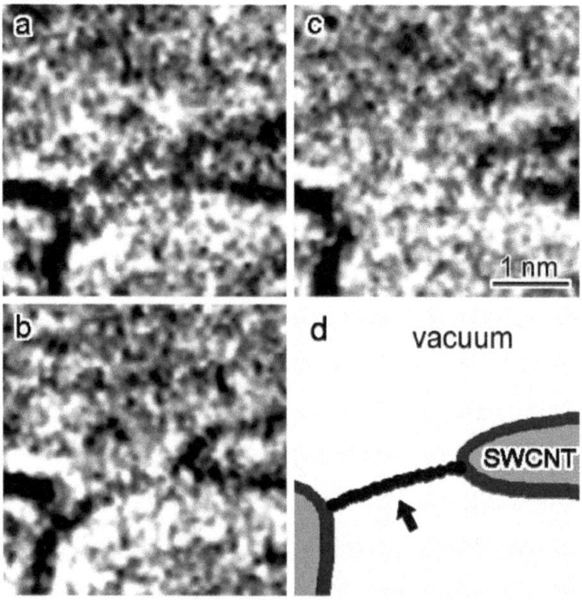

Figure 10.7 (a)–(c) Time-sequential series of high-resolution TEM images of the fracture process of a carbon chain formed between capped single-walled CNTs for tensile deformation. In (c), the carbon chain is completely fractured. (d) Illustration of a carbon chain of (b) formed before fracture. Reprinted with permission from Ref. [28]. Copyright 2005 The American Physical Society.

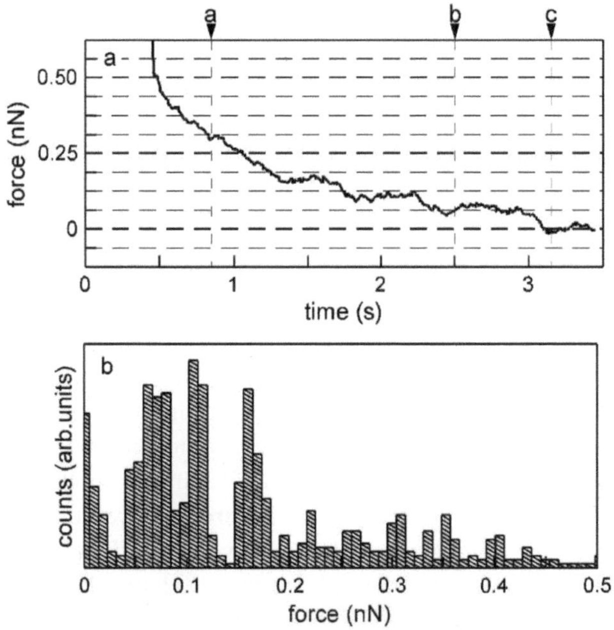

Figure 10.8 (a) Changes in tensile force measured during observation of Fig. 10.7 and (b) its histogram for the value of tensile force. Times indicated by a–c in (a) correspond to those at which Figs. 10.7(a)–(c) were observed. Reprinted with permission from Ref. [28]. Copyright 2005 The American Physical Society.

Tensile deformation of a single-walled CNT has observed with measurement of mechanical properties by *in situ* TEM combined with the functions of scanning probe microscopy [28]. In the final stage of tensile deformation of the single-walled CNT, a carbon chain was formed before the complete fracture of the single-walled CNT, as predicted by theoretical simulations [31]. Figure 10.7 shows a time-sequential series of high-resolution TEM images of the fracture process of a carbon chain formed between capped single-walled CNTs for tensile deformation. The changes in tensile force measured during observation of Fig. 10.7 and its histogram for the value of tensile force are shown in Fig. 10.8. The width of the carbon chain suspended between the ends of the two capped CNT fragments narrows down to a monoatomic scale, as seen in Figs. 10.7(a) and (b). In Fig. 10.8(a), tensile force decreases stepwise with an almost same step height. The stepwise decrease implies that the thinning deformation proceeds by breaking the same type of bonds between carbon atoms. The minimum tensile force corresponding to the monoatomic carbon chain is also revealed to be approximately 0.06 nN, which is about 10^{-2} times smaller than the ideal maximal one calculated by Troiani *et al.* [27]. The decrease may be attributed to several factors such as the tensile deformation rate, electron beam irradiation, and bonding configuration.

10.5 Conclusions

The carbon chains produced in the transmission electron microscope were described in this chapter. High-resolution TEM can be used for not only precise observation at atomic resolution in real time but also synthesis of the monoatomic carbon chain. Furthermore, *in situ* TEM combined with the functions of scanning probe microscopy paves the way for the experimental approach to investigate the electrical and mechanical properties of the isolated carbon chain, although such a challenging approach faces the difficulty in experiment and requires a high order of engineering skills. The bottom-up experimental studies by *in situ* TEM would contribute to the understanding of the structures and properties for the carbon chains and lead to development of potential applications in the future.

References

1. Casari, C. S., and Milani, A. (2018). Carbyne: from the elusive allotrope to stable carbon atom wires. *MRS Communications*, **8**, 207–219.
2. Liu, M., Artyukhov, V. I., Lee, H., Xu, F., and Yakobson B. I. (2013). Carbyne from first principles: chain of C atoms, a nanorod or a nanorope. *ACS Nano*, **7**, 10075–10082.
3. Yang, X., Huang, Y., Cao, B., and To, A. C. (2017). Length and temperature dependence of the mechanical properties of finite size carbine. *Phys. E*, **93**, 124–131.
4. Wang, M., and Lin, S. (2015). Ballistic thermal transport in carbyne and cumulene with micron-scale spectral acoustic phonon mean free path. *Sci. Rep.*, **5**, 18122.
5. Larade, B., Taylor, J., Mehrez, H., and Guo, H. (2001). Conductance, I–V curves, and negative differential resistance of carbon atomic wires. *Phys. Rev. B*, **64**, 075420.

6. Shen, L., Zeng, M., Yang, S.-W., Zhang, C., Wang, X., and Feng, Y. (2010). Electron transport properties of atomic carbon nanowires between graphene electrodes. *J. Am. Chem. Soc.*, **132**, 11484–11486.

7. Dong, Y.-J., Wang, X.-F., Zhai, M.-X., Wu, J.-C., Zhou, L., Han, Q., and Wu, X.-M. (2013). Effects of geometry and symmetry on electron transport through graphene-carbon-chain junctions. *J. Phys. Chem. C*, **117**, 18845–18850.

8. Akdim, B., and Pachter, R. (2011). Switching behavior of carbon chains bridging graphene nanoribbons: effects of uniaxial strain. *ACS Nano*, **5**, 1769–1774.

9. Artyukhov, V. I, Liu, M., Yakobson, B. I. (2014). Mechanically induced metal-insulator transition in carbine, *Nano Lett.*, **14**, 4224–4229.

10. Zeng, M. G., Shen, L., Cai, Y. Q., Sha, Z. D., and Feng, Y. P. (2010). Perfect spin-filter and spin-valve in carbon atomic chains. *Appl. Phys. Lett.*, **96**, 042104.

11. Sorokin, P. B., Lee, H., Antipina, L. Yu., Singh, A. K., and Yakobson, B. I. (2011). Calcium-decorated carbyne networks as hydrogen storage media. *Nano Lett.*, **11**, 2660–2665.

12. Lagow, R. J., Kampa, J. J., Wei, H. C., Battle, S. L., Genge, J. W., Laude, D. A., Harper, C. J., Bau, R., Stevens, R. C., Haw, J. F., and Munson, E. (1995). Synthesis of linear acetylenic carbon: the "sp" carbon allotrope. *Science*, **267**, 362–367.

13. Chalifoux, W. A., and Tykwinski, R. R. (2010). Synthesis of polyynes to model the sp-carbon allotrope carbine. *Nat. Chem.*, **2**, 967–971.

14. Zhao, X., Ando, Y., Liu, Y., Jinno, M., and Suzuki, T. (2003). Carbon nanowire made of a long linear carbon chain inserted inside a multiwalled carbon nanotube. *Phys. Rev. Lett.*, **90**, 187401.

15. Scalese, S., Scuderi, V., Bagiante, S., Simone, F., Russo, P., D'Urso, L., Compagnini, G., and Privitera, V. (2010). Controlled synthesis of carbon nanotubes and linear C chains by arc discharge in liquid nitrogen. *J. Appl. Phys.*, **107**, 014304.

16. Zhang, J., Feng, Y., Ishiwata, H., Miyata, Y., Kitaura, R., Dahl. J. E. P., Carlson, R. M. K., Shinohara, H., and Tománek, D. (2012). Synthesis and transformation of linear adamantane assemblies inside carbon nanotubes. *ACS Nano*, **6**, 8674–8683.

17. Shi, L., Rohringer, P., Suenaga, K., Niimi. Y., Kotakoski, J., Meyer, J. C., Peterlik, H., Wanko, M., Cahangirov, S., Rubio, A., Lapin, Z. J., Novotny, L., Ayala, P., and Pichler, T. (2016). Confined linear carbon chains as a route to bulk carbine. *Nat. Mater.*, **15**, 634–640.

18. Toma, S., Asaka, K., Irita, M., and Saito, Y. (2019). Bulk synthesis of linear carbon chains confined inside single-wall carbon nanotubes by vacuum discharge. *Surf. Interface Anal.*, **51**, 131–135.

19. Asaka, K., Toma, S., and Saito, Y. (2019). Raman spectroscopy and transmission electron microscopy of long linear atomic carbon chains generated by field electron emission accompanied with electrical discharge of carbon nanotube films. *SN Appl. Sci.*, **1**, 493.

20. Jin, C., Lan, H., Peng, L., Suenaga, K., and Iijima, S. (2009). Deriving carbon atomic chains from graphene. *Phys. Rev. Lett.*, **102**, 205501.

21. Chuvilin, A., Meyer, J. C., Algara-Siller, G., and Kaiser, U. (2009). From graphene constrictions to single carbon chains. *New J. Phys.*, **11**, 083019.

22. Song, B., Schneider, G. F., Xu, Q., Pandraud, G., Dekker, C., and Zandbergen, H. (2011). Atomic-scale electron-beam sculpting of neardefect-free graphene nanostructures. *Nano Lett.*, **11**, 2247–2250.

23. Cretu, O., Botello-Mendez, A. R., Janowska, I., Pham-Huu, C., Charlier, J.-C., and Banhart, F. (2013). Electrical transport measured in atomic carbon chains. *Nano Lett.*, **13**, 3487–3493.

24. La Torre, A., Romdhane, F. B., Baaziz, W., Janowska, I., Pham-Huu, C., Begin-Colin, S., Pourroy, G., and Banhart, F. (2014). Formation and characterization of carbon-metal nano-contacts. *Carbon*, **77**, 906–911.

25. La Torre, A., Botello-Mendez, A., Baaziz, W., Charlier, J.-C., and Banhart, F. (2015). Strain-induced metal-semiconductor transition observed in atomic carbon chains. *Nat. Commun.*, **6**, 6636.

26. Lin, Y.-C., Morishita, S., Koshino, M., Yeh, C.-H., Teng, P.-Y., Chiu, P.-W., Sawada, H., and Suenaga, K. (2017). Unexpected huge dimerization ratio in one-dimensional carbon atomic chains. *Nano Lett.*, **17**, 494–500.

27. Troiani, H. E., Miki-Yoshida, M., Camacho-Bragado, G. A., Marques, M. A. L., Rubio, A., Ascencio, J. A., and Jose-Yacaman, M. (2003). Direct observation of the mechanical properties of single-walled carbon nanotubes and their junctions at the atomic level *Nano Lett.*, **3**, 751–755.

28. Asaka, K., and Kizuka, T. (2005). Atomistic dynamics of deformation, fracture, and joining of individual single-walled carbon nanotubes. *Phys. Rev. B*, **72**, 115431.

29. Asaka, K., Karita, M., and Saito, Y. (2011). Joining of multiwall carbon nanotubes for the end-contact configuration by applying electric current. *Mater. Lett.*, **65**, 1832–1834.

30. Casillas, G., Mayoral, A., Liu, M., Ponce, A., Artyukhov, V. I., Yakobson, B. I., and Jose-Yacaman, M. (2014). New insights into the properties and interactions of carbon chains as revealed by HRTEM and DFT analysis. *Carbon*, **66**, 436–441.

31. Yakobson, B. I., Campbell, M. P., Brabec, C. J., and Bernholc, J. (1997). High strain rate fracture and C-chain unraveling in carbon nanotubes. *Comput. Mater. Sci.*, **8**, 341–348.

Chapter 11

Superconductivity of One-Dimensional Carbon

Chi Ho Wong,[a,b] Tao Mengying,[c] and Anatoly Fedorovich Zatsepin[d]

[a]*Department of Industrial and Systems Engineering, The Hong Kong Polytechnic University, Hong Kong, China*
[b]*Research Institute for Advanced Manufacturing, The Hong Kong Polytechnic University, Hong Kong, China*
[c]*Department of Applied Physics, The Hong Kong Polytechnic University, Hong Kong, China*
[d]*Instutite of Physics and Technology, Ural Federal University, Yekaterinburg, Russia*
roych.wong@polyu.edu.hk

With the recent discovery of superconductivity of H_2S at 203 K under pressure [1], high-temperature superconductivity can be achieved not only in correlated electron systems with complex competing or coexisting orders, but also in phonon-mediated classical superconductors that have a high Debye temperature and a large electronic density of states at the Fermi level. This can be found in materials with light atoms and strong covalent bonds.

Emerging Carbyne: Truly One-Dimensional Form of Carbon
Edited by Yahachi Saito

ISBN 978-981-5129-15-1 (Hardcover), 978-1-003-51036-9 (eBook)
www.jennystanford.com

Quasi-1D conductors, which have Van Hove singularities in their electronic density of states, appear to be particularly promising for high temperature superconductivity. In thin carbon nanotubes (CNT), superconductivity at 15 K has been reported, and it is likely the strong curvature of the carbon nanotubes which enhances the phonon-mediated superconductivity. We began our chapter with the introduction of 1D superconductivity followed by a well-known elementary superconducting material consisting of arrays of 4-angstrom carbon nanotubes. Gradually, we proceed to study various 1D carbyne structures and carbon ring, both with and without kinks. Along the way, we optimized the superconducting transition temperature until we reached a maximum of 115 K.

11.1 Superconductivity in 1D System

The Bardeen–Cooper–Schrieffer (BCS) model is the theory of conventional superconductors proposed by John Bardeen, Leon Cooper, and Robert Schrieffer that explains the behavior of certain materials at extremely low temperatures. In the BCS theory, the electrons responsible for electrical conductivity pair up with help of phonon and the electrons move through the lattice with zero resistance, allowing for the flow of electricity without the loss of energy to heat. This phenomenon is known as superconductivity, and it has important applications in areas such as medical imaging, energy transmission, and particle accelerators.

As mentioned, quasi-1D superconductors are potential candidates for high-temperature superconductivity because the one-dimensionality causes Van Hove singularities in the electronic density of states (DOS). Although the Van Hove singularities in 1D superconductors can sometimes lead to very high electronic density of states at the Fermi level, the 1D superconductors significantly increase fluctuations in the phase of the superconducting order parameter, which causes phase slip events in a 1D structure and a non-zero resistance at any finite temperature [1, 2]. This phenomenon is well described by the Langer-Ambegaokar-McCumber-Halperin (LAMH) theory [2, 3] and has been demonstrated in numerous experiments [2, 3]. However, there is one way out of this dilemma: to arrange the 1D elements to form an array of parallel wires or a network with finite Josephson coupling [4]. In this case, the normal state properties such as the electronic density of states largely preserve their 1D character, while superconducting order parameter fluctuations are strongly suppressed [4].

11.2 One-Dimensional Carbon Materials

In 2001, Ping Sheng and their team at the Hong Kong University of Science and Technology discovered that single-walled carbon nanotubes (CNT) exhibit superconductivity [5] at ~15 K. The team discovered these properties in individual carbon nanotubes which were just 0.4 nanometers in diameter. This effectively means that resistance-free current can flow through pure carbon, which could have important implications for various fields such as electronics and mechanics. The researchers found these results to be consistent with the Bardeen-Cooper-Schrieffer (BCS) theory of superconductivity. The results of their discovery validated a prediction made in 1995, suggesting that superconductivity would occur in nanotubes due to enhanced coupling between phonons and electrons.

However, CNT is not the thinnest 1D carbon material. Carbyne in form of a monoatomic linear chain of carbon is the world's thinnest carbon nanowire. Carbyne has attracted considerable interest due to the theoretical prediction of its extremely high tensile strength, Debye temperature and stiffness that breaks all previous records of other known materials so far [6, 7]. The high Debye temperature of the carbyne may be able to activate high-temperature superconductivity based on the BCS theory. However, its fabrication is extremely difficult, although a recent breakthrough has demonstrated the feasibility [8–10]. Linear carbon chains have been predicted to be metallic below 500 K with a structure that consists of repeating double bonds (β-carbyne or cumulene), which transforms above 500 K into the semiconducting α-carbyne (polyyne) with alternating single and triple bonds *via* a Peierls transition [6]. The uncommon occurrence of the metallic phase at a lower temperature presents the possibility of using carbon chains as one-dimensional building blocks in a quasi-one-dimensional bulk single crystalline structure. This structure would have an array of carbyne chains arranged in parallel.

11.3 Scale Factor Approach for 1D Superconductivity of Carbon (CNT, Carbyne)

Basically, the BCS T_c-formula requires a constant DOS as a function of electron energy. However, this is valid only if the Debye temperature is small compared to the range over which the DOS varies significantly. In the case of 1D materials, therefore, fundamental assumptions of the BCS T_c-formula become violated due to the presence of narrow peak-like structures associated with Van Hove singularities in the electronic density of states. Rather than calculating the critical temperature (T_c) of superconductivity directly using Bardeen-Cooper-Schrieffer (BCS) theory, which can be inaccurate for 1D superconducting materials, we employ a phenomenological scale factor approach (vide infra) to calculate the T_c of 1D superconductors based on the results from the ab-initio calculations such as the electronic band diagram, dispersion relationship of phonons, and the density of states of both electrons and phonons through the use of the Generalized Gradient Approximation (GGA) – Perdew-Burke-Ernzerhof (PBE) functional [11, 12].

The scale-factor approach is introduced as follows. The pairing Hamiltonian, $H_{\text{pair}} = \sum_{k\sigma} E_k n_{k\sigma} + \sum_{kl} V_{kl} c^*_{k\uparrow} c^*_{k} c_{l\uparrow} c_{l\downarrow}$, is composed of the single-particle energy E_k relative to the Fermi energy and the interaction term V_{kl}, which scatters the particle from a state with $(l\uparrow, -l\downarrow)$ to $(k\uparrow, -k\downarrow)$. The creation operators, $c^*_{k\uparrow}$ and $c^*_{k\downarrow}$, correspond to spin up and down, respectively, while $n_{k\sigma}$ is the particle number operator, σ is the spin index [13]. We define ψ_G to be the ground state of the BCS wavefunction

$$|\psi_G = \prod_{k=k_1\cdots k_M} \left(u_k + v_k c^*_k c^*_{k\downarrow}\right)|\varphi_0 \tag{11.1}$$

where $|\varphi_0\rangle$ is the vacuum state with no particles present and $|u_k|^2$ refers to the unoccupied probability since $|u_k|^2 + |v_k|^2 = 1$. The energy gap Δ does not depend on k and hence we may write $\Delta = \Delta_k = -\sum_l V_{kl} u_l v_l$ [13]. Multiplying $-\Sigma_k u_k v_k$ to both sides of the above equation leads to

$$-\Delta \sum_k u_k v_k = \sum_{kl} V_{kl} u_k v_k u_l v_l \tag{11.2}$$

The energy gap can then be expressed as $\Delta = \frac{H_{e-ph}}{-\sum_k u_k v_k}$. If the ratio of $-\Sigma_k u_k v_k$ (i.e., the division of transfer function of the pairing states) between two superconductive materials is close to 1, then $\frac{\Delta_A}{\Delta_B} \sim \frac{H_{e-ph(A)}}{H_{e-ph(B)}}$ is satisfied at 0 K. The interaction term arises from the electron-phonon coupling in the expression of [13, 14]

$$H_{e-ph} = \sum_{kk'\sigma\lambda} \left\{ \left[\left(\int d^3 r \phi^*_{\frac{k}{\sigma}}(\mathbf{r}) \phi_{k\sigma}(\mathbf{r}) \nabla_{R_i^0} V\left(\mathbf{r} - \mathbf{R}_i^0\right) \right) \cdot e^{\lambda}(\mathbf{q}) \right] \sqrt{\frac{N}{2M\omega}} c^{\dagger}_{\frac{k}{\sigma}} c_{k\sigma} \left(a_{\lambda}(\mathbf{q}) + a^{\dagger}_{\lambda}(-\mathbf{q}) \right) \right\} \quad (11.3)$$

$$\sum c^{\dagger}_{\frac{k}{\sigma}} c_{k\sigma} = n^{\text{ion}} \left(1 - \frac{1}{\varepsilon} \right) = n_T \nabla \cdot u$$

where n^{ion} is the ionic charge fluctuations induced by ionic displacement and n_τ is the total ionic charge density. The polarization vector $\sum_{\alpha} e_{\alpha}{}^{\lambda}(k) \cdot e_{\alpha}^{\lambda'}(k) = \delta_{\lambda\lambda'}$ is included in the orthogonal cases. $V(r)$ is the lattice potential as a function of space. The electric fields emitted from the face centers, body center or kink structure in different materials is resolved into orthogonal directions before solving the Schrodinger equation of the electrons. We abbreviate $\int d^3 r \phi^*_{k'\sigma}(\mathbf{r}) \phi_{k\sigma}(\mathbf{r}) \nabla_{R_i^0} V\left(\mathbf{r} - \mathbf{R}_i^0\right) \cdot e^{\lambda}(\mathbf{q}) \sqrt{\frac{N}{2M\omega}}$ as $g_{kk'}$, where $\mathbf{R}_i^0$ is the equilibrium position of the ith atom and M is the atomic mass. The dielectric constant $\mathcal{E}$ serves to consider the screening effect. The $a^{\dagger}_{q\lambda} a_{q\lambda} + \frac{1}{2}$ is defined as the quantum number of the phonons, where $\mathbf{q} = \mathbf{k} - \mathbf{k'} + \mathbf{G}$ and $\mathbf{G}$ correspond to a reciprocal lattice vector [13]. The ionic displacement is $u = \frac{1}{\sqrt{NM}} \sum_{k\lambda} Q_{\lambda} e^{\lambda} e^{ik \cdot R_i^0}$. Finally, we obtain $a_{\lambda}(\mathbf{q}) + a^{\dagger}_{\lambda}(-\mathbf{q})$ according to $a_{\lambda}(\mathbf{q}) + a^{\dagger}_{\lambda}(-\mathbf{q}) = Q_{\lambda} \sqrt{2\omega_{\lambda}}$.

The 3D time-independent Schrödinger equation of electrons is solved by the method of separation of variables [13]. The potential energy U in each direction is summed up to 100 unit cells in the Schrödinger equation. The percentage change in the $\frac{\sum_1^{100} U - \sum_1^{99} U}{\sum_1^{100} U}$ is about 0.1% and therefore it can be regarded as an infinitely large system. In other word, the size effect on the electronic band structure and density of states is negligible. For cylindrical or spherical material, the attractive force acting on the electrons is modified due to the change of the effective atomic number $Z_{\text{effective}}$ [15].

$$Z_{\text{effective}} = Z \frac{\sum_r^R U_c(r)}{\sum_r^R U_p(r)} \tag{11.4}$$

The Bloch theorem states that the wavefunctions of electron ψ has the form $\psi(r+R) = e^{ik \cdot R}\psi(r)$ where k is the wave number and R is a lattice vector. The $Z_{\text{effective}}$ is estimated by comparing the attractive potential between the circular U_c and planar U_p shapes. The wavefunction of phonon is solved with help of Hermit polynomial [16]. Due to the low temperatures, considering the lowest order in the Hermit polynomial is a sufficiently precise approximation.

The prediction of the T_c is acquired by computing the scale factor $\frac{T_{c(P)}}{T_{c(Q)}} = \frac{\Delta_P(0)}{\Delta_Q(0)}$, because $\Delta(0) \propto T_c$ per electron [13]. If the T_c of the material Q is known, the T_c of the material P can be predicted according to the above *kl* dependent scale factor. The scale factor is used instead of calculating the T_c directly, because the phonon wavefunction is solved in dimensionless units. However, the $u_l v_l$ depend on Δ and hence another transfer function is required: According to BCS theory, the energy gap can be estimated *via* $\Delta_k = -\frac{1}{2}\sum V_{kl} \frac{\Delta_l}{\left(\Delta_l^2 + E_k^2\right)^{0.5}}$ [13]. The transfer function is derived by calculating the trial energy gap Δ^T, which originates from the electrons at the Fermi level only. In this circumstance the Δ^T is directly proportional to the interaction term since $E_k = 0$. Then the interpreted transfer function, i.e., $u_l^T v_l^T$ as a function of electron energy, will be substituted into $\Delta_k^{\text{corrected}} = -\sum_{kl} V_{kl} u_l^T v_l^T$ in order to correct the energy gap. However, the limitation of using the transfer

Table 11.1 The T_c calculation of conventional superconductors based on the scale factor approach. Reprinted with permission from Ref. [17], Copyright 2023, Springer Nature

Materials	μ	λ	$f(T_D)$	T_c^{ex} (K)	T_c^{scale} (K)	T_c^{Mc} (K)
Al	0.11	0.38	1.0000	1.2	1.2 (Reference)	2.4
Pb	0.10	1.55	1.0227	7.2	6.3	7.0
Sn	0.10	0.72	1.0102	3.7	2.8	5.1
In	0.10	0.69	1.0108	3.4	2.4	3.1
Hg	0.10	1.00	1.0224	4.2	3.9	4.1
Ta	0.10	0.87	1.0101	4.4	3.4	5.8
Li (35 GPa)	0.14	2.75	1.0034	10.0	11.6	13.0
Mo	0.11	0.32	0.9280	0.9	1.0	0.3
Ga	0.11	2.25	1.0157	8.6	9.3	14.1

Table 11.2 The estimated T_c values in various low dimensional superconductors based on the scale factor approach and the empirical method [18–21]

Material	λ_{scale}	μ	R_{shape}	T_c^{scale} **(K)**	T_c^{ex} **(K)**	Scaled by
Al(2D)	0.43	0.12	1.1	1.3	1.3	Al(3D)
Sn(2D)	0.79	0.11	1.1	3.9	4.0	Sn(3D)
Sn(1D)	0.92	0.10	1.3	4.9	5.5	Sn(3D)
Pb(2D hollow sphere)	2.37	0.10	1.5	11.2	11.0	Pb(3D)
Pb(1D)	2.10	0.10	1.4	10.0	11.3	Pb(3D)

function is to satisfy the condition for the Debye energy $\hbar\omega_D \gg \Delta$. Otherwise, the occupational fraction at superconducting states will not drop sharply to zero when the electron energy increases [13].

Before we implement the T_c of the 1D carbon, we have checked carefully that the scale factor does not depend on the lattice structure and curvature. We present the accuracy of the scale factor approach as shown in Tables 11.1 and 11.2. We define $T_c^P = T_c^Q \left| \frac{\lambda_P - \mu_P}{\lambda_Q - \mu_Q} \frac{1}{f(T_D)} \right|$, where $f(T_D)$ is the ratio of the transfer function between two materials P and Q, T_D is the Debye temperature. λ is the electron phonon scattering term, μ is the Coulomb screening. T_c^{ex} and T_c^{Mc} are the T_c values obtained from experimental data and the McMillian formula, respectively. The T_c^{scale} values in Tables 11.1 and 11.2 are obtained by modifying the $g_{kk'}$ only. Table 11.3 shows useful formulae of the scale factor approach.

Table 11.3 Useful formulae in the scale factor approach

$W_{kk'} = \int d^3 r \phi^*_{k'\sigma}(\mathbf{r}) \phi_{k\sigma}(\mathbf{r}) \nabla_{R_i^0} V(\mathbf{r} - \mathbf{R}_i^0)$	$g_{kk'} = (W_{kk'} \cdot e^{\lambda}(\mathbf{q})) \sqrt{\frac{N}{2M\omega}}$
$R_W = \frac{W^A_{kk'\lambda}}{W^B_{kk'\lambda}}$	$R_u = \frac{\langle \nabla \cdot u^A \rangle}{\langle \nabla \cdot u^B \rangle}$
$R_{a(q)} = \frac{\langle a_\lambda(\mathbf{q}) + a^\dagger_\lambda(-\mathbf{q}) \rangle \big|_A}{\langle a_\lambda(\mathbf{q}) + a^\dagger_\lambda(-\mathbf{q}) \rangle \big|_B}$	$R_\omega = \frac{\langle \omega_A \rangle}{\langle \omega_B \rangle}$
$R_{DOS} = \frac{\langle D_{DO}(E_F \to E_{\text{Debye}}) \rangle}{\langle DOS_B(E_F \to E_{\text{Debye}}) \rangle}$	$R_M = \frac{M_A}{M_B}$
$R_{\text{shape}} = \frac{g^A_{kk'}(\omega_p)}{g^B_{kk'}(\omega_p)}$	$R_{kink} = \frac{g^A_{k'}(\omega_p)}{g_{kk'}(\omega_p)}$ Kink $\Leftrightarrow$ change shape

11.3.1 Carbon Nanotube

In the next step we apply this method to calculate the T_c of arrays of 4 Å Angstrom SWCNT (single-wall carbon nanotube), which are formed from hexagonal structure in the presence of curvature. A

SWCNT can be produced by rolling graphene into the shape of a tube. The phonon wavefunction in a SWCNT $\chi\left(\omega_x^{\text{planar}}, \omega_y^{\text{planar}}, \omega_z^{\text{circular}}\right)$ is going to connect with the graphene $\chi\left(\omega_x^{\text{planar}}, \omega_y^{\text{planar}}\right)$ and linear carbon nanowire $\chi\left(\omega_x^{\text{linear}}\right)$ in order to minimize computational cost. The spring constants in the graphene are split into two types of repeating units as shown in Fig. 11.1. The effective spring constants $\left(K_x^{\text{planar}}, K_y^{\text{planar}}\right)$ of the upper and lower repeating units are resolved into the x and y axis, respectively.

To interpret the vibrational frequency of the linear carbon chain, we simulate the dispersion curve and phonon density of states using the GGA functional [11, 12]. This is based on the bond length of graphene, which is approximately 143 pm. We consider the same bond distance of 143 pm and analyze the four nearest carbon atoms along the XY direction.

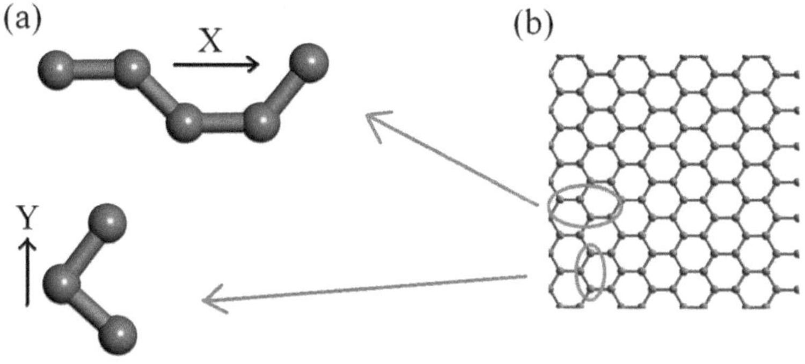

Figure 11.1 (a) Two types of repeating units of graphene. (b) The structure of graphene. Reprinted with permission from Wong *et al.* [17], Copyright 2023, Springer Nature.

The carbon nanowire are grouped to form a linear repeating unit with the known dimensionless spring constant K_x^{linear}. The ratio of the resultant spring constant of the graphene relative to the reference chain can be computed by $\frac{\sqrt{\left(K_x^{\text{planar}} K_x^{\text{planar}} + K_y^{\text{planar}} K_y^{\text{planar}}\right)}}{K_x^{\text{linear}}}$. According to the classical mass-spring approach, it's possible to interpret the vibrational frequency of graphene. This means we can estimate the approximated phonon wavefunction in graphene [16].

The approximated wavefunction of the SWCNT is determined in combination with the concept of surface phonon softening due to the curvature. About 13 atoms along the armchair path or chirality number (4, 2) form a loop in the SWCNT, which means that the tilt angle between the adjacent atoms is 360/13 = 27.6 degrees. Another relative spring constant K_z^{circular} of the SWCNT can be found by resolving the vector into radial and tangential components [14, 15] and eventually the approximated wavefunction of the SWCNT can be interpreted. As a results, R_{shape} and R_{kink} (Table 11.3) can be found where ω_p is not obtained from the phonon dispersion. The electronic band diagram and the electronic density of states of the SWCNT are also computed by the GGA functional without making extra connection to the graphene and carbon nanowire.

To determine the potential energy in the Schrodinger equation of electrons, one can simply project the electric fields onto the orthogonal directions once more [16]. The (5, 0) SWCNT is managed in a similar manner. Finally, the theoretical T_c values of the (4, 2) and (5, 0) SWCNT arrays can be determined. As a test, the prediction of T_c in the SWCNT is still accurate if we start from the original BCS pairing Hamiltonian without taking the mentioned approximations. The accuracy of the transfer function is acceptable for SWCNTs, because the Debye temperature is much higher than the superconducting transition temperature. If the Debye energy $\hbar\omega_D$ is infinitely large, the BCS occupational probability will drop quickly to zero beyond the Fermi level, which more or less matches the pattern of the transfer function. Assuming that the Δ^T of the (4, 2) SWCNT is very large, the average offset [13] between $\sum_{E_F}^{E_F+3\Delta^T} u_l v_l$ and $\sum_{E_F}^{E_F+\Delta^T} u_l v_l$ is still less than 4% when we sum over the survival energies of the paired electrons. The accuracy of the Δ^T may not perfect, but the error of the transfer function should be significantly decreased once the Debye frequency is very large.

11.3.2 Monoatomic Carbon Chain

To predict the T_c of various monoatomic carbon chains with different structural arrangements in the form of kinks, the electronic density of states of the carbon chain is simulated by the GGA functional [11].

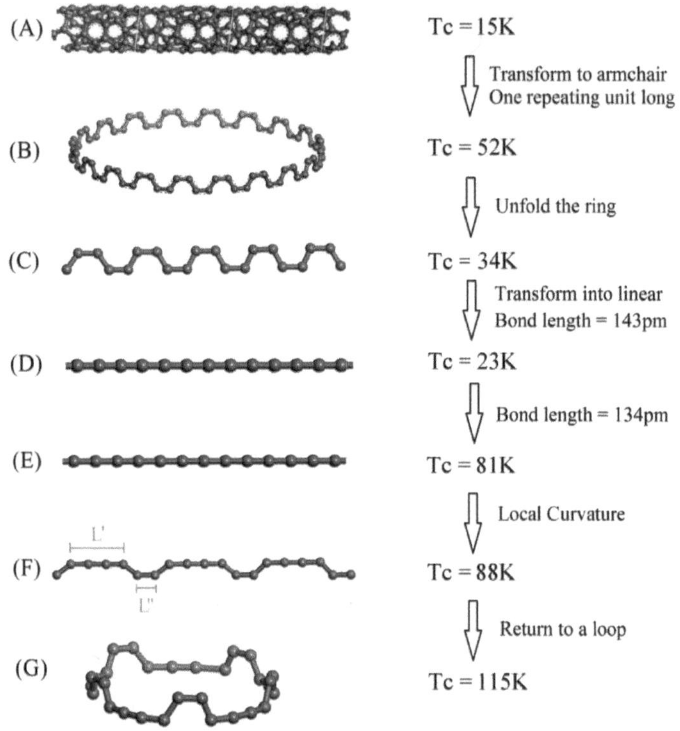

Figure 11.2 The process of designing a 115 K superconductor is illustrated in this diagram. Different types of carbon materials are shown and labeled as Sample A to G, respectively. Sample A depicts an infinitely long (4, 2) SWCNT, with only 3 repeating units being shown. The kink angles of Sample B and C are 60 degrees but are reduced to 10 degrees in Sample F. Sample D and E represent infinitely long linear carbon chains in cumulene phase, with the trivial kink angle of 0 degrees. The bond lengths of Sample A, B, C, and D are 143 pm, while the bond distances of Sample E, F, and G are 134 pm. Each sample in the simulation contains about 100 atoms except Sample A, D and F. The ratio of branch length $R_B = L'/L''$ in Sample F equals to 3. Reprinted with permission from Wong *et al.* [17], Copyright 2023, Springer Nature.

By using the technique mentioned above to address the coupling of electric fields to electrons in the kink-structural carbon chain, we can solve for the time-independent Schrodinger equation of the electrons. We will draw a parallel between the kink structural carbon chain and linear carbon chain with the same bond distance. Then the phonon wavefunction and the T_c of the kink-structural carbon chain and become interpretable. We will make use of the surface phonon softening g [18–20] to estimate the T_c when it is returned to circular shape. The predicted T_c of the Sample B, C, D, E, F and G in Fig. 11.2 will be obtained by utilizing a scale factor of Al, Ta, Hg, Mo, Ga, Pb, In, Sn or the (4, 2) SWCNT respectively, to ensure all predicted T_c values are almost identical, despite it is scaled by different materials, different lattice structures and different curvatures. The isolated carbon chain refers to the lateral chain-to-chain separation of 1340 pm. We do not substitute the physical quantities into the conclusive equation of BCS theory directly [13], i.e., $\Delta = \frac{\hbar\omega_D}{\sinh\left[1/DOS(E_F)V\right]}$, because we suspect that this conclusive formula may not work in some 1D materials with high Debye frequency. The formula is obtained by considering the finite integral of the electron energy up to the Debye frequency. If the 1D material contains extremely narrow peaks in the DOS, the difference between the DOS(E_F) and DOS(E_F + dE) may become very large, and hence the formula likely becomes inapplicable.

11.4 Toward Phonon-Mediated High-Temperature Superconductivity in Carbon Chains

We investigated how different parameters, such as bond angle and bond length, as well as the separation in bulk structures formed by arrays of these low-dimensional elements, affect the superconducting transition temperature. Figure 11.2 shows the T_c of the various carbon structures we have investigated. The theoretical T_c of the (4, 2) SWCNT arrays is ~15 K. Here a scale factor derived from the gap and T_c values of either Al, Ta, Hg, Mo, Ga, Pb, In or Sn in combination with the intermediate step of the sample D was used. It is consistent to the experimental T_c of the 4 angstrom SWCNT (Sample A) at 15 K [5]. On the other hand, the theoretical T_c of the (5, 0) SWCNT is ~9 K with help of the scale factors again as shown in Tables 11.4 and 11.5. R_w, R_{DOS}, R_{kink}, R_{shape}, R_ω, R_M, R_u, $R_{a(q)}$ and $f(T_D)$ are defined in Table 11.3. By comparing Tables 11.4 and 11.5, modifying $g_{kk'}$ is accurate enough to provide a good reliability when considering the effect of weak curvature and dimensionality in the carbon nanotubes [14].

Table 11.4 The T_c values of SWCNT by the modification of $g_{kk'}$ and surface phonon softening [5]

Material	R_W	R_{DOS}	R_{shape}	R_{kink}	λ_{scale}	μ	$f(T_D)$	T_c^{scale} (K)	T_c^{ex} (K)
(4, 2) SWCNT	3.9	0.13	1.69	0.85	3.82	0.14	1.0062	16.3	15.0
(5, 0) SWCNT	4.3	0.08	1.20	0.85	1.85	0.13	1.0157	7.6	7–15

Table 11.5 The T_c calculations of SWCNT based on the original Hamiltonian [5]

Material	R_W	R_{DOS}	R_ω	R_M	R_u	$R_{a(q)}$	λ_{scale}	μ	$f(T_D)$	T_c^{scale} (K)
(4, 2) SWCNT	3.9	0.13	1.7	1	1.8	0.98	3.64	0.14	~1	13.9
(5, 0) SWCNT	4.3	0.08	1.6	1	1.6	0.99	2.29	0.13	~1	9.5

The (4, 2) and (5, 0) SWCNT with the Debye temperature of 1000 K are used. We compare the accuracy between the McMillian method and scale factor approach in the 1D regime. We estimate the electron-phonon coupling of the (4, 2) and (5, 0) SWCNT from $\lambda_{Mc} = 2\int \frac{\alpha^2(\omega)F(\omega)}{\omega}d\omega = \frac{DOS(E_F)\langle g^2\rangle}{M\langle\omega^2\rangle}$, respectively, where α is the electron-phonon scattering term. $\langle g^2\rangle$ is the average over the Fermi surface of the square of the electronic matrix element and M is the mass of atom [22]. However, the λ_{Mc} values are much smaller than the λ_{scale} values (see Tables 11.5 and 11.6). The margin between the λ_{scale} and λ_{Mc} values is likely increasing when going from the 3D to the 1D regime. The reason is also found in the Van Hove singularities in $\lambda_{Mc} = \frac{DOS(E_F)\langle g^2\rangle}{M\langle\omega^2\rangle}$. Even worse, there remain significant uncertainties in values of λ_{Mc} obtained in this way, which are caused by the presence of the transmission coefficients of tunneling and by inelastic effects across the barrier region [23]. In addition, the McMillian method fails to predict the T_c of SWCNT because the Van Hove Singularities [18–20] in the electronic DOS are not taken into consideration. Therefore, it is wiser to calculate the electron-phonon coupling of 1D materials by the original form of the Hamiltonian [21, 24, 25]. The α of SWCNT is obtained by resolving the repeating unit into orthogonal directions.

Table 11.6 The comparison between the T_c of the SWCNT obtained from the McMillian T_c formula and experimental data [5]

SWCNT	$\lambda_{Mc} = 2\int \alpha^2(F(\omega)/\omega)d\omega$	μ	$T_c^{Mc}(K)$	$T_c^{ex}(K)$
(4, 2)	0.62	0.14	6.1	15
(5, 0)	0.31	0.13	0.2	7–15

In experimental work on superconducting SWCNT arrays grown in the zeolite matrices containing a mixture of nanotubes with different chiralities with mostly (4, 2) and (5, 0) SWCNT [5], the measured onset T_c at 15 K is thus likely attributed to the (4, 2) SWCNT, in contrast to what has been proposed previously [26], and the sharper downturns of resistance at ~7 K at zero magnetic field [5] is likely triggered by the onset of superconducting fluctuations in the (5, 0) nanotubes, which dramatically enhances the transverse Josephson coupling within the array, and thus triggers a transition towards a three-dimensional bulk superconducting state [22] within the array. The electronic energy band diagrams of the SWCNT, using a built-in coordinate system, are metallic, as shown in Fig. 11.3. At first glance, this seems to contradict earlier work on individual nanotubes, where a bandgap was found for (4, 2) nanotubes [26]. However, the difference originates from the arrangement of the nanotubes within a hexagonal array with weak transverse coupling. The 0.12 eV indirect band gap of the (4, 2) SWCNT in the separation of ~1.5 nm computed by "Quantum ESPRESSO" is suppressed by lateral coupling as shown in Fig. 11.4. The valence and conduction bands are clearly overlapped if the (4, 2) SWCNTs are spaced by ~0.7 nm. The center-to-center distance of ~0.7 nm corresponds to the wall-to-wall distance of ~0.3 nm, which is still much larger than single bond length, and therefore we do not expect any extra covalent bond to form laterally across the tubes [27]. Based on a linear interpolation of two band edges as a function of tube-to-tube separations, the metal-semiconductor transition point may occur at a center-to-center distance of approximately 1.13 nm.

When rolling graphene into nanotubes, it's important to consider both the radial and tangential components of the atomic spring constants. Since there are no nearest neighbors along the radial axis, the out-of-plane vibration of the carbon is usually ignored. This means that the tangential vibration is weakened by a sinusoidal factor. The slower lattice vibration allows longer electron-phonon scattering times, causing a T_c enhancement. For instance, the T_c of 1D Pb nanowires (or Sn nanowires) is enhanced by a factor of ~1.6 compared to the bulk T_c by this surface phonon softening effect [18, 19]. Sample B is produced by shortening the armchair SWCNT to one repeating unit in the larger diameter of 2.3 nm that shows a remarkable effect on T_c which is increased to 52 K because the DOS of sample B is 10 times larger than Sample A in Fig. 11.5, where the DOS of sample B and C are assumed to be identical.

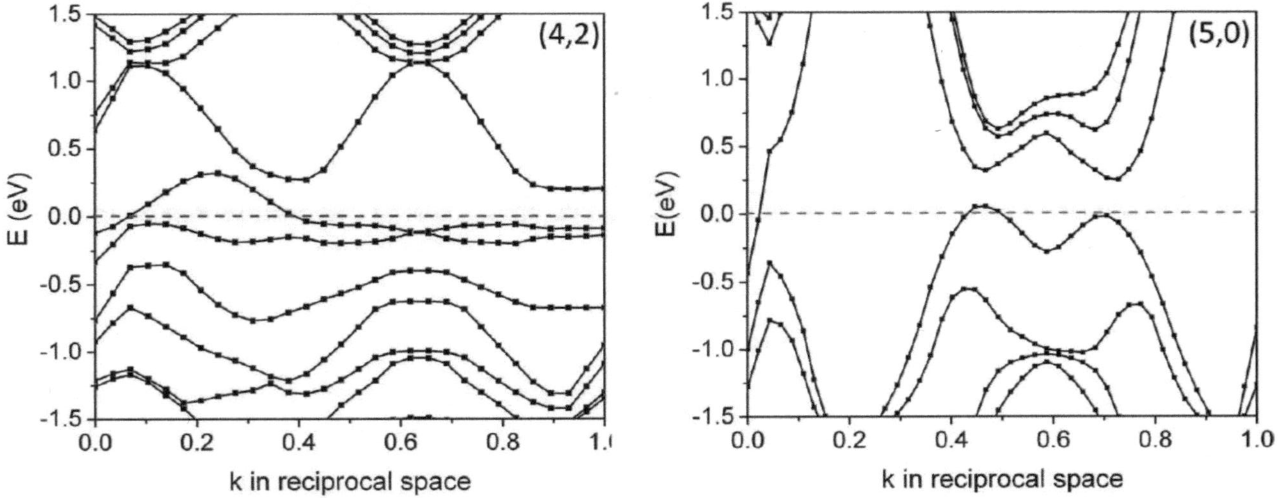

Figure 11.3 The energy band diagrams for the hexagonal arrays of (4, 2) and (5, 0) SWCNT indicate a metallic character. The SWCNTs are separated by approximately 0.7 nm, center to center. The Fermi-level was adjusted to 0 eV for improved readability. Reprinted with permission from Wong *et al*. [17], Copyright 2023, Springer Nature.

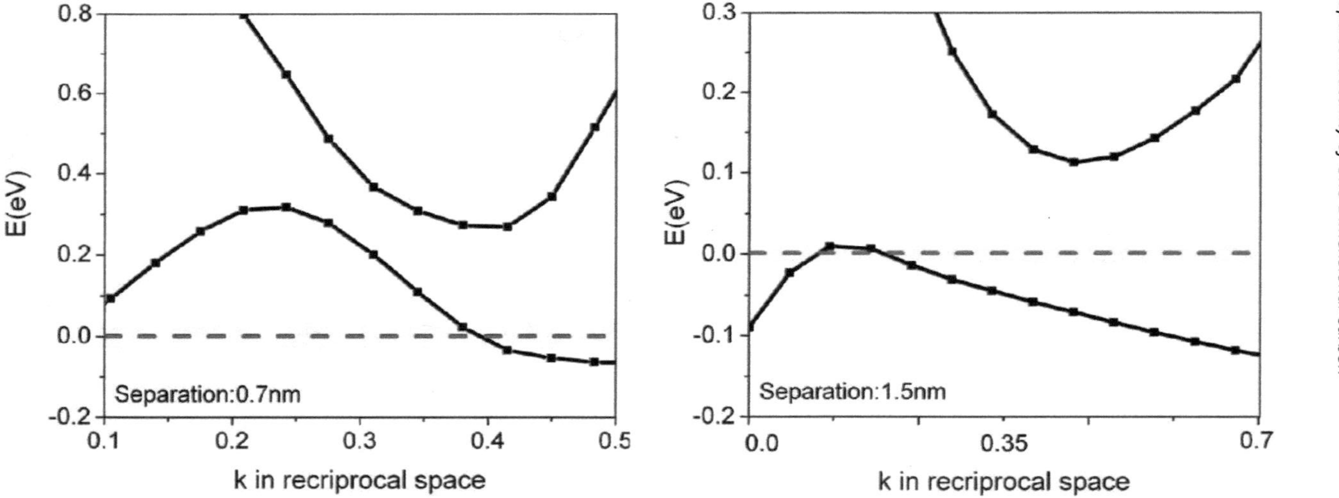

Figure 11.4 The local band structures of (4, 2) SWCNTs are impacted by the center-to-center separation, which is approximately 0.7 nm and 1.5 nm, respectively. The interaction between the tubes triggers a transition from semiconductor to metal. Reprinted with permission from Wong *et al*. [17], Copyright 2023, Springer Nature.

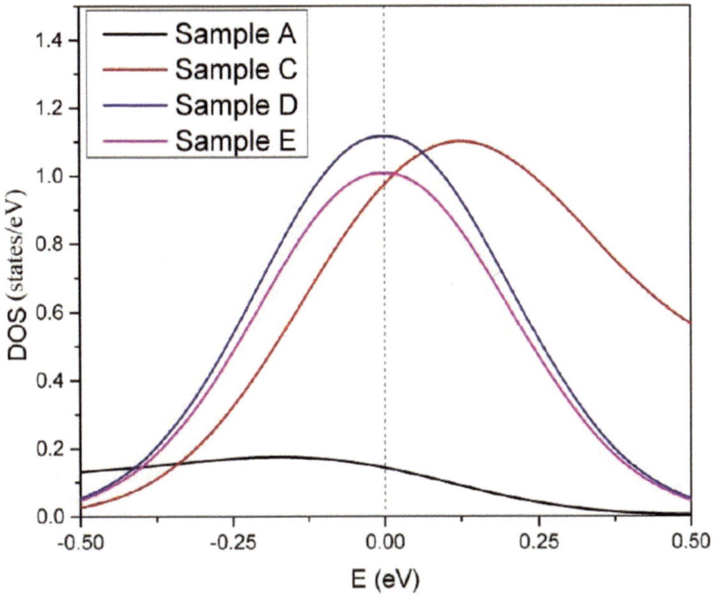

Figure 11.5 The electronic density of states (DOS) per atom was measured for several samples. Sample A had a much lower DOS at the Fermi level compared to the other samples. The DOS for Sample B and C were assumed to be identical due to their weak global curvature. Reprinted with permission from Wong *et al.* [17], Copyright 2023, Springer Nature.

We acknowledge that one may question the validity of our scale factor approach when the SWCNT is shortened to a carbon ring. However, it's worth noting that the Debye temperature of the carbon chain is even higher than that of the SWCNT [28, 29], and therefore using the scale factor to predict the T_c of the carbon nanowire is even more accurate. In the absence of phonon softening due to curvature, the electron-phonon coupling is lowered [14, 18, 20], and therefore unfolding the Sample B to Sample C lowers T_c by 1.5 times. Removal of the kinks leads to Sample D with T_c = 23 K. This can also be explained by the loss of local curvature, which weakens the electron-phonon coupling. However, the curvature is not the only parameter we can tune in order to optimize T_c [6, 7]. Indeed, if we decrease the bond length from 143 pm to 134 pm, as demonstrated in Sample E, the predicted T_c is enhanced tremendously from 23 K to 81 K. The calculation based on the scale factor approach is illustrated in Table 11.7.

Table 11.7 The T_c^{scale} calculation in various materials. The reference material is Al with $\lambda_{\text{ref}} = 0.38$. The C-chains are separated by 1340 pm. The compressed lithium is shown here as an example to estimate λ_{scale} [30, 31]

Material	R_W	R_{DOS}	R_ω	R_M	R_u	$R_{a(q)}$	$\lambda/\lambda_{\text{ref}}$	λ_{scale}	μ	$f(T_D)$	T_c^{scale} (K)
Al	1	1	1	1	1	1	1	0.38	0.11	1.0000	1.2
Li (35 GPa)	1.59	0.64	2.82	0.26	6.94	0.89	7.34	2.79	0.14	1.0034	11.8
Sample D	3.92	1.59	4.06	0.44	3.57	0.83	14.0	5.25	0.15	1.0022	22.6
Sample E	8.81	1.61	5.13	0.44	5.10	1.01	48.4	18.4	0.15	1.0004	81.1

The ratio of branch length R_B in Sample F equals to 3. Figure 11.6 demonstrates that the DOS is shifted to lower energies when the kink angle is increased from 10 to 50 degrees, except for the case of 60 degrees. The sharpness of the DOS distribution decreases with increasing kink angle owing to a denser particle concentration. It stimulates us to investigate the dependence of T_c of Sample F as a function of the kink angle. We observe that T_c becomes optimized at 10 degrees (Fig. 11.7). Two effects are competing. On one hand, the electron-phonon coupling benefits dramatically from the local curvature [18, 20]. On the other hand, the electron-phonon coupling is weakened by the loss of electronic density of states [13]. The effect of the local curvature is not sufficient to compensate the effect of the DOS for large kink angles, and hence T_c becomes significantly reduced beyond 10 degrees. As the carbon ring (Sample G) consists of ~100 atoms only, the tilt angle between the adjacent atoms is

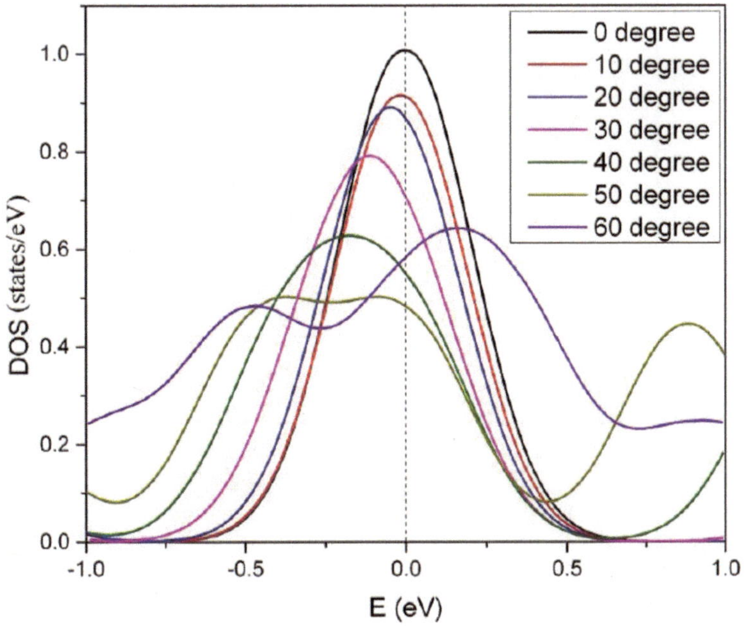

Figure 11.6 The electronic density of states (DOS) per atom in Sample F is shown as a function of kink angle. The Fermi level is adjusted to 0 eV and coincides with the Van Hove singularity in the DOS at a kink angle of zero degree. Reprinted with permission from Wong *et al.* [17], Copyright 2023, Springer Nature.

about 360/100 = 3.6 degrees which is ~8 times smaller than the (4, 2) SWCNT. The few degrees of curvature do not amend the electronic band diagram and the electronic DOS remarkably [14]. Modifying the lattice vibrations is sufficient to dramatically tune T_c of such weakly-curved low-dimensional carbon-based superconductors [14]. As a result, we use the electronic band diagram and the electronic DOS obtained from the linear carbon chain to model a carbon ring in the following. If Sample F is finally bent to form a ring structure of diameter of 4.2 nm (Sample G), T_c becomes optimized and reaches a value as large as 115 K. The T_c is thus enhanced by a factor of 1.3.

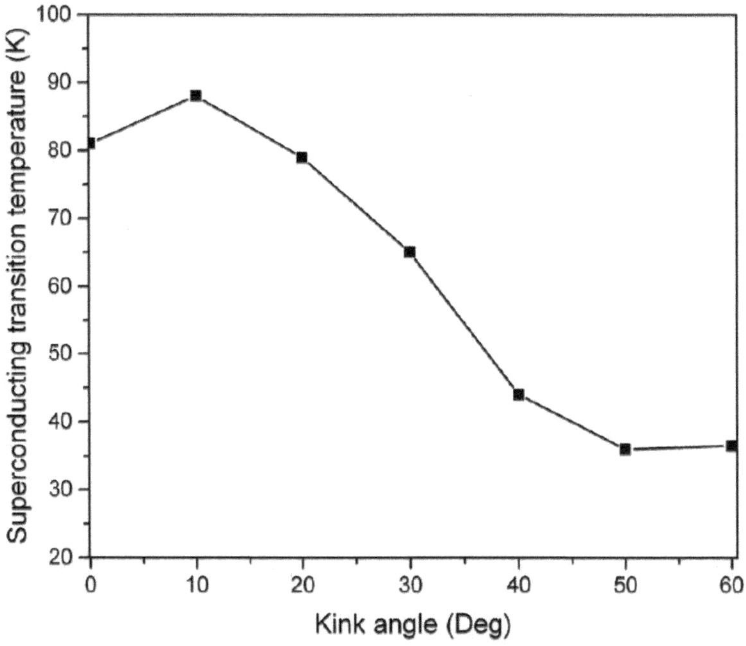

Figure 11.7 The T_c of the Sample F as a function of kink angle. The T_c is increased to 88 K after the kink is bent to 10 degrees. However, a dramatic reduction of T_c occurs at larger kink angles. Reprinted with permission from Wong *et al.* [17], Copyright 2023, Springer Nature.

Another interesting phenomenon is observed in the coupled linear carbon nanowires (Sample E) as shown in Fig. 11.8. The T_c remains constant for chain separations exceeding 1.2 nm. However, the T_c is suppressed due to the lateral coupling, because the DOS(E_F)

of the coupled nanowires are significantly reduced for inter-chain distances of less than 0.7 nm (Fig. 11.9). However, there may exist an opportunity to increase T_c in the presence of strong chain-to-chain couplings in 1D chains of materials with larger atomic numbers, because the larger atomic number of the element, the stronger the Coulomb attraction the material generates.

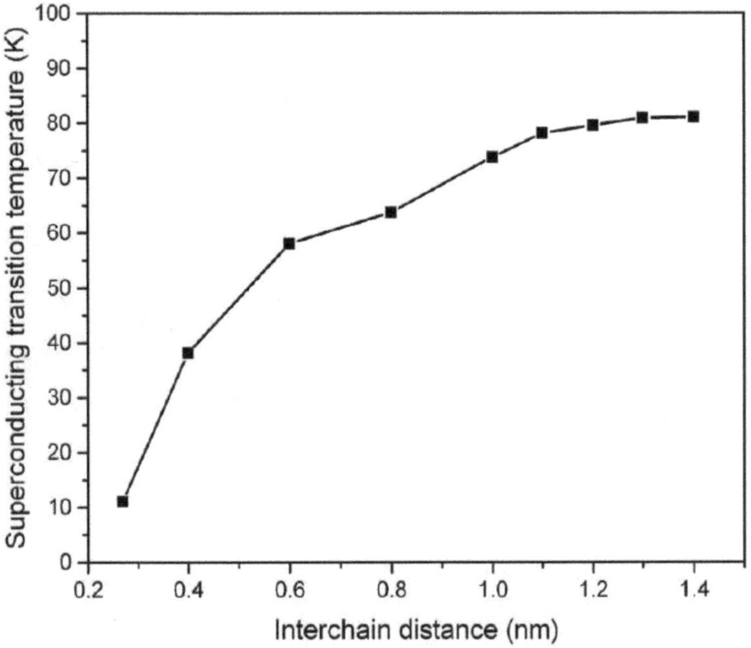

Figure 11.8 The T_c of the laterally coupled carbon chains (Sample E) as a function of chain-to-chain separation. Reprinted with permission from Wong *et al.* [17], Copyright 2023, Springer Nature.

Simulating coupled chains is crucial because the low dimensionality of the isolated kink structural carbon ring can lead to significant phase slip events along the chain. As a result, it causes finite resistance at all temperatures in the superconducting state [2] unless the rings are coupled laterally *via* the Josephson effect [18]. According to recent theoretical and experimental work [4, 5, 32, 33], it is possible for cylindrical superconducting nanowire arrays to undergo a dimensional crossover to a three-dimensional bulk superconducting state triggered by the transverse Josephson effect.

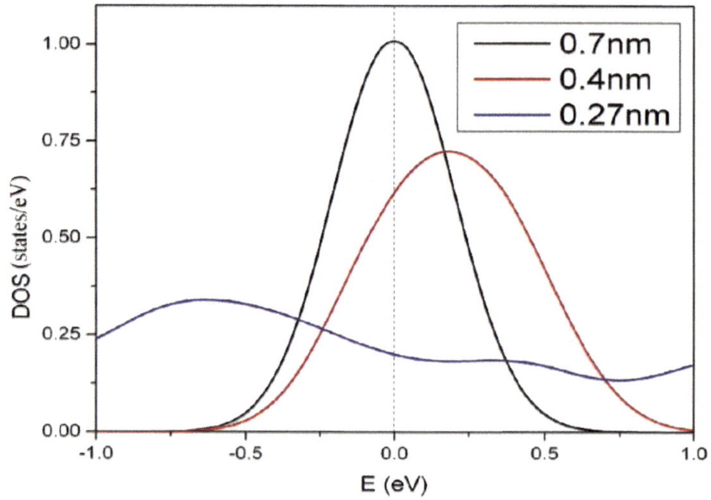

Figure 11.9 The electronic density of states of the coupled arrays of Sample E at various lateral coupling strengths. The sharpness of the DOS peaks rapidly decays at close chain density within the array. Reprinted with permission from Wong *et al.* [17], Copyright 2023, Springer Nature.

This dimensional transition should not be affected by the curvature of the nanowire arrays, which only influences the critical temperature at which Cooper pairs form. The coherence length ζ of the kink structural carbon ring is about 12 nm as obtained from $\zeta = \frac{\hbar v_F}{\pi \Delta}$, where v_F is Fermi velocity [13]. If the carbon rings are arranged on top of each other with a ring-to-ring distance along the axial direction of 1.2 nm, a Berezinskii-Kosterlitz-Thouless-like (BKT) transition [34] is likely able to trigger the dimensional crossover from 1D to 3D [4], because the coherence length exceeds the ring-to-ring separation. As a result, the global phase coherence in the coupled kink structural carbon ring is expected to establish a true bulk superconducting zero resistance state at finite temperature [4, 18]. However, the strength of phase fluctuations causing phase slip events will depend crucially on the Josephson coupling between the rings, and it is unclear whether there would remain a large separation between T_c (where the Cooper pairs form in the individual rings) and the temperature (T_{BKT}) below which global phase coherence and thus zero resistance occurs. We compare the electron-phonon coupling of the sample F obtained from the well accepted formula, $\lambda_{Mc} = 2\int \alpha^2 \frac{F(\omega)}{\omega} d\omega$, with help of the phonon data in Fig. 11.10. The Debye temperature of the sample F is ~1500 K. We only show the phonon data of sample F for clarity. The α^2 contains the average square electron-phonon matrix [22]. The T_c^{Mc} estimated from the McMillian method can still reach more than 40 K, as shown in Table 11.8. The renormalized factor $(1 + \lambda)$ is valid in the strong coupling regime.

Table 11.8 The T_c attained from the McMillian method. The T_c^{scale} is obtained from λ_{scale} instead of λ_{Mc}

Material	$\lambda_{Mc} = 2\int \alpha^2 (F(\omega)/\omega) d\omega$	$\lambda/(1 + \lambda)$	μ	$\mu/(1 + \lambda)$	$T_c^{Mc}(K)$	$T_c^{\text{scale}}(K)$
Sample E	1.51	0.60	0.15	0.059	41.9	81
Sample F	1.53	0.61	0.15	0.059	42.6	88

Table 11.9 Phonon frequencies of the various carbon structures. Reprinted with permission from Wong *et al.* [17], Copyright 2023, Springer Nature

Sample	**A**	**B**	**C**	**D**	**E**	**F**	**G**
$\langle\omega\rangle$ (THz)	36	5.5	12	21	27	26	15

A large discrepancy between T_c^{Mc} and T_c^{scale} is observed in the SWCNT array. If the diameter is reduced to one atom thickness only, as in the case of a carbon chain, we expect the error to increase further. In view of this, the formation of Cooper pairs in sample E at 81 K is likely more reasonable. The average phonon frequencies of the samples are shown in Table 11.9. The absolute value of the phonon frequency is not important in the scale factor approach, which requires a ratio of phonon frequencies only.

Table 11.10 The T_c of the kink structural carbon ring for various ring diameters

Total Number of Atoms in Sample G	**Superconducting Transition Temperature/K**
100	115
130	96
170	90

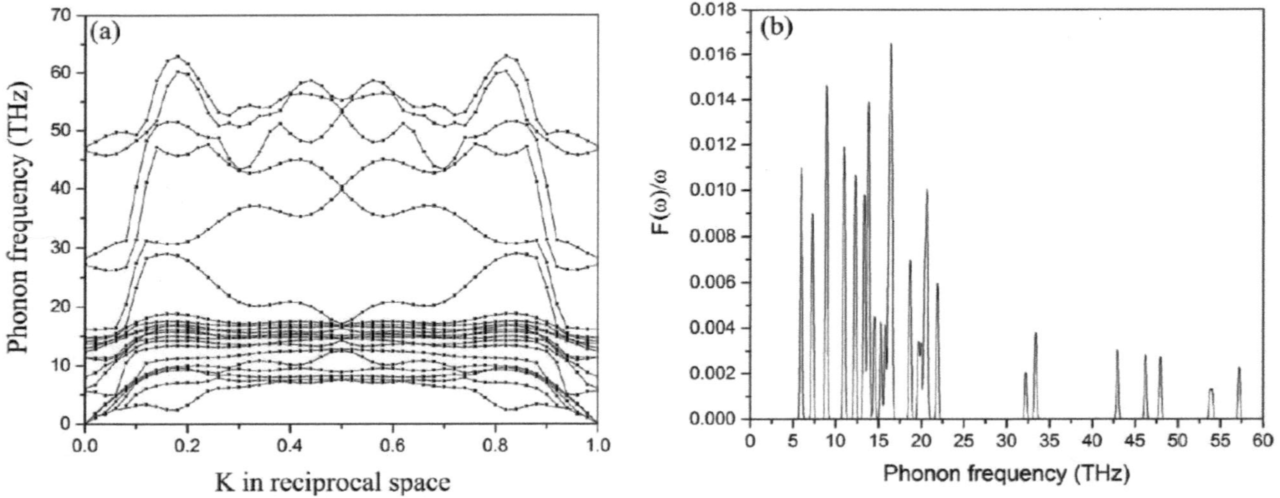

Figure 11.10 Phonon dispersion relation (a) and phonon spectrum (b) of Sample F obtained from "Quantum ESPRESSO". Reprinted with permission from Wong *et al*. [17], Copyright 2023, Springer Nature.

Table 11.10 shows the reduction of T_c for large diameter rings. The smaller ring provides a higher T_c due to the curvature-induced phonon softening. All our attempts to increase the T_c beyond our predicted 115 K failed so far. For example, reducing the radius of the kink structural carbon ring would reinforce the phonon softening. However, the electron-phonon coupling is weaker if the nanowire becomes too short [35]. Our 115 K ($\lambda \sim 25$) prediction thus appears to be the optimal T_c value, although there may exist more complex carbon structure that could push this high temperature superconductor limit in carbon based structures further. In the end it will be the experimental realization of such a carbon ring array structure which may or may not confirm our theoretical prediction.

11.5 Conclusion

We have investigated quasi-1D carbon structures as potential high temperature superconductors and have concluded that they are promising candidates. Starting with carbon nanotubes, an initial calculation yielded a T_c of 13.9 K, in good agreement with experiments. We then proceeded to reduce the nanotube to a ring, open the ring to form chains, optimize bond length and kink structure, and finally form a new type of carbon ring that reached a T_c value of 115 K. These results demonstrate the potential of quasi-1D carbon structures as novel high temperature superconductors.

References

1. Drozdov, A. P., Eremets, M. I., Troyan, I. A., Ksenofontov, V., & Shylin, S. I. Conventional superconductivity at 203 Kelvin at high pressures in the sulfur hydride system. *Nature* **525**, 73–76 (2015).
2. Altomare, F., & Chang, A. M. *One-Dimensional Superconductivity in Nanowires*, Wiley-VCH, 2013.
3. Joshi, D. G., & Bhattacharyay, A. Revisiting the Langer-Ambegaokar-McCumber-Halperin theory of resistive transitions in onedimensional superconductors with exact solutions. *J. Phys. Cond. Mat.* **23**, 342203 (2011).
4. He, M., *et al.* 1D to 3D dimensional crossover in the superconducting transition of the quasi-one-dimensional carbide superconductor Sc_3CoC_4. *J. Phys. Cond. Mat.* **27**, 075702 (2015).

5. Tang, Z. K., *et al.* Superconductivity in 4 angstrom single-walled carbon nanotubes. *Science* **292**, 2462–2465 (2001).
6. Liu, X., Zhang, G., & Zhang, Y. W. Tunable mechanical and thermal properties of one-dimensional carbyne chain: phase transition and microscopic dynamics. *J. Phys. Chem. C* **119**, 24156–24164 (2015).
7. Wong, C. H., Buntov, E. A., Rychkov, V. N., Guseva, M. B., & Zatsepin, A. F. Simulation of chemical bond distributions and phase transformation in carbon chains. *Carbon* **114**, 106–110 (2017).
8. Guseva, M., Babaev, V., Novikov, N., inventors; *Tetracarbon*, United States Patent 6,355,350 B, 1996 January 18.
9. Guseva, M., Babaev, V., Novikov, N., inventors; *Tetracarbon*, United States Patent 6,454,797 B, 2002 September 24.
10. Guseva M., Babaev V., Novikov, N., inventors; *Tetracarbon*, United States Patent 6,555,224 B, 2003 April 29.
11. Perdew, J. P., *et al.* Atoms, molecules, solids, and surfaces: applications of the generalized gradient approximation for exchange and correlation. *Phys. Rev. B* **46**, 6671–6687 (1992).
12. Becke, A. D. Density-functional exchange-energy approximation with correct asymptotic behavior. *Phys. Rev. A* **38**, 3098–3100 (1988).
13. Tinkham, M. *Introduction to Superconductivity*, Dover Publications, 2004.
14. Wong, C. H., & Lortz, R. Edge effect and significant increase of the superconducting transition onset temperature of 2D superconductors in flat and curved geometries. *Physica C* **521–522**, 50–54 (2016).
15. Benedict, L. X., Crespi, V. H., Louie, S. G., & Cohen, M. L. Static conductivity and superconductivity of carbon nanotubes: relations between tubes and sheets. *Phys. Rev. B* **52**, 14935–14940 (1995).
16. Christman, J. R. *Fundamentals of Solid State Physics*, John Wiley & Sons, 1988.
17. Wong, C. H., Lortz, R., Buntov, E. A., Kasimova, R. E., & Zatsepin, A. F. A theoretical quest for high temperature superconductivity on the example of low-dimensional carbon structures. *Sci. Rep.* **7**(1), 15815 (2017).
18. He, M., *et al.* "Giant" enhancement of the upper critical field and fluctuations above the bulk T_c in superconducting ultrathin lead nanowire arrays. *ACS Nano* **7**, 4187–4193 (2013).
19. Zhang, Y., *et al.* Dramatic enhancement of superconductivity in single-crystalline nanowire arrays of Sn. *Sci. Rep.* **6**, 32963 (2016).

20. Xu, Y. L., Gong, H. C., Fan, Q. S., & Zhou, S. M. Hydrothermal fabrication and superconductivity of isotropic Pb hollow microspheres. *Mater. Lett.* **130**, 57–60 (2014).

21. Khukhareva, I. S. The superconducting properties of thin aluminum films. *Soviet Physics JETP* **16**, 828–832 (1963).

22. Allen, P. B., & Dynes, R. C. Transition temperature of strong-coupled superconductors reanalyzed. *Phys. Rev. B* **12**, 905–922 (1975).

23. Poole, C. P. Jr. *Handbook of Superconductivity*, Ch. 9, Academic Press, New York, 1999, pp.478–483.

24. Sun, S. J., Lin, K. H., Ju, S. P., & Li, J. Y. Electronic and structural properties of ultrathin tungsten nanowires and nanotubes by density functional theory calculation. *J. App. Phys.* **116**, 133704 (2016).

25. McMillian, W. L. Transition temperature of strong-coupled superconductors. *Phys. Rev.* **167**, 331–344 (1968).

26. Machon, M., Reich, S., & Thomsen, C. Ab initio calculations of the optical properties of 4-Å-diameter single-walled nanotubes. *Phys. Rev. B* **66**, 155410 (2002).

27. Hu, M., *et al.* Compressed carbon nanotubes: a family of new multifunctional carbon allotropes. *Sci. Rep.* **3**, 1331 (2013).

28. Wang, M., & Lina, S. Ballistic thermal transport in carbyne and cumulene with micron-scale spectral acoustic phonon mean free path. *Sci. Rep.* **5**, 18122 (2015).

29. Hone, J. Phonons and thermal properties of carbon nanotubes, in *Carbon Nanotubes: Synthesis, Structure, Properties, and Applications* (Dresselhaus, M. S., *et al.*, eds), Springer, Berlin, 2001, pp. 273–286.

30. Shi, L., & Papaconstantopoulos, D. A. Theoretical predictions of superconductivity in alkali metals under high pressure. *Phys. Rev. B.* **73**, 184516 (2006).

31. Shimizu, K., Ishikawa, H., Takao, D., Yagi, T., & Amaya, K. Superconductivity in compressed Lithium at 20 K. *Nature* **419**, 597–599 (2002).

32. Wong, C. H., Wu, P. H., & Lortz, R. Phase fluctuations in two coaxial quasi-one-dimensional superconducting cylindrical surfaces serving as a model system for superconducting nanowire bundles. *Physica C* **534**, 45–49 (2017).

33. Bergk, B., *et al.* Superconducting transitions of intrinsic arrays of weakly coupled one-dimensional superconducting chains: the case of the extreme quasi-1D superconductor $Tl_2Mo_6Se_6$. *New J. Phys.* **13**, 103018 (2011).

34. Kosterlitz, J. M. The critical properties of the two-dimensional XY model. *J. Phys. C: Solid State Phys.* **7**, 1046–1060 (1974).

35. Zhong, G., Chen, X. J., & Lin, H. Q. Prediction of superconductivity in potassium-doped benzene. *arXiv* 1501, 00240 (2015).

Index